AF614716

Methods in Molecular Biology™

Series Editor
John M. Walker
School of Life Sciences
University of Hertfordshire
Hatfield, Hertfordshire, AL10 9AB, UK

For other titles published in this series, go to
www.springer.com/series/7651

Membrane Transporters in Drug Discovery and Development

Methods and Protocols

Edited by

Qing Yan

PharmTao, Santa Clara, California, USA

Editor
Qing Yan
PharmTao
PO Box 5672
Santa Clara, CA 95056-5672
USA
qyan@pharmtao.com

ISSN 1064-3745 e-ISSN 1940-6029
ISBN 978-1-60761-699-3 e-ISBN 978-1-60761-700-6
DOI 10.1007/978-1-60761-700-6
Springer New York Dordrecht Heidelberg London

Library of Congress Control Number: 2010923408

Printed on acid-free paper

Humana Press is part of Springer Science+Business Media (www.springer.com)

Preface

Studies of membrane transporters have had a great impact on our understanding of human diseases and the design of effective drugs. About 30% of current clinically marketed drugs are targeting membrane transporters or channels. This book provides various practical methodologies of the ongoing research on membrane transporters, especially applications of transporter technologies in drug discovery and development. To provide readers the most up-to-date information, new and useful fields and methodologies are embraced in this volume, including pharmacogenomics, nutrigenomics, systems biology, bioinformatics, nuclear magnetic resonance (NMR), imaging, and quantitative real-time PCR. Transporter studies in drug discovery and development for various diseases are discussed, such as neuropsychiatric disorders, cardiovascular diseases, ophthalmic diseases, cancer, and diabetes.

Pharmacogenomics and systems biology studies of membrane transporters are useful in drug discovery and in predicting drug responses in the clinic. In this volume, the current status of these emerging fields in relevance to transporter studies is reviewed and the key issues are discussed.

Bioinformatics is frequently used in transporter studies and has become indispensable for all kinds of research methods. Commonly used bioinformatics methods, such as databases and tools, are collected in this book to facilitate transporter research. Because of heterogeneous sources and tremendous amount of data, data integration has become one of the most important issues in transporter studies. A brief introduction of data integration methodology offered here can help researchers manage their data for further knowledge discovery. The discussion of data modeling approaches can help with the understanding of necessary computational decision support for transporter studies in pharmacogenomics and systems biology.

Bioinformatics analyses may allow definition of phylogenetic relationships in transport protein superfamilies. These novel programs can be used to analyze the major superfamilies of secondary carriers that export hydrophobic and amphipathic compounds including drugs. Applications of these programs are introduced in this volume, including phylogenetic clustering of the families and the analysis of phylogenetic relationships of sequence-divergent drug exporters.

Applications of both in vitro and in silico techniques are helpful for the understanding of transporter behavior. For example, in vitro techniques include mammalian cell-based transporter assays, and in silico techniques include ligand-based transporter modeling. The combination of these in vitro and in silico approaches may assist in faster discovery of ligands for transporter-targeted drug delivery.

Many transporters can be differentially up-regulated in cancer cells compared to normal tissues. Such differential expression of transporters may provide good targets for improving drug delivery, with more focused distribution to the target tissues and enhanced bioavailability. Methodologies to analyze these activities of transporters in cancer are discussed in this volume.

The analysis of gene expression is important for the understanding of interactions between transporters and drugs in the treatment of cancer. Quantitative real-time PCR (qRT-PCR) is a popular technique for assessing gene expression levels with many advantages over microarrays. Gene expression analysis by qRT-PCR has potential as a diagnostic tool for predicting drug responses in cancer patients. In this volume, applications of the qRT-PCR technique are described and compared for analyzing transporter expression levels in various drug-resistant cancer cell lines.

Resistance to multiple drugs is a serious obstacle to chemotherapy treatment of human cancers. Many clinically useful drugs have limited uptake in the intestine and cannot access the brain. For example, it is difficult to directly quantitate drug binding to P-glycoprotein or measure drug transport rates. Fluorescence spectroscopic approaches are very useful in overcoming these problems. Detailed description is provided on using fluorescence tools to quantitate the affinity of binding of various drugs and measure the drug transport activity in real time.

The blood–brain barrier (BBB) is the physiological barrier that regulates the passage of substances into and out of the brain. A useful in vitro BBB model is introduced for the transport studies of the nutrition, physiology, and pharmacology of the brain.

Vesicular monoamine transporters (VMATs) are an important target for biological research in neuropsychiatric disorders. Different structurally related but pharmacologically distinct VMATs have been identified. Various PCR methods are described in this volume for genotyping polymorphisms in these transporters implicated in neuropsychiatric disorders. Human vesicular acetylcholine transporter (hVAChT) in cholinergic nerve terminals stores acetylcholine (ACh) in synaptic vesicles. Methods for characterizing equilibrium binding and transport by VAChT are also discussed.

ATP-binding cassette (ABC) transporters have been linked to Stargardt's disease, fundus flavimaculatus, cone–rod dystrophy, retinitis pigmentosa, and age-related macular degeneration. Genetic pathways involved in the pathogenesis of ABCR-related ophthalmic conditions, as well as diagnostic and therapeutic objectives for these diseases, are discussed in this volume.

Skeletal muscle is crucial in regulating whole body glucose homeostasis. Severe dysfunction in insulin-mediated glucose uptake features insulin-resistant states and type II diabetes. Intravital imaging of protein translocation in skeletal muscle is a very useful technique for the study of insulin signaling. Such analysis may help elucidate the molecular mechanisms of both normal and insulin-resistant conditions.

Glucose transporters have been identified in parasites including *Leishmania*, *Trypanosoma*, and *Plasmodium* that can cause leishmaniasis, African sleeping sickness, and malaria. Hexose transporters are essential for the infectious stage of these parasites and can be important targets for development of novel anti-parasitic drugs.

Because structural and functional studies are essential in understanding the interactions of drugs and the functional transporter proteins, methodologies and protocols are provided from various points of view to tackle structure–function problems. These methods include nuclear magnetic resonance (NMR), site-directed mutagenesis, and the use of *Xenopus* oocytes. These systems can help in the mechanistic and structural characterization of transport proteins and provide insight in signal transduction pathways. Methods to study specific transporters are also provided. For example, the plasma membrane calcium/calmodulin-dependent ATPase (PMCA) plays an important role in signal transduction, especially in the nitric oxide signaling pathway. Different methods in determining PMCA activity are described, such as patch clamp and enzyme assay.

Readers are encouraged to explore integrated views and comprehensive methodologies from different chapters of this book, as these methods are not isolated techniques but rather complementary to each other. One technology may have several other methods and techniques involved. For example, studying the expression system of *Xenopus* oocytes uses fluorescence microscopy. To measure intracellular pH, which is important for understanding the role of membrane transporters in cellular processes, NMR and fluorescent techniques may also be used.

This is an updated and supplemental volume of a previously published book *Membrane Transporters: Methods and Protocols* (Methods in Molecular Biology Series, Volume 227, published by Humana Press, 2003), with emphasis on transporter methodologies in drug discovery and development. This volume strives to deliver to readers not only a collection of practical protocols that can be used immediately in the lab but also critical surveys of key topics by leading researchers in this field. Readers can develop their own workable schemes for their personal studies based on these powerful methodologies. Biomedical researchers in various fields who are interested in membrane transporters and drug development will find the book useful, including chemists, biochemists, molecular biologists, geneticists, physiologists, microbiologists, immunologists, bioinformatics researchers, pharmaceutical scientists, toxicologists, bioengineers, and clinical researchers.

I would like to thank all authors for sharing their valuable experience and insights with the research community at large. I would also like to thank the series editor Dr. John Walker for help with reviewing the chapters.

Qing Yan

Contents

Contributors

RIHAM ABOU-LEISA • *Cardiovascular Medicine Research Group, University of Manchester, Manchester, United Kingdom*

SURESH V. AMBUDKAR • *Laboratory of Cell Biology, Center for Cancer Research, National Cancer Institute, NIH, DHHS, Bethesda, MD*

PRAVEEN M. BAHADDURI • *Genzyme Corporation, Waltham, MA*

FLORENCE M. BAUDOIN-STANLEY • *Cardiovascular Medicine Research Group, University of Manchester, Manchester, United Kingdom*

DENNIS J. BOBILYA • *Department of Molecular, Cellular, and Biomedical Sciences, University of New Hampshire, Durham, NH*

STEFAN BRÖER • *Research School of Biology, Australian National University, Canberra, Australia*

ANNA MARIA **Calcagno** Laboratory of Cell Biology, Center for Cancer Research, National Cancer Institute, NIH, DHHS, Bethesda, MD

ELIZABETH CARTWRIGHT • *Cardiovascular Medicine Research Group, University of Manchester, Manchester, United Kingdom*

JOSEPH R. CASEY • *CIHR Membrane Protein Research Group, Departments of Physiology and Biochemistry, University of Alberta Edmonton, AB, Canada*

JONATHAN S. CHEN • *Division of Biological Sciences, University of California San Diego, La Jolla, CA*

GABRIEL A. COOK • *Department of Chemistry and Biochemistry, University of California San Diego, La Jolla, CA*

SEAN EKINS • *Collaborations in Chemistry, Jenkintown, PA*

RUTH M. EMPSON • *Department of Physiology, Otago School of Medical Sciences, University of Otago, Dunedin, New Zealand*

PARUL KHARE • *Department of Chemistry and Biochemistry, Neuroscience Research Institute, University of California, Santa Barbara, CA*

SCOTT M. LANDFEAR • *Department of Molecular Microbiology and Immunology, Oregon Health and Science University, Portland, OR*

HANS P.M.M LAURITZEN • *Research Division, Joslin Diabetes Center and Harvard Medical School, Boston, MA*

RONGHUA LIU • *Department of Molecular and Cellular Biology, University of Guelph, Guelph, ON, Canada*

FALK W. LOHOFF • *Translational Research Laboratories, Department of Psychiatry, Center for Neurobiology and Behavior, University of Pennsylvania School of Medicine, Philadelphia, PA*

FREDERICK B. LOISELLE • *CIHR Membrane Protein Research Group, Departments of Physiology and Biochemistry, University of Alberta Edmonton, AB, Canada*

JOSE L. MARQUEZ • *Division of Biological Sciences, University of California San Diego, La Jolla, CA*

AUDRA A. MCKINZIE • *Transporter Biology Group, Discipline of Pharmacology and Bosch Institute, University of Sydney, Sydney, NSW, Australia*

KEISUKE MITSUOKA • *Department of Membrane Transport and Biopharmaceutics, School of Pharmaceutical Sciences, Kanazawa University, Kanazawa, Japan*

TAMER M. A. MOHAMED • *Cardiovascular Medicine Research Group, University of Manchester, Manchester, United Kingdom*

ANUPRAO MULAKALURI • *Department of Chemistry and Biochemistry, Neuroscience Research Institute, University of California, Santa Barbara, CA*

TAKEO NAKANISHI • *Department of Membrane Transport and Biopharmaceutics, School of Pharmaceutical Sciences, Kanazawa University, Kanazawa, Japan; The Program in Experimental Therapeutics, Marlene and Stewart Greenebaum Cancer Center (UMGCC) and the Division of Hematology and Oncology, Department of Medicine, University of Maryland School of Medicine, Baltimore, MD*

LUDWIG NEYSES • *Cardiovascular Medicine Research Group, University of Manchester, Manchester, United Kingdom*

DELVAC OCEANDY • *Cardiovascular Medicine Research Group, University of Manchester, Manchester, United Kingdom*

STANLEY J. OPELLA • *Department of Chemistry and Biochemistry, University of California San Diego, La Jolla, CA*

STANLEY M. PARSONS • *Department of Chemistry and Biochemistry, Neuroscience Research Institute, University of California, Santa Barbara, CA*

JAMES E. POLLI • *Department of Pharmaceutical Sciences, University of Maryland, Baltimore, MD*

CHRIS J. ROOME • *Department of Physiology, Otago School of Medical Sciences, University of Otago, Dunedin, New Zealand*

DOUGLAS D. ROSS • *The Department of Membrane Transport and Biopharmaceutics, School of Pharmaceutical Sciences, Kanazawa University, Kanazawa, Japan; Program in Experimental Therapeutics, Marlene and Stewart Greenebaum Cancer Center (UMGCC) and the Division of Hematology and Oncology, Department of Medicine, University of Maryland School of Medicine, Baltimore, MD*

RENAE M. RYAN • *Transporter Biology Group, Discipline of Pharmacology and Bosch Institute, University of Sydney, Sydney, NSW, Australia*

MILTON H. SAIER • *Division of Biological Sciences, University of California San Diego, La Jolla, CA*

FRANCES J. SHAROM • *Department of Molecular and Cellular Biology, University of Guelph, Guelph, ON, Canada*

ERIC I. SUN • *Division of Biological Sciences, University of California San Diego, La Jolla, CA*

PETER W. SWAAN • *Department of Pharmaceutical Sciences, University of Maryland, Baltimore, MD*

ROBERT J. VANDENBERG • *Transporter Biology Group, Discipline of Pharmacology and Bosch Institute, University of Sydney, Sydney, NSW, Australia*

BALPREET VINEPAL • *Department of Molecular and Cellular Biology, University of Guelph, Guelph, ON, Canada*

COREY WESTERFELD • *Department of Ophthalmology, Massachusetts Eye and Ear Infirmary, Harvard Medical School, Boston, MA*

AUBREY R. WHITE • *Department of Chemistry and Biochemistry, Neuroscience Research Institute, University of California, Santa Barbara, CA*

QING YAN • *PharmTao, Santa Clara, CA*

MING REN YEN • *Division of Biological Sciences, University of California San Diego, La Jolla, CA*

Chapter 1

Membrane Transporters and Drug Development: Relevance to Pharmacogenomics, Nutrigenomics, Epigenetics, and Systems Biology

Qing Yan

Abstract

The study of membrane transporters may result in breakthroughs in the discovery of new drugs and the development of safer drugs. Membrane transporters are essential for fundamental cellular functions and normal physiological processes. These molecules influence drug absorption and distribution and play key roles in drug therapeutic effects. A primary goal of current research in drug discovery and development is to fully understand the interactions between transporters and drugs at both the system levels in the human body and the individual level for personalized therapy. Systematic studies of membrane transporters will help in not only better understanding of diseases from the systems biology point of view but also better drug design and development. The exploration of both pharmacogenomics and systems biology in transporters is necessary to connect individuals' genetic profiles with systematic drug responses in the human body. Understanding of gene–diet interactions and the effects of epigenetic changes on transporter gene expression may help improve clinical drug efficacy. The integration of pharmacogenomics, nutrigenomics, epigenetics, and systems biology may enable us to move from disease treatment to disease prevention and optimal health. The key issues in such integrative understanding include the correlations between structure and function, genotype and phenotype, and systematic interactions among transporters, other proteins, nutrients, drugs, and the environment. The exploration in these key issues may ultimately contribute to personalized medicine with high efficacy but less toxicity.

Key words: Transporters, membrane proteins, pharmacogenomics, systems biology, drugs, drug development, polymorphisms, genotype, phenotype, pathway, nutrigenomics, epigenetics, toxicity.

Q. Yan (ed.), *Membrane Transporters in Drug Discovery and Development*, Methods in Molecular Biology 637,
DOI 10.1007/978-1-60761-700-6_1,

1. Introduction

1.1. Membrane Transporters: Essential for Normal Physiological Functions

Membrane transporters or channels are targets of about 30% of current clinically marketed drugs. Transporters are proteins that span the lipid bilayer and form a transmembrane channel lined with hydrophilic amino acid side chains. Membrane transporters play crucial roles in fundamental cellular functioning and normal physiological processes of archaebacteria, prokaryotes, and eukaryotes (1). These molecules are critical during the formation of electrochemical potentials, uptake of nutrients, removal of wastes, endocytotic internalization of macromolecules, and oxygen transport in respiration (2, 3).

About one-third of all the proteins of a cell are embedded in biological membranes and about one-third of these proteins function to catalyze the transport of molecules across the membrane (4, 5). Membrane transporters have been grouped into different types according to their different functions. Some transporters are called "uniporters," as they mediate the unidirectional translocation of a single substrate. When two substrates are transported in opposite directions in a firmly coupled process, transporters function as "antiporters." There are also "symporters" that are involved in the connected co-transport of two separate substrates in the same direction. Substrates of transporters move across the lipid bilayer through the transmembrane channels and increase the rate of transmembrane passage. Multiple α-helices constitute transmembrane domains (TMD), which form the secondary structure of transporters. During the process of solute translocation across the membrane, transporters undergo conformational changes.

Based on mechanisms and energetics, membrane transporters can be categorized into two broad classes: passive transporters and active transporters. Passive transporters include ion channels, such as the Na^+ channel, and facilitated diffusion such as glucose transporter. Primary active transporters, such as H^+-ATPase and Na^+K^+ATPase, make use of ATP, light, or substrate oxidation as energy resources. Secondary active transporters, such as Na^+/amino acid symporters and H^+/peptide transporters, use ion gradients as their energy source. In addition to transport mode and energy coupling, phylogenetic grouping reveals structure, function, mechanism, and substrate specificity, providing a reliable secondary basis for classification (6). The tertiary basis for classification is substrate specificity and polarity of transport, which are more readily altered during the evolutionary history.

Many primary active transporters contain an ATP-binding cassette (ABC) and belong to the ABC superfamily that comprises proteins with very diverse functions (7). More than 50 human transporters have been identified in this superfamily, including the transporter associated with antigen processing (TAP) and P-glycoprotein multidrug transporter (Pgp). Generally, the ABC superfamily members transport various kinds of substrates, including ions, sugars, amino acids, phospholipids, toxins, and different drugs.

Transporter malfunction can cause disorders in different systems of the human body, which also demonstrates their important roles in normal physiological processes. For example, glucose galactose malabsorption, characterized by severe diarrhea, is associated with defects in the Na^+-dependent glucose transporter (SGLT1) (8). Loss of transporters for Lys, Arg, and Cys from intestinal and renal brush borders can cause cystinuria and kidney stones. Mutations in the transporter protein SLC3A1 (also known as rBAT) have been determined to be the cause of type I cystinuria (9). The genetic disease cystic fibrosis is caused by the dysfunction of cystic fibrosis transmembrane conductance regulator (CFTR) (10).

Genetic polymorphisms can also cause physiological disorders. Polymorphisms are allelic variants in genes that exist stably in the population, typically with an allele frequency above 1%. Polymorphism within the promoter region of the serotonin transporter gene (5-HTT) is considered as a potential genetic risk factor for Alzheimer's disease (AD) (11). Polymorphisms of the dopamine transporter (DAT) and *N*-acetyltransferase 2 (NAT2) have been found to be significantly associated with Parkinson's disease (12–14).

With new technologies such as microarray and bioinformatics, it will be possible to catalog all transporter genes. Features of transporters, including genetic sequences, tissue distribution and functions, as well as influences of polymorphisms, can also be analyzed. Systematic studies of membrane transporters will help in not only better understanding of diseases from the systems biology point of view but also better drug design and development.

1.2. Pharmaceutical Relevance of Transporters

Another key role of membrane transporters is the effect they have on drug therapeutics. Transporters are important in the absorption of oral medications across the gastrointestinal tract. For example, dipeptide transporters are H^+-coupled, energy-dependent transporters. These transporters are crucial in the oral absorption of β-lactam antibiotics, angiotensin-converting enzyme (ACE) inhibitors, renin inhibitors, and an antitumor drug, bestatin (15). Active drug transporter Pgp has been found to be involved in apafant and digoxin absorption (16).

Membrane transporters also influence drug distribution. Nucleosides and their analogs including antivirals and antineoplastics depend on specific transporters to reach their target sites. Transporters for amino acids, monocarboxylic acids, organic cations, hexoses, nucleosides, and peptides are involved in drug transformation across the blood–brain barrier (17). Without these transporters, hydrophilic compounds cannot cross the barriers. Recently, regulating the activity of efflux transporters has been suggested to improve the brain entry of certain substrates (18). In addition to drug entrance, membrane transporters are also crucial for drug exit from the body. For example, diverse secretary and absorptive transporters in the renal tubule enable renal disposition of drugs (19).

The development of the biology of transporters is of particular pharmaceutical relevance (20). Structural modification and specific transporter targeting are considered promising strategies for drug design with increased bioavailability and tissue distribution. For example, an approach has been explored to enhance therapeutic efficacy, by pharmacological modulation of Pgp functions to improve drug bioavailability to the body and drug targets (21). The intestinal peptide transporter can be used to increase the bioavailability of several classes of peptidomimetic drugs, especially ACE inhibitors and β-lactam antibiotics (22). The strategy of using the breast cancer resistance protein (BCRP) and Pgp inhibitor GF120918 has been found to significantly increase the bioavailability of topotecan (23).

The study of membrane transporters may result in breakthroughs in the discovery of new drugs. The antiepileptic drug tiagabine, a γ-aminobutyric acid (GABA) uptake inhibitor, came from the research on amino acid transporters (24). Neurotransmitter transporters are suggested to be the "fruitful targets" for central nervous system (CNS) drug discovery. In addition, multiple drug-resistant (MDR) genes, which are implicated in native and acquired resistance to antineoplastic agents, have drawn intensive interest (25, 26).

Membrane transporters are also important for the development of safer drugs. For example, polymorphisms in the serotonin transporter (5-HTTLPR) gene have been found to cause an increased risk of adverse events during the treatment of mental disorders, such as depression, using selective serotonin-reuptake inhibitors (27). Studies of these factors may help with the development of personalized medicine.

The use of transporters in designing drugs is not limited to humans but can be extended to all kinds of therapeutics. The world's three best-selling veterinary antiparasitic drugs (i.e., parasiticides) act on ligand-gated ion channels (28).

Although the role of membrane transporters in drug effects has attracted much recent interest, the relevant transporters are still unknown for most drugs. A primary goal of current research in drug discovery and development is to fully understand the interactions between transporters and drugs at both the system levels in the human body and the individual level for personalized therapy. The exploration in emerging disciplines including pharmacogenomics, nutrigenomics, and systems biology may help us achieve this goal.

2. Pharmacogenomics, Nutrigenomics, Epigenetics, and Systems Biology of Membrane Transporters

2.1. Pharmacogenomics and Systems Biology

With the impending identification of most human genes, molecular biology is moving from the structural phase toward the functional phase (29). As a fast growing scientific discipline, pharmacogenomics is translating functional genomics into clinical medicine (30). Pharmacogenomics studies the genetic basis of the individual variations in response to drug therapy (31). It involves the analysis of gene expression variations related to drug response. The goal of pharmacogenomics is to achieve optimal therapy for the individual patient, using genetic and genomic principles to facilitate drug discovery and development and to improve drug therapy (32). The development of pharmacogenomics can have great impact on every phase of biomedicine, from clinical laboratory tests to personalized (or individually tailored) medicine (33–36).

The word "pharmacogenomics" has been used interchangeably with "pharmacogenetics." The term "pharmacogenetics" was first introduced by Friedrich Vogel in 1959 (37). At that time, this field was primarily concentrated on genetic polymorphisms in drug-metabolizing enzymes and how the differences affect drug effects (38). Today, people use the term "pharmacogenomics" to represent the entire spectrum of genes that determine drug behavior and sensitivity, although the two words are used with similar meanings in most occasions.

Pharmacogenomics can establish the correlation between specific genotypes and certain phenotypes in the therapeutic context. However, the exploration of this correlation should not be limited to single genes or single nucleotide polymorphisms (SNPs), as these molecules interact with each other in their network structure. Systems biology that provides the understanding at the systems level (39), including complicated pathways and interactions, can help explore what roles biological molecules play in the underlying mechanisms of diseases. In order to connect individuals' genetic profiles with systematic drug responses in the human body, the exploration of both pharmacogenomics and systems biology is needed. The study at the systems level is fundamental for the insight in the key issues in pharmacogenomics of membrane transporters, including associations between genetic structure and function, gene and drug, and genotype and phenotype.

2.2. Key Issues in Pharmacogenomics and Systems Biology of Transporters

The study of pharmacogenomics and systems biology in membrane transporters may contribute significantly to our understanding of interindividual variability in response to numerous therapeutic agents at the systems level. For example, polymorphisms of DAT1 have been found to play a role in response to methylphenidate, which was used to treat attention-deficit hyperactivity disorder in children (40).

Membrane transporters also have significant roles in systematic interactions and pathways. For example, the ABC transporter MRP4 interacts with enzyme SULT2a1 in an integrated pathway that mediates the elimination of sulfated steroid and bile acid metabolites from the liver (41). The export pathway of macrophage migration inhibitory factor (MIF) involves another ABC transporter, ABCA1 (42).

Pgp (also called MDR) functions as an efflux pump that translocates substrates from the inner side of the membrane to the outer side. Sequence variations in a Pgp transporter may have functional importance for drug absorption and elimination, as well as clinical relevance to drug resistance response. A significant correlation has been observed between a polymorphism in exon 26 (C3435T) of MDR-1 and the expression levels and function of MDR-1 (43). Individuals homozygous for this polymorphism have been found to have significantly reduced duodenal MDR-1 expression and increased digoxin plasma levels. This polymorphism has been suggested to influence the absorption and tissue concentrations of other substrates of MDR-1.

Serotonin transporter (5-HTT) is another example of polymorphisms with impact on drug efficacy. Serotonin (5-HT) is a neurotransmitter that plays important roles in many physiological processes. The malfunction of serotonin may cause severe depression. The protein 5-HTT is critical in the termination of serotonin neurotransmission. This transporter is the target for selective serotonin-reuptake inhibitors. A functional polymorphism in the transcriptional control region upstream of the coding sequence of 5-HTT has been reported (44). It has been observed that this polymorphism influences responses to antidepressants such as fluvoxamine and paroxetine (45–48). Polymorphisms in the promoter of serotonin transporter have been found to affect responses to a 5-HT(3) antagonist that relieves symptoms in women with diarrhea-predominant irritable bowel syndrome (D-IBS) (49).

In the following sections some key issues will be discussed, which have triggered great interest and have to be solved before we can achieve the ultimate goal of pharmacogenomics and systems biology.

2.2.1. The Structure–Function Association

The objective of studying transporter genetic structures is to find out how they affect functional consequences, which may be used later in therapeutics. One of the most important issues in pharmacogenomics and systems biology of membrane transporters is to elucidate the relationships between the structural and the functional properties of transporters at both molecular and systems levels. For example, the function of nucleotide-binding domains (NBDs) of CFTR is hydrolyzing ATP to regulate channel gating (50). The CFTR regulatory (R) domain phosphorylation controls the functional channel activity (51). Sequence structural variation may have clinical consequences. For example, studies have shown that allelic variants in the promoter region of the serotonin transporter have an association with the risk for alcohol dependence (52).

The complexity of the structure–function relationships may be clarified through molecular cloning of transporter subtypes. Transporter subtypes can have a similar function but different tissue distribution, regulation, and specificity toward a drug. For example, the multiple drug resistance-associated protein MRP1 has the highest levels in tissues of testes, skeletal muscle, heart, kidney, and lung, but low levels in the liver and intestine (53–55). However, another protein in the same subfamily, MRP2 (also called canalicular multiple organic anion transporter [cMOAT]), is abundant in the liver, kidney, and intestine (54). The structural analysis may also provide perception into gene regulation and evolution, such as the example that vesicular choline transporter is contained entirely in the first intron of the choline acetyltransferase gene (56, 57).

Understanding the correlations between transporter structure and function will enable a better description of transport mechanisms. Through the insight of transport mechanisms, we can better understand how the transporter proteins may be altered in diseases and regulated by therapeutic agents. The identification of structural elements is necessary to explain the direction of translocation and subcellular localization. A more complete understanding of transporter structure requests the elucidation of the secondary and three-dimensional (3D) structure. Available 3D information can be found at Protein Data Bank (PDB) (58), and transporter 3D data are collected at Membrane Transporter Database Portal (59).

Another systematic approach in structure–function studies is to elucidate the role of a transporter in the whole genome and the relationship of a transporter gene to other genes nearby on the chromosome. **Table 1.1** is a sample list of chromosome loca-

Table 1.1
Chromosome locations of some transporters in the ABC superfamily

Chromosome location	Transporter
1p22	ABCA4
1q21–q23	ATP1A2
1q25–q32	PMCA4; ATP2B4
2q24	BSEP
3q13.3–q21	PEPT2
3p26–p25	ATP2B2; PMCA2
3q22–q23	ATP1B3
3q27	MRP5
4q22	ABCP
5q31–q34	DTD
6p21.3	TAP1; TAP2; NPT3; NPT4
7pter–7qter	ZNT3
7q21.1	MDR1; MDR2; ABCB4
7q31	CFTR
9q22–q31	ABC1
10q23–q24	MRP2
12q11–q12	ABCD2; ALDR
12q12	ALD1; hBNaC2
12q21–q23	ATP2B1; PMCA1
13q12.1–q12.3	ATP1AL1
13q14.3	ATP7B
13q32	MRP4
13q33–34	PEPT1
14q24.3	ABCD4; PMP69
16p12	SERCA1
16p13.1	MRP6
16p13.12–13	MRP1
16p13.3	ABCA3
17p	ATP1B2
17q21–q23	MRP3
18q21	FIC1
19q12–q13.2	ATP1A3
19q13.1	CSNU3; SLC7A10; ATPGG
20q11.2	ZNT4
Xq12–q13	ATP7A
Xq13.1–q13.3	ABCB7
Xq28	CRTR; ABCD1; PMCA3g

tions of transporters in the ABC superfamily (59). Such information indicates where the transporters – and maybe the whole family – are in the whole genome, how they are related spatially in the genome, and what the nearby transporters are. This information may also give us some hints about potential interactions between transporter genes, which may help us understand the structure–function relationship at the systems level. **Figure 1.1** is a screen shot showing the information of pathways of transports. This information is retrieved from the Membrane Transporter Database Portal (59).

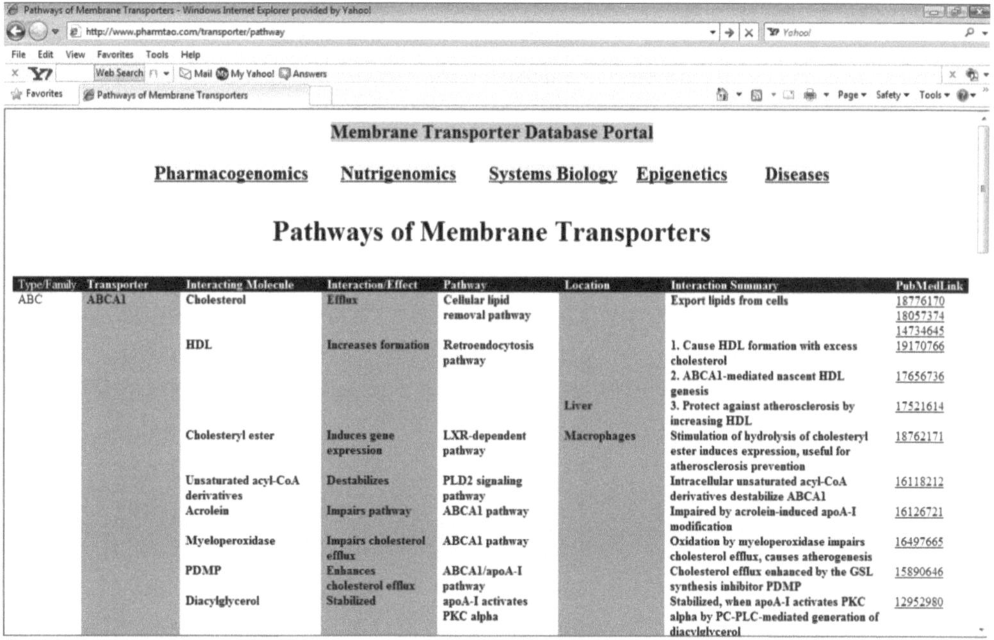
Membrane Transporter Database Portal

Pharmacogenomics Nutrigenomics Systems Biology Epigenetics Diseases

Pathways of Membrane Transporters

Type/Family	Transporter	Interacting Molecule	Interaction/Effect	Pathway	Location	Interaction Summary	PubMedLink
ABC	ABCA1	Cholesterol	Efflux	Cellular lipid removal pathway		Export lipids from cells	18776170 18057374 14734645
		HDL	Increases formation	Retroendocytosis pathway		1. Cause HDL formation with excess cholesterol	19170766
						2. ABCA1-mediated nascent HDL genesis	17656736
					Liver	3. Protect against atherosclerosis by increasing HDL	17521614
		Cholesteryl ester	Induces gene expression	LXR-dependent pathway	Macrophages	Stimulation of hydrolysis of cholesteryl ester induces expression, useful for atherosclerosis prevention	18762171
		Unsaturated acyl-CoA derivatives	Destabilizes	PLD2 signaling pathway		Intracellular unsaturated acyl-CoA derivatives destabilize ABCA1	16118212
		Acrolein	Impairs pathway	ABCA1 pathway		Impaired by acrolein-induced apoA-I modification	16126721
		Myeloperoxidase	Impairs cholesterol efflux	ABCA1 pathway		Oxidation by myeloperoxidase impairs cholesterol efflux, causes atherogenesis	16497665
		PDMP	Enhances cholesterol efflux	ABCA1/apoA-I pathway		Cholesterol efflux enhanced by the GSL synthesis inhibitor PDMP	15890646
		Diacylglycerol	Stabilized	apoA-I activates PKC alpha		Stabilized, when apoA-I activates PKC alpha by PC-PLC-mediated generation of diacylglycerol	12952980

Fig. 1.1. Pathways of membrane transporters (from MEMBRANE Transporter Database Portal).

2.2.2. The Genotype–Phenotype Correlation

The correlation between genotype and phenotype plays a crucial role in the translation of pharmacogenomics into clinical medicine. Genetic polymorphisms can often alter the kinetics of transporters, modulate tissue distribution, and change subsequent drug disposition (60). When we study a transporter gene, one of the first things we need to know is where the transporter is located in tissues. Tissue distribution is one of the most critical phenotypic information about transporters. Such information is especially important in understanding transporter pharmacogenomics at the system level. The information about the abundance of transporters in different tissues can be used in designing new drugs that need to target on certain tissues. To determine which drug transporters are candidates for genotyping, this information may also be helpful. In addition, such information may assist in guiding drug therapy in individual patients on the basis of the drug transporter genotype and phenotype. **Table 1.2** presents some examples of transporters expressed in the liver, kidney, intestines, and brain (59).

Another key part of phenotype is disease. When the data of human genome sequence and genetic variability become available,

Table 1.2
Transporters in tissues of the liver, kidney, intestines, and brain

Liver	Kidney	Intestines	Brain
AE1	AE2	4F2HC	ALDR
ANT2	ASNA1	ACATN	ASCT1
BGT-1	CNT1	ATP2A3	ATP1AL1
BSEP	ENT1	CAT1	CAT4
CAT1	FATP4	CNT1	CNT1
CAT2	GLUT2	CNT2	CNT2
CNT1	GLUT5	CTR-1	DAT1
CNT2	GLUT6	CTR-2	EAAT1
CTR-1	KCC1	EAAT3	EAAT2
CTR-2	KCC3	ENT1	EAAT3
EAAT2	KCC4	GLUT2	ENT1
EAAT5	LAT-2	GLUT5	ENT2
ENT1	LAT-3	GLUT6	GAT-1
FATP4	MCT4	KCC1	GAT-3
G6PT	MCT5	LAT-2	GLCR2
GLCR2	MDR1	MCT7	GLUT1
GLUT2	MRP1	MDR1	GLUT3
GLUT6	MRP3	MRP1	GLUT6
LAT-2	NCX1	MRP3	GLYT1
LST-1	NHE1	NBC	GLYT2
MCT7	NHE2	NCCT	HTT
MDR1	NHE3	NHE1	KCC1
MDR2	NHE6	NHE2	KCC3
MNK	NKCC1	NHE3	KCC4
MRP1	NKCC2	NHE6	LAT-1
MRP2	NPT1	NTCP2	LAT-2
MRP3	NPT2	OCT1	MCT2
NHE1	NTCP1	OCTN2	MCT6
NHE6	NTCP2	ORCTL2	MCT7

(continued)

Table 1.2 (continued)

Liver	Kidney	Intestines	Brain
OAT2	OAT1	PEPT1	MRP1
OATP1	OAT2	PGT	NAT1
OCT1	OAT3	PMCA1	NHE1
OCTN1	OATP1	rBAT	NHE5
OCTN2	OCT1	RFC	NHE6
PEPT1	OCT2	SATT	OAT1
PGT	OCTN1	SBC2	OATP1
PMCA4	OCTN2	SDCT1	PROT
PMP34	ORCTL2	SGLT1	WHITE1
PMP70	ORCTL3	SGLT2	ZNT-1
TAUT	ORCTL4	TSC	ZNT-3
UGT1	TAUT	ZNT-1	ZNT-4

whatever is buried in these data contains the entire range of phenotypic variation. For example, at least five distinct disease phenotypes, including retinitis pigmentosa, cone-rod dystrophy, and Stargardt macular dystrophy, have been related to mutations of a photoreceptor-specific transporter ABCR (61).

Genetic variations may influence the systematic interaction of transporters with other proteins and result in certain disease phenotypes. For example, Tokuhiro et al. (62) investigated the genetic contribution of the cytokine gene cluster in chromosome 5q31 to susceptibility to rheumatoid arthritis. They found that a SNP affects the transcriptional efficiency of a transporter SLC22A4 through changing its affinity for the transcription factor RUNX1. They also discovered that polymorphisms of both genes are associated with susceptibility to rheumatoid arthritis. This example indicates that studies at the systems level are necessary to understand the correlation between genotype–phenotype, because multiple interacting genes are usually involved in a complex phenotypic trait. **Table 1.3** presents examples of genetic variations, including nucleotide variations, positions, and resulting amino acid changes in membrane transporter genes and their correlations with malfunctions or diseases (59).

2.2.3. The Gene–Drug Interaction

Correlations between the sequence variation genotype and disease phenotype might affect drug targets and the correlated drug-response phenotype. The essential gene–drug interaction has been considered "extremely important" in drug development and clinical medicine (78, 79). This interaction composes the

Table 1.3
Transporter variations and associations with diseases: examples of the genotype–phenotype correlation

Transporter	Nucleotide change	Nucleotide position	Amino acid change	Variation-related malfunction/disease	Ref.
ABC1/ ABCA1	A→G; A→G	1730; 2744	Arginine → glutamine; asparagine → serine	Tangier disease and familial high-density lipoprotein deficiency	(63, 64)
ABCR/ ABCA4	G→C; T→C	2588; 1622	gly863 → ala; leu541 → pro	Stargardt disease and age-related macular degeneration	(65, 66)
ABCB4/ MDR3	C→A		Alanine 546 → aspartic acid	Intrahepatic cholestasis of pregnancy	(67)
ABCD1/ ALD	G→A; C→G	1258; 1551	Glutamic acid → lysine; arg389→gly	Adrenoleuko-dystrophy/adrenomyelo-neuropathy	(68, 69)
ATP7A	A→G	2462	Exon skipping and activation of a cryptic splice acceptor site	Occipital horn syndrome	(70)
ATP7B	GAC→AAC; CGG→CTG		asp765 → asn; arg778 → leu	Wilson disease	(71, 72)
SLC4A1/ AE1	CCC→CGC		pro327 → arginine	Spherocytic hemolytic anemia because of Band 3 Tuscaloosa	(73)
SLC22A5	G→A	1196	Arginine →glutamine	Primary carnitine deficiency	(74)
SLC26A4	T→C; G→T	707; 626	leu236 → pro; gly → val	Pendred syndrome; enlarged vestibular aqueduct	(75, 76)
TAP1	A→G; A→G	1069; 1982	Isoleucine-333 → valine; asp-637 → glycine	Involved in antigen processing	(77)

central part of pharmacogenomics. Pharmacogenomics represents studies on the essential gene–drug interactions through genetic mechanisms and the functional pharmacological context, in cooperation with clinical studies. When drugs enter the human body, their fate is affected by uptake, binding, distribution, biotransformation, and excretion.

On a molecular basis, the efficacy of a drug is influenced by the alterations in receptor affinity, transporters, or protein binding. An example of transporter gene–drug interaction is that the functional polymorphism in 5-HTT affects antidepressant responses to fluvoxamine and paroxetine (45–48). **Table 1.4** lists some examples of transporter variations and their impact on certain drug/substrate effects (59).

The gene–drug interactions can be illustrated from several aspects. These may include the responding genes to specific drugs, the expression level of these genes, the sensitivity of a cell to a drug, as well as pharmacological characteristics of the drug action influenced by multiple interacting genes at systems levels. For example, the polymorphism of MDR-1 could significantly increase digoxin plasma levels in patients and affect the absorption and tissue concentrations of other substrates of MDR-1 (43). In a clinical trial of bupropion, the altered dopamine function had significant effect on prolonged smoking abstinence and relapse during the treatment phase (85). The altered dopamine function was caused by the interaction between dopamine transporter SLC6A3 and receptor DRD2 with polymorphisms. This example demon-

Table 1.4
Effect of transporter variations on drugs/substrates: examples of gene–drug interactions

Transporter gene with variations	Drug/substrates	Effects	Ref.
Plasmodium falciparum chloroquine resistance transporter gene (pfcrt)	Chloroquine (CQ)	Resistance	(80, 81)
Serotonin transporter (5-HTT)	Fluvoxamine; paroxetine	Poor response	(45–48)
MDR1 (P-glycoprotein)	Digoxin; fexofenadine	Enhanced efflux of digoxin; enhanced in vivo activity	(43, 82)
Organic anion transporting polypeptide-C (OATP-C) (gene SLC21A6)	Estrone sulfate and estradiol 17beta-d-glucuronide	Reduced uptake	(83)
Dopamine transporter (DAT1)	Methylphenidate	Nonresponse	(84)

strates that genetic polymorphisms can affect systematic interactions between proteins and result in functional and drug-response changes.

2.3. Nutrigenomics and Epigenetics: Interactions Between Transporters and the Environment

As personalized medicine is the aim of pharmacogenomics, personalized nutrition is the goal of nutrigenomics. Nutrigenomics studies the interaction of food and the human genotype that may contribute to disease prevention and the maintenance of good health (86). Pharmacogenomics and nutrigenomics are two closely aligned disciplines that may enable us to move from disease treatment to disease prevention.

Interactions between diets and membrane transporters play important roles in various health conditions such as diabetes, obesity, and the process of weight loss. For example, high-fat diet may inhibit the expression of cholesterol transporter proteins including ATP-binding cassette subfamily A member 1 (ABCA1), ATP-binding cassette subfamily G member 5 (ABCG5), and ABCG8 (87). ABCG5 and ABCG8 are proteins that regulate biliary cholesterol secretion in response to cholate and diosgenin. Genetic variants in the glucose transporter type 2 have been associated with higher intakes of sugars (88). Polymorphisms of the uncoupling protein 1 (UCP-1) gene have effects on body fat accumulation and body weight with different diets (89, 90).

Understanding of these gene–diet interactions is crucial for the elucidation of disease mechanisms and for the discovery and development of more effective drugs. For instance, silencing of the transporter solute carrier family 27 member 5 (SLC27A5, fatty acid transporter) has been found to reverse diet-induced non-alcoholic fatty liver disease and improve hyperglycemia (91). Such proteins may be potential drug targets. **Table 1.5** shows some examples of how the interactions between transporters and diets can affect health and diseases (59).

Many studies have investigated the effects of polymorphisms in transporters on pharmacokinetics (PK) and pharmacodynamics (PD) of clinically important drugs. However, it has recently been reported that epigenetic mechanisms, such as DNA methylation, play crucial roles in the expression of these transporter genes (98). Understanding the mechanisms of epigenetic changes on transporter gene expression may help improve clinical drug efficacy. For example, it has been discovered that doxorubicin-selected cancer cells may overexpress the ABCG2 drug transporter. Such effects may be mediated through epigenetic changes (99). Elucidation of the mechanisms underlying the interactions between transporter genes and the environment, including nutrients and epigenetic factors, will enable treatment of diseases at the system level. The study of systems biology of transporters should embrace nutrigenomics and epigenetics to achieve a thorough understanding of health and diseases.

Table 1.5
Nutrigenomics of transporters: examples of gene–diet interactions

Transporter	Nutrients/diets	Interactions	Ref.
ABCA1	High-fat diet	A high-fat diet inhibits gene expression	(87)
ABCG1	Unsaturated fatty acids	Unsaturated fatty acids can inhibit the stimulatory effects of oxysterols and retinoids on the expression	(92)
ATP1A2	Overfeeding	The polymorphism affects the changes of skeletal muscle metabolic properties after overfeeding	(93)
CFTR	Food intake	It is associated with stature, food intake, and energy homeostasis	(94)
SLC15A1	Diet that induces obesity	With diet-induced obesity, intestinal absorption of dipeptides is reduced	(95)
SLC2A2	Sugars	Genetic variant is associated with higher intakes of sugars	(88)
SLC23A1	Vitamin C	It mediates the uptake of vitamin C	(96)
SLC27A5	Diets that induce non-alcoholic fatty liver disease	Silencing of this protein may reverse diet-induced non-alcoholic fatty liver disease	(91)
SLC5A6	Biotin	It is responsible for biotin uptake and transport	(97)
UCP1	A high-fat diet	Polymorphism has effects on body fat accumulation	(89, 90)

3. Conclusion

In this chapter, we have briefly reviewed the definition, current status, and key issues in pharmacogenomics, nutrigenomics, epigenetics, and systems biology of transporters. Here, the key issues are identified to help us clarify the complex field. However, a comprehensive understanding of these issues will be needed. Some subtopics are also included in these points, such as the protein–protein interaction between transporters and between transporters and other proteins.

The key issues in studying pharmacogenomics and systems biology and transporters cannot be separated but are tightly connected and interlinked, as shown in **Fig. 1.2**. The correlation between genetic structure and observable normal functions can be represented as normal phenotypes. Altered genetic structure may cause malfunctions at the molecular level, which would influence the downstream gene–gene interactions, pathways, and networks at the cellular level. Such changes may then lead to tis-

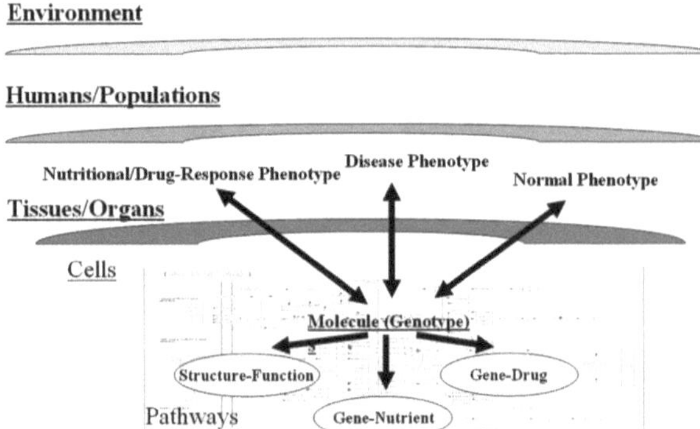

Fig. 1.2. The comprehensive diagram of the key correlations in pharmacogenomics, nutrigenomics, and systems biology.

sue or organ disorders that are disease phenotypes reflected as symptoms of the whole body. In addition, varied genetic structure and altered functions may influence the gene–nutrient and gene–drug interactions, which ultimately affect nutritional or drug-response phenotypes. On the other hand, interactions among genes, foods, drugs, and the environment at higher levels may also affect the structure and function of genes at the molecular level, which would in turn change downstream reactions and phenotypes, forming a feedback loop. The understanding of such interwoven network may be the ultimate key to accurately identifying drug targets and to avoid adverse reactions. In these correlations, genotype–phenotype is the broadest concept that covers the whole biomedical process of drug therapy. However, identification of the other correlations is also important to provide detailed cause–effect relations in this process.

Here, "pharmacogenomics" is not limited to studies in structural genomics, but really requires insights into functional genomics, proteomics (the study of gene expression at the protein level), disease pathogenesis, systems biology, as well as pharmacology and toxicology. Pharmacogenomics is, in fact, a multidisciplinary field with interlinking and overlapping of various knowledge domains (100). We emphasize the integration of pharmacogenomics with systems biology because the approach to individualized drug could be "extremely difficult" if only genomics is considered (101). Diseases and drug responses are very complex phenotypes with various types of genes and multiple pathways involved.

The correlations between allelic variants in transporters and functional consequences will elucidate the impact of genetic variations on therapeutic susceptibility and safety. Studies on functional and systematic interactions between transporters and other proteins, between transporters and drugs, and between transporters and the environment will improve our perception of health, complex diseases, and drug effects. These may ultimately contribute to the development of personalized medicine with high efficacy but less toxicity. The integration of pharmacogenomics, nutrigenomics, epigenetics, and systems biology may enable us to move from disease treatment to disease prevention and optimal health. This is also the overall goal of medical studies.

References

1. Clayton, R.A., White, O., Ketchum, K.A., and Venter, J.C. (1997) The first genome from the third domain of life. *Nature* **29**, 459–462.
2. Lehninger, A.L., (1993) *Principles of Biochemistry*. Worth Publishing, New York, pp. 1–1013.
3. Lodish, H., Baltimore, D., Berk, A., Zipursky, S.L., Matsudaira, P., and Darnell, J. (1995) *Molecular Cell Biology*. Scientific American Books, New York, pp. 1–1417.
4. Paulsen, I.T., Sliwinski, M.K., and Saier, M.H. Jr. (1998) Microbial genome analyses: global comparisons of transport capabilities based on phylogenies, bioenergetics and substrate specificities. *J. Mol. Biol.* **277**, 573–592.
5. Paulsen, I.T., Sliwinski,M.K., Nelissen, B., Goffeau, A., and Saier, M.H. Jr. (1998) Unified inventory of established and putative transporters encoded within the complete genome of Saccharomyces cerevisiae. *FEBS Lett.* **430**, 116–125.
6. Saier, M.H. Jr. (2000) A functional-phylogenetic classification system for transmembrane solute transporters. *Microbiol. Mol. Biol. Rev.* **64**, 354–411.
7. Higgins, C.F. (1992) ABC transporters: from microorganisms to man. *Curr. Opin. Cell Biol.* **8**, 67–113.
8. Martin, M.G., Lostao, M.P., Turk, E., Lam, J., Kreman, M., and Wright, E.M. (1997) Compound missense mutations in the sodium/D-glucose cotransporter result in trafficking defects. *Gastroenterology* **112**, 1206–1212.
9. Palacin, M., Bertran, J., and Zorzano, A. (2000) Heteromeric amino acid transporters explain inherited aminoacidurias. *Curr. Opin. Nephrol. Hypertens.* **9**, 547–553.
10. Sheppard, D.N. and Welsh, M.J. (1999) Structure and function of the CFTR chloride channel. *Physiol. Rev.* **79**, S23–S45.
11. Hu, M., Retz, W., Baader, M., Pesold, B., Adler, G., Henn, F.A., Rosler, M., and Thome, J. (2000) Promoter polymorphism of the 5-HT transporter and Alzheimer's disease. *Neurosci. Lett.* **294**, 63–65.
12. Le Couteur, D.G., Leighton, P.W., McCann, S.J., and Pond, S. (1997) Association of a polymorphism in the dopamine-transporter gene with Parkinson's disease. *Mov. Disord.* **12**, 760–763.
13. Kim, J.W., Kim, D.H., Kim, S.H., and Cha, J.K. (2000) Association of the dopamine transporter gene with Parkinson's disease in Korean patients. *J. Korean Med. Sci.* **15**, 449–451.
14. Tan, E.K., Khajavi, M., Thornby, J.I., Nagamitsu, S., Jankovic, J., and Ashizawa, T. (2000) Variability and validity of polymorphism association studies in Parkinson's disease. *Neurology* **55**, 533–538.
15. Lee, V.H. (2000) Membrane transporters. *Eur. J. Pharm. Sci.* **11**, S41–S50.
16. Leusch, A., Volz, A., Muller, G., Wagner, A., Sauer, A., Greischel, A., and Roth, W. (2002) Altered drug disposition of the platelet activating factor antagonist apafant in mdr1a knockout mice. *Eur. J. Pharm. Sci.* **16**, 119–128.
17. Tamai, I. and Tsuji, A. (2000) Transporter-mediated permeation of drugs across the blood-brain barrier. *J. Pharm. Sci.* **89**, 1371–1388.
18. Sugiyama, Y., Kusuhara, H., and Suzuki, H. (1999) Kinetic and biochemical analysis of carrier-mediated efflux of drugs through the blood-brain and blood-cerebrospinal fluid barriers: importance in the drug delivery to the brain. *J. Control Release.* **62**, 179–186.
19. Inui, K.I., Masuda, S., and Saito, H. (2000) Cellular and molecular aspects of drug transport in the kidney. *Kidney Int.* **58**, 944–958.
20. Sadee, W., Drubbisch, V., and Amidon, G.L. (1995) Biology of membrane transport proteins. *Pharm. Res.* **12**, 1823–1837.
21. Silverman, J.A. (1999) Multidrug-resistance transporters. *Pharm. Biotechnol.* **12**, 353–386.
22. Oh, D.M., Han, H.K., and Amidon, G.L. (1999) Drug transport and targeting. Intestinal transport. *Pharm. Biotechnol.* **12**, 59–88.
23. Kruijtzer, C.M., Beijnen, J.H., Rosing, H., ten Bokkel Huinink, W.W., Schot, M., Jewell, R.C., Paul, E.M., and Schellens, J.H. (2002)

Increased oral bioavailability of topotecan in combination with the breast cancer resistance protein and P-glycoprotein inhibitor GF120918. *J. Clin. Oncol.* **1**, 2943–2950.
24. Iversen, L. (2000) Neurotransmitter transporters: fruitful targets for CNS drug discovery. *Mol. Psychiatry.* **5**, 357–362.
25. Rund, D., Azar, I., and Shperling, O. (1999) A mutation in the promoter of the multidrug resistance gene (MDR1) in human hematological malignancies may contribute to the pathogenesis of resistant disease. *Adv. Exp. Med. Biol.* **457**, 71–75.
26. Gerlach, J.H., Kartner, N., Bell, D.R., and Ling, V. (1986) Multidrug resistance. *Cancer Surv.* **5**, 25–46.
27. Smits, K., Smits, L., et al. (2007) Serotonin transporter polymorphisms and the occurrence of adverse events during treatment with selective serotonin reuptake inhibitors. *Int. Clin. Psychopharmacol.* **22**, 137–143.
28. Raymond, V. and Sattelle, D.B. (2002) Novel animal-health drug targets from ligand-gated chloride channels. *Nat. Rev. Drug Discov.* **1**, 427–436.
29. Kennedy, G.C. (2000) The impact of genomics on therapeutic drug development. *EXS.* **89**, 1–10.
30. Evans, W.E. and Relling, M.V. (1999) Pharmacogenomics: translating functional genomics into rational therapeutics. *Science* **286**, 487–491.
31. Shi, M.M., Bleavins, M.R., and de la Iglesia, F.A. (1999) Technologies for detecting genetic polymorphisms in pharmacogenomics. *Mol. Diagn.* **4**, 343–351.
32 Sadee, W. (2002) Pharmacogenomics: the implementation phase. *AAPS Pharm. Sci.* **4**, E5.
33. Emilien, G., Ponchon, M., Caldas, C., Isacson, O., and Maloteaux, J.M. (2000) Impact of genomics on drug discovery and clinical medicine. *QJM.* **93**, 391–423.
34. Hess, P. and Cooper, D. (1999) Impact of pharmacogenomics on the clinical laboratory. *Mol. Diagn.* **4**, 289–298.
35. Roses, A.D. (2000) Pharmacogenetics and pharmacogenomics in the discovery and development of medicines. *Novartis Found Symp.* **229**, 63–66.
36. March, R. (2000) Pharmacogenomics: the genomics of drug response. *Yeast* **17**, 16–21.
37. Vogel, F. (1959) Moderne probleme der Humangenetik. *Ergeb. Inn. Med. Kinderheilkd.* **12**, 52–125.
38. Nebert, D.W. (1997) Polymorphisms in drug-metabolizing enzymes: what is their clinical relevance and why do they exist? *Am. J. Hum. Genet.* **60**, 265–271.
39. Kitano H. (2002) Systems biology: a brief overview. *Science* **295**, 1662–1664.
40. Winsberg, B.G. and Comings, D.E. (1999) Association of the dopamine transporter gene (DAT1) with poor methylphenidate response. *J. Am. Acad. Child Adolesc. Psychiatry* **38**, 1474–1477.
41. Assem, M., Schuetz, E.G., Leggas, M., Sun, D., et al. (2004) Interactions between hepatic Mrp4 and Sult2A as revealed by the constitutive androstane receptor and Mrp4 knockout mice. *J. Biol. Chem.* **279**, 22250–22257.
42. Flieger, O., Engling, A., Bucala, R., Lue, H., Nickel, W., Bernhagen, J. (2003) Regulated secretion of macrophage migration inhibitory factor is mediated by a non-classical pathway involving an ABC transporter. *FEBS Lett.* **11**, 78–86.
43. Hoffmeyer, S., Burk, O., von Richter, O., Arnold, H.P., Brockmoller, J., Johne, A., Cascorbi, I., Gerloff, T., Roots, I., Eichelbaum, M., Brinkmann, U. (2000) Functional polymorphisms of the human multidrug-resistance gene: multiple sequence variations and correlation of one allele with P-glycoprotein expression and activity in vivo. *Proc. Natl. Acad. Sci. USA* **97**, 3473–3478.
44. Lesch, K.P., Bengel, D., Heils, A., Sabol, S.Z., Greenberg, B.D., Petri, S., Benjamin, J., Muller, C.R., Hamer, D.H., Murphy, D.L. (1996) Association of anxiety-related traits with a polymorphism in the serotonin transporter gene regulatory region. *Science* **274**, 1527–1531.
45. Smeraldi, E., Zanardi, R., Benedetti, F., Di Bella, D., Perez, J., Catalano, M. (1998) Polymorphism within the promoter of the serotonin transporter gene and antidepressant efficacy of fluvoxamine. *Mol. Psychiatry* **3**, 508–511.
46. Kim, D.K., Lim, S.W., Lee, S., Sohn, S.E., Kim, S., Hahn, C.G., Carroll, B.J. (2000) Serotonin transporter gene polymorphism and antidepressant response. *Neuroreport.* **11**, 215–219.
47. Pollock, B.G., Ferrell, R.E., Mulsant, B.H., Mazumdar, S., Miller, M., Sweet, R.A., Davis, S., Kirshner, M.A., Houck, P.R., Stack, J.A., Reynolds, C.F., Kupfer, D.J. (2000) Allelic variation in the serotonin transporter promoter affects onset of paroxetine treatment response in late-life depression. *Neuropsychopharmacology* **23**, 587–590.
48. Zanardi, R., Benedetti, F., Di Bella, D., Catalano, M., Smeraldi, E. (2000) Efficacy of paroxetine in depression is influenced by a functional polymorphism within the pro-

moter of the serotonin transporter gene. *J. Clin. Psychopharmacol.* **20**, 105–107.
49. Camilleri, M., Atanasova, E., Carlson, P.J., Ahmad, U., Kim, H.J., Viramontes, B.E., McKinzie, S., Urrutia, R. (2002) Serotonin-transporter polymorphism pharmacogenetics in diarrhea-predominant irritable bowel syndrome. *Gastroenterology* **123**, 425–432.
50. Anderson, M.P., Berger, H.A., Rich, D.P., Gregory, R.J., Smith, A.E., Welsh, M.J. (1991) Nucleoside triphosphates are required to open the CFTR chloride channel. *Cell* **67**, 775–784.
51. Berger, H.A., Anderson, M.P., Gregory, R.J., Thompson, S., Howard, P.W., Maurer, R.A., Mulligan, R., Smith, A.E., Welsh, M.J. (1991) Identification and regulation of the cystic fibrosis transmembrane conductance regulator-generated chloride channel. *J. Clin. Invest.* **88**, 1422–1431.
52. Lichtermann, D., Hranilovic, D., Trixler, M., Franke, P., Jernej, B., Delmo, C.D., Knapp, M., Schwab, S.G., Maier, W., Wildenauer, D.B. (2000) Support for allelic association of a polymorphic site in the promoter region of the serotonin transporter gene with risk for alcohol dependence. *Am. J. Psychiatry* **157**, 2045–2047.
53. Hipfner, D.R., Deeley, R.G., Cole, S.P. (1999) Structural, mechanistic and clinical aspects of MRP1. *Biochim. Biophys. Acta* **1461**, 359–376.
54. Borst, P., Evers, R., Kool, M., Wijnholds, J. (1999) The multidrug resistance protein family. *Biochim. Biophys. Acta* **1461**, 347–357.
55. Konig, J., Nies, A.T., Cui, Y., Leier, I., Keppler, D. (1999) Conjugate export pumps of the multidrug resistance protein (MRP) family: localization, substrate specificity, and MRP2-mediated drug resistance. *Biochim. Biophys. Acta* **1461**, 377–394.
56. Bejanin, S., Cervini, R., Mallet, J., Berrard, S. (1994) A unique gene organization for two cholinergic markers, choline acetyltransferase and a putative vesicular transporter of acetylcholine. *J. Biol. Chem.* **269**, 21944–21947.
57. Erickson, J.D., Varoqui, H., Schafer, M.K., Modi, W., Diebler, M.F., Weihe, E., Rand, J., Eiden, L.E., Bonner, T.I., Usdin, T.B. (1994) Functional identification of a vesicular acetylcholine transporter and its expression from a "cholinergic" gene locus. *J. Biol. Chem.* **269**, 21929–21932.
58. Protein Data Bank (PDB): http://www.rcsb.org/pdb/ (accessed in May 2009).
59. Membrane Transporter Database Portal: http://www.pharmtao.com/transporter (accessed in May 2009).
60. Sissung, T. M., Gardner, E. R., et al. (2008) Pharmacogenetics of membrane transporters: a review of current approaches. *Methods Mol. Biol.* **448**, 41–62.
61. Lewis, R.A., Shroyer, N.F., Singh, N., Allikmets, R., Hutchinson, A., Li, Y., Lupski, J.R., Leppert, M., Dean, M. (1999) Genotype/Phenotype analysis of a photoreceptor-specific ATP-binding cassette transporter gene, ABCR, in Stargardt disease. *Am. J. Hum. Genet.* **64**, 422–434.
62. Tokuhiro, S., Yamada, R., Chang, X., Suzuki, A., et al. (2003) An intronic SNP in a RUNX1 binding site of SLC22A4, encoding an organic cation transporter, is associated with rheumatoid arthritis. *Nat Genet.* **35**, 341–348.
63. Brooks-Wilson, A., Marcil, M., Clee, S. M., Zhang, L.-H., Roomp, K., van Dam, M., Yu, L., Brewer, C., Collins, J. A., Molhuizen, H. O. F., Loubser, O., Ouelette, B. F. F., and 14 others. (1999) Mutations in ABC1 in Tangier disease and familial high-density lipoprotein deficiency. *Nat. Genet.* **22**, 336–345.
64. Bodzioch, M., Orso, E., Klucken, J., Langmann, T., Bottcher, A., Diederich, W., Drobnik, W., Barlage, S., Buchler, C., Porsch-Ozcurumez, M., Kaminski, W. E., Hahmann, H. W., Oette, K., Rothe, G., Aslanidis, C., Lackner, K. J., Schmitz, G. (1999) The gene encoding ATP-binding cassette transporter 1 is mutated in Tangier disease. *Nat. Genet.* **22**, 347–351.
65. Allikmets, R., Shroyer, N. F., Singh, N., Seddon, J. M., Lewis, R. A., Bernstein, P. S., Peiffer, A., Zabriskie, N. A., Hutchinson, A., Dean, M., Lupski, J. R., Leppert, M. (1997) Mutation of the Stargardt disease gene (ABCR) in age-related macular degeneration. *Science* **277**, 1805–1807.
66. Rivera, A., White, K., Stohr, H., Steiner, K., Hemmrich, N., Grimm, T., Jurklies, B., Lorenz, B., Scholl, H. P. N., Apfelstedt-Sylla, E., Weber, B. H. F. (2000) A comprehensive survey of sequence variation in the ABCA4 (ABCR) gene in Stargardt disease and age-related macular degeneration. *Am. J. Hum. Genet.* **67**, 800–813.
67. Dixon, P. H., Weerasekera, N., Linton, K. J., Donaldson, O., Chambers, J., Egginton, E., Weaver, J., Nelson-Piercy, C., de Swiet, M., Warnes, G., Elias, E., Higgins, C. F., Johnston, D. G., McCarthy, M. I., Williamson, C. (2000) Heterozygous MDR3 missense mutation associated with intrahepatic cholestasis of pregnancy: evidence for a defect in protein trafficking. *Hum. Mol. Genet.* **9**, 1209–1217.

68. Cartier, N., Sarde, C.-O., Douar, A.-M., Mosser, J., Mandel, J.-L., Aubourg, P. (1993) Abnormal messenger RNA expression and a missense mutation in patients with X-linked adrenoleukodystrophy. *Hum. Mol. Genet.* **2**, 1949–1951.
69. Krasemann, E. W., Meier, V., Korenke, G. C., Hunneman, D. H., Hanefeld, F. (1996) Identification of mutations in the ALD-gene of 20 families with adrenoleukodystrophy/adrenomyeloneuropathy. *Hum. Genet.* **97**, 194–197.
70. Kaler, S. G., Gallo, L. K., Proud, V. K., Percy, A. K., Mark, Y., Segal, N. A., Goldstein, D. S., Holmes, C. S., Gahl, W. A. (1994) Occipital horn syndrome and a mild Menkes phenotype associated with splice site mutations at the MNK locus. *Nat. Genet.* **8**, 195–202.
71. Figus, A., Angius, A., Loudianos, G., Bertini, C., Dessi, V., Loi, A., Deiana, M., Lovicu, M., Olla, N., Sole, G., De Virgiliis, S., Lilliu, F., and 21 others. (1995) Molecular pathology and haplotype analysis of Wilson disease in Mediterranean populations. *Am. J. Hum. Genet.* **57**, 1318–1324.
72. Kim, E. K., Yoo, O. J., Song, K. Y., Yoo, H. W., Choi, S. Y., Cho, S. W., Hahn, S. H. (1998) Identification of three novel mutations and a high frequency of the arg778-to-leu mutation in Korean patients with Wilson disease. *Hum. Mutat.* **11**, 275–278.
73. Jarolim, P., Palek, J., Rubin, H. L., Prchal, J. T., Korsgren, C., Cohen, C. M. (1991) Band 3 Tuscaloosa: pro327-to-arg327 substitution in the cytoplasmic domain of erythrocyte band 3 protein associated with spherocytic hemolytic anemia and partial deficiency of protein 4.2. (Abstract) *Blood* **78** (suppl.): 252a.
74. Wang, Y., Korman, S. H., Ye, J., Gargus, J. J., Gutman, A., Taroni, F., Garavaglia, B., Longo, N. (2001) Phenotype and genotype variation in primary carnitine deficiency. *Genet. Med.* **3**, 387–392.
75. Van Hauwe, P., Everett, L. A., Coucke, P., Scott, D. A., Kraft, M. L., Ris-Stalpers, C., Bolder, C., Otten, B., de Vijlder, J. J. M., Dietrich, N. L., Ramesh, A., Srisailapathy, S. C. R., Parving, A., Cremers, C. W. R. J., Willems, P. J., Smith, R. J. H., Green, E. D., Van Camp, G. (1998) Two frequent missense mutations in Pendred syndrome. *Hum. Mol. Genet.* 7, 1099–1104.
76. Usami, S., Abe, S., Weston, M. D., Shinkawa, H., Van Camp, G., Kimberling, W. J. (1999) Non-syndromic hearing loss associated with enlarged vestibular aqueduct is caused by PDS mutations. *Hum. Genet.* **104**, 188–192.
77. Colonna, M., Bresnahan, M., Bahram, S., Strominger, J. L., Spies, T. (1992) Allelic variants of the human putative peptide transporter involved in antigen processing. *Proc. Nat. Acad. Sci. USA* **89**, 3932–3936.
78. Nebert, D.W. (1999) Pharmacogenetics and pharmacogenomics: why is this relevant to the clinical geneticist? *Clin. Genet.* **56**, 247–258.
79. Dirckx, C., Donati, M.B., Iacoviello, L. (2000) Pharmacogenetics: a molecular sophistication or a new clinical tool for cardiologists? *Ital. Heart J.* **1**, 662–666.
80. Mockenhaupt, F.P., Eggelte, T.A., Till, H., Bienzle, U. (2001) Plasmodium falciparum pfcrt and pfmdr1 polymorphisms are associated with the pfdhfr N108 pyrimethamine-resistance mutation in isolates from Ghana. *Trop. Med. Int. Health* **6**, 749–755.
81. Basco, L.K., Ringwald, P. (2001) Analysis of the key pfcrt point mutation and in vitro and in vivo response to chloroquine in Yaounde, Cameroon. *J. Infect. Dis.* **183**, 1828–1831.
82. Kim, R.B., Leake, B.F., Choo, E.F., Dresser, G.K., Kubba, S.V., Schwarz, U.I., Taylor, A., Xie, H.G., McKinsey, J., Zhou, S., Lan, L.B., Schuetz, J.D., Schuetz, E.G., Wilkinson, G.R. (2001) Identification of functionally variant MDR1 alleles among European Americans and African Americans. *Clin. Pharmacol. Ther.* **70**, 189–199.
83. Tirona, R.G., Leake, B.F., Merino, G., Kim, R.B. (2001) Polymorphisms in OATP-C: identification of multiple allelic variants associated with altered transport activity among European- and African-Americans. *J. Biol. Chem.* **276**, 35669–35675.
84. Winsberg, B.G., Comings, D.E. (1999) Association of the dopamine transporter gene (DAT1) with poor methylphenidate response. *J. Am. Acad. Child Adolesc. Psychiatry* **38**, 1474–1477.
85. Lerman, C., Shields, P.G., Wileyto, E.P., Audrain, J., Hawk, L.H. Jr., Pinto, A., Kucharski, S., Krishnan, S., Niaura, R., Epstein, L.H. (2003) Effects of dopamine transporter and receptor polymorphisms on smoking cessation in a bupropion clinical trial. *Health Psychol.* **22**, 541–548.
86. Ghosh, D., Skinner, M. A., et al. (2007) Pharmacogenomics and nutrigenomics: synergies and differences. *Eur. J. Clin. Nutr.* **61**, 567–574.
87. de Vogel-van den Bosch, H. M., de Wit, N. J., et al. (2008) A cholesterol-free, high-fat diet suppresses gene expression of cholesterol transporters in murine small intestine. *Am. J. Physiol. Gastrointest. Liver Physiol.* **294**, G1171–1180.

88. Eny, K. M., Wolever, T. M., et al. (2008) Genetic variant in the glucose transporter type 2 is associated with higher intakes of sugars in two distinct populations. *Physiol. Genomics* **33**, 355–360.
89. Kim, K. S., Cho, D. Y., et al. (2005) The finding of new genetic polymorphism of UCP-1 A-1766G and its effects on body fat accumulation. *Biochim. Biophys. Acta.* **1741**, 149–155.
90. Shin, H. D., Kim, K. S., et al. (2005) The effects of UCP-1 polymorphisms on obesity phenotypes among Korean female subjects. *Biochem Biophys. Res. Commun.* **335**, 624–630.
91. Doege, H., Grimm, D., et al. (2008) Silencing of hepatic fatty acid transporter protein 5 in vivo reverses diet-induced non-alcoholic fatty liver disease and improves hyperglycemia. *J. Biol. Chem.* **283**, 22186–22192.
92. Uehara, Y., Miura, S., et al. (2007) Unsaturated fatty acids suppress the expression of the ATP-binding cassette transporter G1 (ABCG1) and ABCA1 genes via an LXR/RXR responsive element. *Atherosclerosis* **191**, 11–21.
93. Ukkola, O., Joanisse, D. R., et al. (2003) Na+-K+-ATPase alpha 2-gene and skeletal muscle characteristics in response to long-term overfeeding. *J. Appl. Physiol.* **94**, 1870–1874.
94. Mekus, F., Laabs, U., et al. (2003) Genes in the vicinity of CFTR modulate the cystic fibrosis phenotype in highly concordant or discordant F508del homozygous sib pairs. *Hum. Genet.* **112**, 1–11.
95. Hindlet, P., Bado, A., et al. (2009) Reduced intestinal absorption of dipeptides via PepT1 in mice with diet-induced obesity is associated with leptin receptor down-regulation. *J. Biol. Chem.* **284**, 6801–6808.
96. Varma, S., Campbell, C. E., et al. (2008) Functional role of conserved transmembrane segment 1 residues in human sodium-dependent vitamin C transporters. *Biochemistry* **47**, 2952–2960.
97. Luo, S., Kansara, V. S., et al. (2006) Functional characterization of sodium-dependent multivitamin transporter in MDCK-MDR1 cells and its utilization as a target for drug delivery. *Mol. Pharm.* **3**, 329–339.
98. Hirota, T., Takane, H., et al. (2008) Epigenetic regulation of genes encoding drug-metabolizing enzymes and transporters; DNA methylation and other mechanisms. *Curr. Drug Metab.* **9**, 34–38.
99. Calcagno, A. M., Fostel, J. M., et al. (2008) Single-step doxorubicin-selected cancer cells overexpress the ABCG2 drug transporter through epigenetic changes. *Br. J. Cancer* **98**, 1515–1524.
100. Sadee, W. (1998) Genomics and drugs: finding the optimal drug for the right patient. *Pharm. Res.* **15**, 959–963.
101. Nebert, D.W., Jorge-Nebert, L., Vesell, E.S. (2003) Pharmacogenomics and "individualized drug therapy": high expectations and disappointing achievements. *Am. J. Pharmacogenomics* **3**, 361–370.

Chapter 2

Bioinformatics for Transporter Pharmacogenomics and Systems Biology: Data Integration and Modeling with UML

Qing Yan

Abstract

Bioinformatics is the rational study at an abstract level that can influence the way we understand biomedical facts and the way we apply the biomedical knowledge. Bioinformatics is facing challenges in helping with finding the relationships between genetic structures and functions, analyzing genotype–phenotype associations, and understanding gene–environment interactions at the systems level. One of the most important issues in bioinformatics is data integration. The data integration methods introduced here can be used to organize and integrate both public and in-house data. With the volume of data and the high complexity, computational decision support is essential for integrative transporter studies in pharmacogenomics, nutrigenomics, epigenetics, and systems biology. For the development of such a decision support system, object-oriented (OO) models can be constructed using the Unified Modeling Language (UML). A methodology is developed to build biomedical models at different system levels and construct corresponding UML diagrams, including use case diagrams, class diagrams, and sequence diagrams. By OO modeling using UML, the problems of transporter pharmacogenomics and systems biology can be approached from different angles with a more complete view, which may greatly enhance the efforts in effective drug discovery and development. Bioinformatics resources of membrane transporters and general bioinformatics databases and tools that are frequently used in transporter studies are also collected here. An informatics decision support system based on the models presented here is available at http://www.pharmtao.com/transporter. The methodology developed here can also be used for other biomedical fields.

Key words: Bioinformatics, pharmacogenomics, systems biology, data modeling, data integration, object oriented, Unified Modeling Language, computational, decision support, transporters, drug development, databases.

Q. Yan (ed.), *Membrane Transporters in Drug Discovery and Development*, Methods in Molecular Biology 637,
DOI 10.1007/978-1-60761-700-6_2, © Springer Science+Business Media, LLC 2010

1. Bioinformatics and Membrane Transporter Studies

With a history of less than 40 years, bioinformatics is a rapidly growing area that applies computational approaches to solve biological problems. Similar to the composition of the word itself, "bioinformatics" is an independent field developed from the union of computer science and molecular biology. Rather than a simple combination of computer science and biology, bioinformatics should be an "organic" integration of the two. This merge was started in 1970s, when it was found that RNA secondary structure might be predicted with computational techniques (1, 2). At that time, people began to build databases of nucleic acids (3) and proteins (4). Algorithms and programs were developed to translate DNA sequences into protein sequences (5, 6) and to detect patterns including restriction enzyme recognition sites (7, 8).

Various bioinformatics approaches have been used in transporter studies. For example, the sequence similarity searching tool BLAST (Basic Local Alignment Search Tool) has been used extensively in the analysis of transport systems in different organisms (9) and in the identification of transporter genes (10). The database PROSITE has been used for functional analysis in transporter genes (11, 12).

Table 2.1 lists some bioinformatics resources designed specifically for membrane transporter studies (Websites accessed in May 2009). **Table 2.2** summarizes some general bioinformatics databases and tools that are frequently used in

Table 2.1
Data sources for membrane transporter and ion channel studies

Category	Databases and tools	Links
Transporter Portal	Human Membrane Transporter Database Portal	http://www.pharmtao.com/transporter
Transporter Classification	Transport Classification Database (TCDB)	http://tcdb.ucsd.edu/index.php
Genomic Comparisons	TransportDB	http://www.membranetransport.org/
Membrane Proteins in Different Species	Human membrane protein library (HMPL)	http://wardlab.cbs.umn.edu/human/
	Functional Genomics of Plant Transporters (PlantsT)	http://plantst.genomics.purdue.edu/

(continued)

Table 2.1 (continued)

Category	Databases and tools	Links
	Aramemnon : Plant membrane protein database	http://aramemnon.botanik.uni-koeln.de/
	Arabidopsis Membrane Protein Library	http://wardlab.cbs.umn.edu/arabidopsis/
	Rice Membrane Protein Library (RMPL)	http://wardlab.cbs.umn.edu/rice/
	Yeast membrane protein library (YMPL)	http://wardlab.cbs.umn.edu/yeast/
	Schizosaccharomyces pombe membrane protein library (SpMPL)	http://wardlab.cbs.umn.edu/pombe/
	Drosophila membrane protein library (DMPL)	http://wardlab.cbs.umn.edu/fly/
	C. elegans membrane protein library (CeMPL)	http://wardlab.cbs.umn.edu/worm/
Ion Channels	Ligand-Gated Ion Channel database	http://www.ebi.ac.uk/compneur-srv/LGICdb/LGICdb.php
	Ion Channel Diseases	http://neuromuscular.wustl.edu/mother/chan.html
	Voltage-gated potassium channel database (VKCDB)	http://vkcdb.biology.ualberta.ca/
	ChannelDB	http://www.modelersworkspace.org/channeldb/ChannelDB.html
ABC Transporters	ABCISSE database	http://www1.pasteur.fr/recherche/unites/pmtg/abc/database.iphtml
	Human ATP-Binding Cassette Transporters	http://nutrigene.4t.com/humanabc.htm
	ABC Transporter Genes Database	http://www.humanabc.bio.titech.ac.jp/
	P-type ATPases database	http://biobase.dk/~axe/Patbase.html
	Arabidopsis ABC superfamily	http://www.arabidopsis.org/info/genefamily/ABC_proteins.html
	Archaeal and Bacterial ABC transporter database	http://www-abcdb.biotoul.fr/
Specific Diseases	Wilson Disease Mutation Database	http://www.wilsondisease.med.ualberta.ca/database.asp

Table 2.2
General bioinformatics databases and tools for membrane transporter studies

Category	Database/tool example	URL
Nucleotide and Protein Portal	Entrez	http://www.ncbi.nlm.nih.gov/Entrez/
	European Bioinformatics Institute (EBI)	http://www.ebi.ac.uk/
Homology search	BLAST	http://www.ncbi.nlm.nih.gov/BLAST/
Multiple alignment	Clustal W	http://www.ebi.ac.uk/clustalw/
DNA	dbEST	http://www.ncbi.nlm.nih.gov/dbEST/index.html
Gene-oriented cluster	UniGene	http://www.ncbi.nlm.nih.gov/UniGene/index.html
Sequence variation	dbSNP	http://www.ncbi.nlm.nih.gov/SNP/index.html
	International HapMap Project	http://www.hapmap.org/
Human Gene Mutation Database	HGMD	http://www.uwcm.ac.uk/uwcm/mg/hgmd0.html
Motif analysis	Pfam	http://pfam.sanger.ac.uk/
	ProfileScan	http://hits.isb-sib.ch/cgi-bin/PFSCAN
Exon finding and gene annotation	GenScan	http://genes.mit.edu/GENSCAN.html
Secondary structure prediction	PredicProtein	http://cubic.bioc.columbia.edu/predictprotein/submit_def.html
Transmembrane region detection	TMPred	http://www.ch.embnet.org/software/TMPRED_form.html
Structure (3D) database	PDB	http://www.rcsb.org/pdb/
3D structure prediction	Geno3D	http://geno3d-pbil.ibcp.fr/
Pathway and cellular regulation	KEGG	http://www.genome.ad.jp/kegg/kegg2.html
	Reactome	http://www.reactome.org/
	Human Protein Reference Database (HPRD)	http://www.hprd.org/
	Pathguide	http://www.pathguide.org/
Disorders	OMIM	http://www.ncbi.nlm.nih.gov/omim/
	Genes and Disease Map	http://www.ncbi.nlm.nih.gov/disease/Transporters.html
Literature	PubMed	http://www.ncbi.nlm.nih.gov/PubMed/

transporter studies (Websites accessed in May 2009). These bioinformatics databases and tools are linked and integrated in a comprehensive database portal for transporters at http://www.pharmtao.com/transporter (accessed in May 2009).

As we enter the transition era from structural to functional genomics and proteomics, especially with the overwhelming variety and volume of data, bioinformatics becomes increasingly important and indispensable for other biomedical sciences. Different from most traditional biomedical sciences that are grounded in the observation of the physical world, bioinformatics is the rational study at an abstract level that can influence the way we understand biomedical facts and the way we apply the biomedical knowledge. At this stage, bioinformatics is facing challenges in helping with finding the relationships between genetic structures and functions, analyzing genotype–phenotype associations, and understanding gene–environment interactions at the systems level.

One of the most important issues in bioinformatics is data integration. This includes the integration of data from heterogeneous resources, from various data types, and enterprise-wide data integration among different groups and departments. There can be valuable knowledge buried in various unorganized data, and the process of data integration can help "unveil" the hidden knowledge. This chapter briefly introduces methodologies on how to extract useful information from various data so that the information can be applied directly in research and development projects.

With the volume of data and the high complexity, computational decision support is essential for integrative transporter studies in pharmacogenomics, nutrigenomics, epigenetics, and systems biology. For the development of such a decision support system, object-oriented (OO) models can be constructed using the Unified Modeling Language (UML). The modeling methods for decision support in transporter studies of these emerging fields will be described in detail, such as how to construct UML diagrams based on biomedical models at different system levels. The methodology developed here can also be used for other biomedical fields.

2. Data Integration Methods in Membrane Transporter Studies

Data integration is not only just for simple data access but also for knowledge discovery and decision support. The two words "data" and "information" are often used interchangeably. In fact, they are quite different. The term "data" implies a collection of discrete elements, such as a file. Data are rarely clean and may have different formats. Some data are from multiple competing sources. Some data have missing and incomplete fields. In addition, data formats and contents may change over time. When data are cleaned, structured, merged, aggregated, derived, sorted, and displayed, they become "information."

Usually a database system provides an area to collect, integrate, and store data to perform the actions to enrich and enhance the value of the data. A database system offers a platform to transform data into information and is often useful for decision support purposes. The data integration methods introduced here can be used when researchers try to organize and integrate public and in-house data. This kind of work has become crucial in the routine lab work, in order to organize and even publish one's research results.

Figure. 2.1 shows the data transformation and integration process. This is also a process that standardizes names and values, resolves inconsistencies in representation of data, and integrates common values together. The "equal" values of data from disparate sources that represent the same biomedical facts are also resolved in this process. This transformation process is repeated over and over again, during the original development of the target database, when adding new sources to the existing database, and when distributing data from the system to users.

Here we focus on introducing the data consolidation approach in data integration, although other approaches can be used, such as the method of federation (13). The consolidation approach is based on constructing a database with a single large data model, e.g., when we need to extract data from various data sources and centralize these extracted data at one place. The major

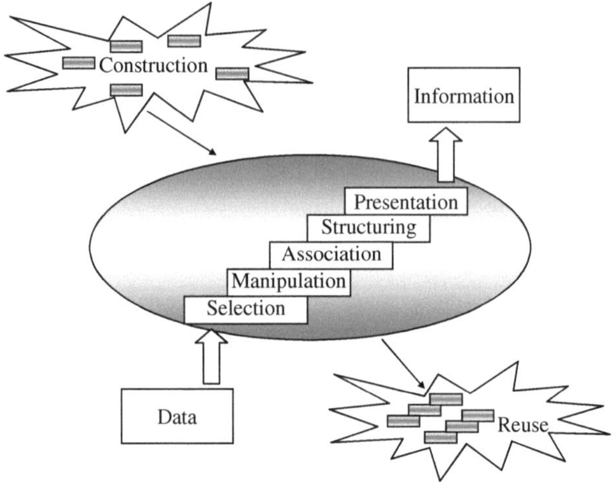

Fig. 2.1. Data integration: the process from data to information.

benefits of this approach include enforcement of the standardization of heterogeneous data.

The data integration and transformation process begins with the selection of data sources, as shown in **Fig. 2.1**. Data are selected through the screening of all available sources and choosing the ones that can best fulfill the requirements. For transporter studies, data sources can be from the tools and databases listed in **Tables 2.1** and **2.2**. The selected data are then manipulated and transformed. Data consolidation is a procedure that analyzes and merges data from disparate sources or systems into a single, integrated data structure. This is achieved through identifying data that is common across the various source files, and investigating the rules that manage the usage of the data. For example, polymorphism data about transporter genes can be retrieved from several sources, such as databases dbSNP and OMIM (Online Mendelian Inheritance in Man). It is necessary to integrate data from these different sources into a common data structure, which includes variation types such as deletion or point mutation. Such data structure can be constructed based on the data models developed during the requirement analysis phase, which will be introduced in detail in the next section (*see* **Section 3.2**).

The process of data consolidation also includes identifying data elements that have common biomedical meaning even though the names are different (synonyms) or those that have the same name but represent different biomedical facts (homonyms). Failing to properly identify these in the source files may result in disparate data that fail to provide the true integration points. For example, failure to identify the synonyms of one transporter gene may cause these different names to be regarded as different genes, which may lead to unclean data and serious data redundancy, even repeated experiments. The classification of transporter genes can provide a comprehensive view of all transporters and help elucidate the directions of transporter research (14). Transporter gene classification is also one of the most important parts involved in data consolidation.

During the data manipulation process, data are also cleaned. In this step, redundant data are removed, and outdated data are updated. Data cleaning can be done together with data consolidation and conversion. For example, sometimes the synonyms of a transporter gene are recorded as different records with redundant gene sequences. These redundancies can be removed with the identification of the synonyms. When the data are manipulated through consolidation, conversion, and cleaning, the result can be presented with a structuring of the information. Such data presentation may include tables and graphs, which can also be used in the decision support process.

3. Data Modeling for Informatics Support of Transporter Pharmacogenomics and Systems Biology

3.1. Informatics Support in Pharmacogenomics and Systems Biology

Pharmacogenomics is multi-disciplinary involving molecular biology and human genetics, genomics, bioinformatics, physiology, pharmacology, and internal medicine (13, 15). As discussed in **Chapter 1**, the emerging fields of nutrigenomics, epigenetics, and systems biology add more dimensions to this complexity. It is even difficult for experts from these different domains to communicate with each other. These multi-level characteristics and domain knowledge barriers bring great challenges to decision making in clinics and labs. Comprehensive and integrative informatics methodologies are needed to break these barriers and to improve the information flow for better communication.

For example, genetic variations have been suggested to be useful for decision making about drug treatment in clinics (16). To achieve this goal effectively, information of genetic variations needs to be processed and provided by a computational system to support the clinical decision making. In addition, the application of high-throughput technologies and the analysis of patient genetic profiles require powerful informatics support. With the amount of available data rapidly increasing, the need for strong information technology support becomes increasingly urgent.

A bioinformatics decision support system (DSS) is a system that provides information to assist biomedical experts in making decisions and doing their job more effectively in both laboratory research and clinical practice. Such systems can help record, store, analyze, and mine the data. Here a decision is an irreversible choice among alternative ways to allocate valuable resources.

To build such informatics systems, intercommunication and interoperation between different biomedical databases are becoming critical issues. To solve these problems, bioinformatics is demanding a common literacy with mutual intelligibility that can be widely accepted (17).

This goal has been difficult to achieve and has been considered to be the first obstacle in biological knowledge modeling and encoding (18). These problems are central to providing informatics support for transporter pharmacogenomics and systems biology studies because this is a complex area requiring heterogeneous data sources. In addition, the special features of pharmacogenomics, such as genetic polymorphisms and genotype–phenotype correlations, are difficult to process using the traditional passive and static flat files or even relational models.

To solve these problems and construct useful informatics support systems, advanced and integrative data models, such as using the object-oriented (OO) methodology, need to be built. A data model covers the scope of the system development including relationships, attributes, and definitions. Data modeling provides a formal methodology for documenting users' data needs. The models we use for building pharmacogenomics and systems biology computational systems will have a profound influence upon how a scientific or clinical problem is attacked, how a solution is shaped, and how a result is interpreted.

3.2. Unified Modeling Language (UML)

To construct an accurate and usable model for transporter pharmacogenomics and systems biology, it is necessary to capture the important concepts and relationships of the domain knowledge, and convert such understanding into physical data structures for decision support systems (*see* **Section 2**). Building a biomedical model from the domain requirement analysis is necessary for approaching the complexity. This model is scientific in nature and consists of accurate description and illustration of the fundamental factors and processes of our understanding of this particular area of science.

From this scientific model, important and repetitively occurring concepts and factors can be abstracted and identified. These concepts and factors can be viewed as objects in the sense of object-oriented (OO) methodology. In real-world terms, an object can be defined as a concept, abstraction, or a thing with crisp boundaries and meanings for the problem at hand (19). In software terms, an object is an intelligent piece of a program that can encapsulate code and data. These objects, together with the interrelationships among them, should be able to represent the outputs as well as the inner workings of the biomedical model. These objects are also our building blocks for constructing sophisticated information systems.

Based on the biomedical model and abstracted concepts, OO models can be constructed using the Unified Modeling Language (UML). Object-oriented methods offer a unifying paradigm for the three traditional phases of software development: analysis, design, and implementation (20). This unification leads to a smooth transition from one phase to the next. UML is an object-oriented design language for specifying, visualizing, constructing, and documenting the objects of a system (21). It is a standard modeling language that has been widely accepted in computer science and used extensively in the business world. The application of UML can help the transformation from the logical model to the physical mode smoothly.

In recent years, UML has been adopted by more and more biomedical systems, especially in the medical imaging field (22). It was used in the implementation of brain computer interface (BCI) systems (23). In the Biomedical Research Integrated Domain Group (BRIDG) project, declarative and procedural knowledge were represented with the UML class, activity, and state diagrams (24).

UML has been suggested as a useful language for cell and biochemistry modeling (25). The UML approach was adopted in systematic modeling, capturing, and disseminating proteomics experimental data (26). It has been used in the integration of microarray gene expression, proteomics, and metabolomics data in the Chemical Effects in Biological Systems (CEBS) through building the Systems Biology Object Model (SysBio-OM) (27).

To overcome the barriers between different knowledge domains and capture the essence of different disciplines for a coherent decision support system in pharmacogenomics and systems biology, a methodology for model construction can be used (13). This approach is from domain requirement analysis and biomedical models ⇒ concept abstraction and object design ⇒ OO UML models. The application of UML in this approach helps decompose the complexity and make the system comprehensible for data analysis in pharmacogenomics and systems biology. A DSS built based on this approach can represent the most important correlations and key issues in transporter studies discussed in **Chapter 1**. Information from the requirement analysis and the biomedical model will feed the data modeling directly.

To build the DSS for transporter pharmacogenomics and systems biology studies, the first phase is system design, which starts from understanding the biomedical needs and data requirements of the system users. Analyzing and designing are critical for the success of the system construction. The analysis process identifies what the problem is and what a system needs to do. The design process provides a logical solution so that the system satisfies the requirements. Once the requirements are specified, UML diagrams can be used to analyze, design, and develop applications. The following sections focus on this design phase.

The phase after the design is system implementation (13). A physical database system can be developed according to the data model designed in the previous phase. During this phase, data analysis and integration are performed to determine the best and cleanest source of data (*see* **Section 2**). The step after the implementation is system application. In this phase, data access tools are used to build reports and support data mining.

3.2.1. Biomedical Models at the Molecular and Systems Level

The biomedical system we are dealing with is overwhelmingly complex. It is necessary to decompose it into understandable chunks to comprehend and manage the complexity. To do this, models can be constructed to describe, abstract, and represent essential aspects of the system. As mentioned earlier, the first step for model construction in pharmacogenomics and systems biology is domain requirement analysis and building biomedical models. The systematic overview of the knowledge domain based on biomedical models is necessary to clearly identify the target domain first. A biomedical model describes the natural scenarios and reflects the understanding and thinking process of domain users for whom the software system is built. It helps capture the most important issues that the users are concerned about. Biomedical modeling before data modeling can help lay the ground for further concept identification.

Biomedical models represent the problems in the biomedical system in an intuitive way using the language of the field. The requirement of a biomedical model includes that it should show the biological objects, associations, and processes as they are understood by biomedical experts. These models then need to be translated into data models that are the foundations of further software development. Based on these biomedical models, use cases can be described and concepts can be abstracted (*see* **Section 3.2.2**). The step of concept abstraction can lead to the data modeling phase with the creation of series of diagrams.

Figure 2.2 shows a biomedical model describing the structure–function, gene–drug, and genotype–phenotype correlations of transporters at the molecular level. Two typical transporters are used as examples in the model to illustrate different aspects. G1 is used to represent common transporter families such as those in the ABC superfamily, for example, multidrug-resistance protein (MRP). MRPs are organic anion transporters that transport anionic drugs such as methotrexate, and neural drugs conjugated to acidic ligands such as sulfate (28, 29). Compounds can be transported by MRPs in complexes with glutathione (GSH). G2 is used to represent other types such as the families of ion channels. For example, intermediate conductance Ca^{2+}-activated K^{+} channel (IKCa1) modulates calcium influx by regulating the membrane potential and the driving force for calcium entry. These two kinds of transporters are also used to represent the correlations at two levels, i.e., protein and nucleotide levels.

Here the modeling part of G1 is used to focus on the description of protein structure and functions. The topologies of these genes include transmembrane domains (TMDs). More detailed topology of transporter proteins is also described, such as the nucleotide-binding domain (NBD), and a "signature" motif that defines the NBDs of ABC transporters (30).

Elevated levels of G1 can confer resistance to drugs. For example, overexpression of MRP2 was found to result in resistance to cisplatin, etoposide, doxorubicin, and epirubicin (31).

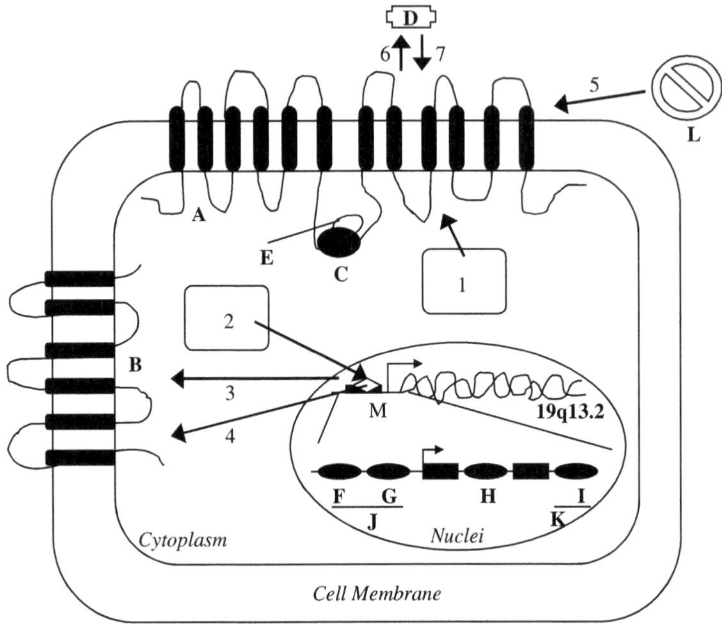

Fig. 2.2. The biomedical model of correlations of transporters at the molecular level. *A*: G1; *B*: G2; *C*: nucleotide-binding domain (NBD); *D*: Drug; *E*: "Signal" motif; *F*: Enhancer; *G*: Promoter; *H*: Intron; I: Silencer; *J*: 5'-region; *K*: 3'-region; *L*: Inhibitors; *M*: Regulatory unit. *1*: Biochemical pathway; *2*: PKC signaling pathway; *3*: increase expression; *4*: mutation activates expression; *5*: inhibit transporter; *6*: drug efflux; *7*: drug influx.

Because of the potential involvement of these drug pumps in the clinical phenotypes such as drug resistance, inhibitors are also important in describing transporters' functions. For example, high-affinity substrates can be potent competitive inhibitors, such as leukotriene C4 and S-decylglutathione for MRP1 (32).

In the modeling part of G2, the structure–function correlation at the nucleotide level is emphasized to represent the common mechanisms in human genes. In the genome, IKCa1 is located at chromosome 19q13.2. IKCa1 can be upregulated through the stimulation of PKC pathway (e.g., in T cells), which can trigger transcriptional activation of the IKCa1 promoter. The regulatory regions of a gene include enhancer, promoter, and silencer. These regulatory units are located in 5′- and 3′-gene flanking regions and in introns. The locations of these regulatory elements and the nucleotide sequences can describe their structure features. Their corresponding functional characteristics can be described in the effect on gene transcriptional activity and tissue and stage specificities.

The major correlations, especially gene—drug and genotype–phenotype interactions, are described at the systems level in **Fig. 2.3**. Abnormal function of transporter proteins can cause abnormal phenotypes, i.e., diseases. Transporters can also be

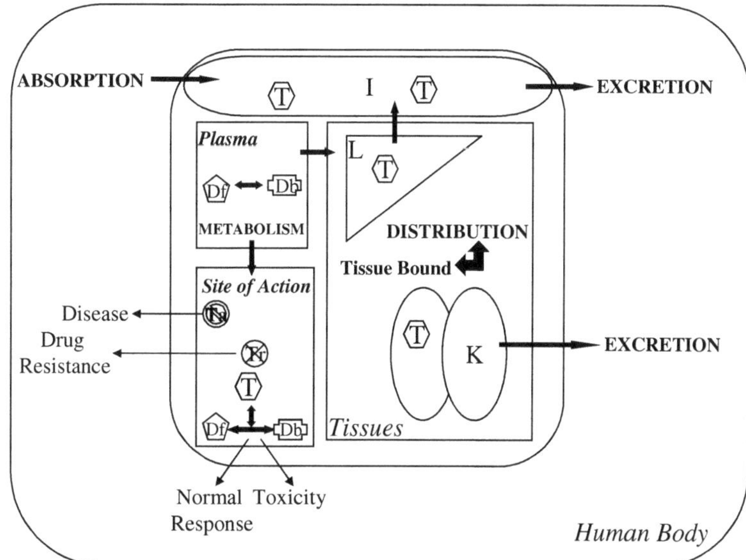

Fig. 2.3. The biomedical model of correlations of transporters at the systems level. *I*: Intestines; *L*: Liver; *K*: Kidney; *Df*: Free drug; *Db*: Bound drug; *T*: transporter; *Ta*: Altered transporters (that cause diseases); *Tr*: Transporters that are responsible for drug resistance.

involved in drug-response phenotypes of resistance, toxicity, or normal responses. The diagram illustrates the processes involved in drug transport and the effect of transporter actions on the bioavailability of drugs. Drug availability can be controlled by drug absorption and excretion, as shown in the diagram. Besides absorption and excretion, the interaction processes between drugs and the human body also include the distribution and metabolism. In addition, transporters are distributed in different tissues.

3.2.2. Use Case Diagrams

In UML terminology, user requirements are expressed in terms of *use cases*. A use case is a process that fulfills certain requirements of a system user (33). A process describes a sequence of events, actions, and transactions needed to complete something of usefulness to a user, from start to finish. Use case diagrams describe the main processes in a system and the interactions between the processes (use cases) and the external systems or actors. An actor is an entity outside the system that in some way participates in the story of the use case. Actors are represented by the role they play in the use case, such as a pharmacologist or a bioinformatician. A use case diagram defines the system boundaries, as well as the users that will utilize the system. Use cases comprise all the system functions identified during the prior requirement analysis and biomedical modeling. The processes in these use cases can be simple data query and retrieving, as well as more complicated knowledge discovery. In the later case, the system is also functioning as a data mining tool.

In a use case diagram, a rectangle with rounded corners represents the application or system (*see* **Fig. 2.4**). Use cases are shown in ovals, with the name of the use case written inside the oval. Actors are represented in stick figures, with the name written under the figure. A line links the actors and the use cases they interact with.

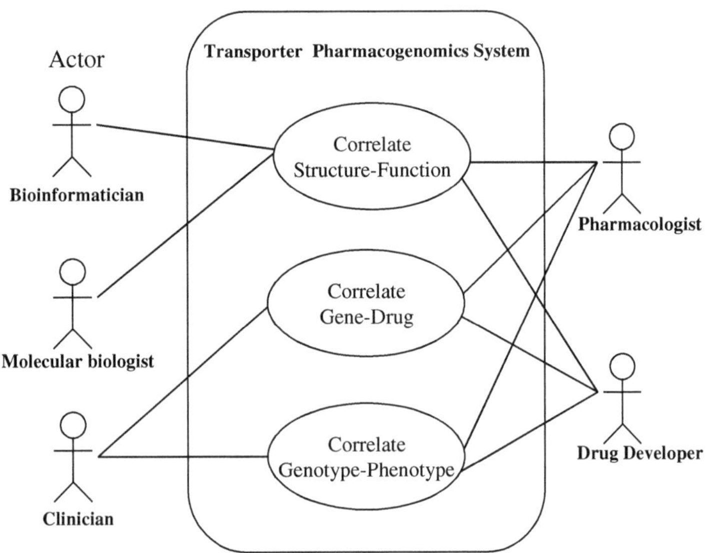

Fig. 2.4. The use case diagram of the transporter pharmacogenomics decision support system.

Figure. 2.4 shows the use case diagram of the transporter pharmacogenomics and systems biology decision support system (34). The actors in the diagram include molecular biologists, bioinformatician, pharmacologists, drug developers, and clinicians. The use cases that the system supports include, but not limited to, "Correlate Structure–Function," "Correlate Gene–Drug," and "Correlate Genotype—Phenotype."

Pharmacologists and drug developers (the actors) may be interested in finding the gene–drug interactions. For example, genetic alterations in transporter genes BCRP and MXR are shown to be associated with resistance to mitoxantrone in breast cancer cell lines (35). It may be interesting to determine if other transporter genes are involved in the resistance. The actors can also categorize the genes and drugs with known interactions, which might help predict new interactions. For instance, to answer the question "For a new drug, what genes may interact with it?" analysis of the interaction patterns in drugs with similar structures and functions might be helpful. In this data mining process, the information of the structure–function correlation is also important.

The study of genotype—phenotype correlation may be helpful to pharmacologists and drug developers to get some feedback about the use of drugs. This correlation information can assist more accurate drug targeting in the drug design process. For example, the identification of potential gene markers in the drug-resistance phenotype may provide clues for these actors to design new drugs targeting the markers to reverse or overcome the resistance. The software system will be a very useful tool in these decision-making processes (34).

3.2.3. Concept Abstraction and Class Diagrams

In UML, a class describes a set of objects that share the same attributes, methods, and relationships (33). A class diagram illustrates classes and the relationships between classes. In a class diagram (*see* **Fig. 2.5**), a large rectangle is divided into three horizontal compartments. The name of the class is written in the top section. The second section records the attributes of the class. The lower section usually includes operations and methods. The associations needed to record relationships are also added. An association is a relationship between objects that designate some meaningful and interesting connection. Associated classes are linked by lines. In some cases an association can be an object that has its

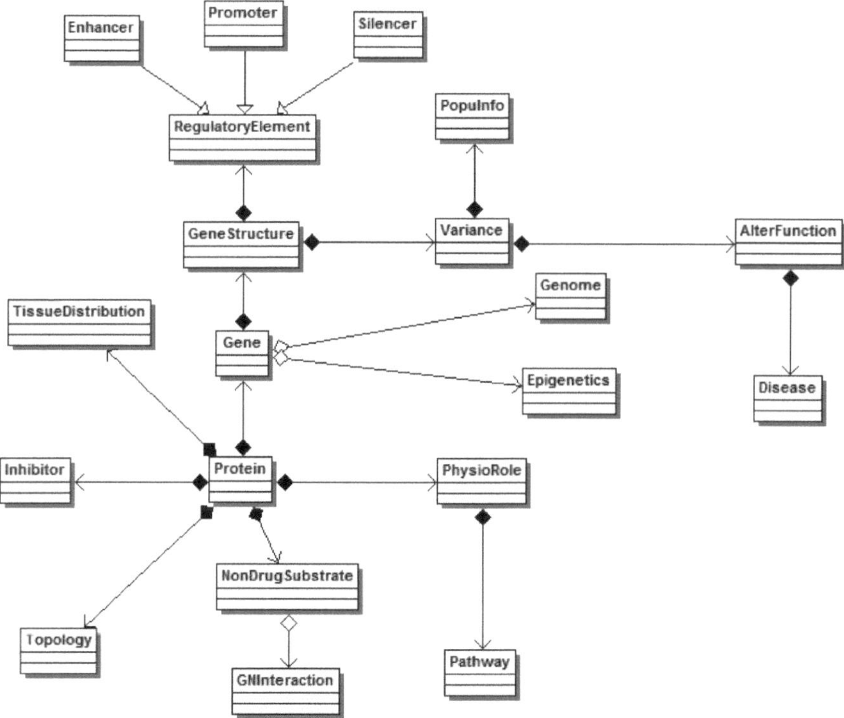

Fig. 2.5. The class diagram of the structure–function correlation.

own attributes. The representation of an association class is useful when there is a relationship between many objects and many other objects, while the attributes are the characteristics of the connection itself but not any of those classes being linked. Such an association class can be connected to the regular association line with a dotted line (*see* **Fig. 2.6**). A dotted line stands for a dependency. A dependency implies that one of the elements will change when the other changes.

A class hierarchy can be formed when the subclasses inherit attributes and associations of a super class. The subclasses can have particular characteristics of their own. This inheritance relationship between super classes and subclasses is illustrated with an open arrowhead (*see* **Fig. 2.5**).

Another type of association is aggregation, which refers to the part–whole relationship. An aggregation relationship is represented by a small diamond at the end of the association line that runs between part classes and the whole classes, heading toward the whole class (*see* **Fig. 2.5**). In an aggregation relationship, the part/child class instance can outlive its parent (the whole class). Such an aggregation relationship can be represented

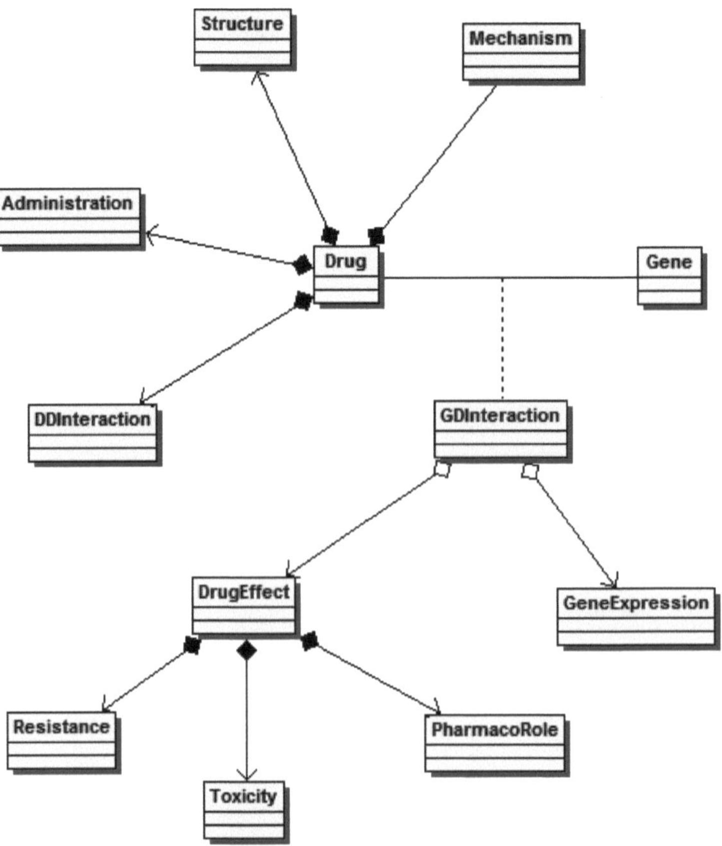

Fig. 2.6. The class diagram of gene–drug interactions.

with an unfilled diamond shape on the parent class's association end. The composition aggregation relationship means the child class's instance is dependent on the parent class's instance lifecycle, which is represented with the filled diamond shape.

3.2.3.1. Data Modeling of the Structure–Function Correlation

As shown in **Fig. 2.5**, gene, protein, and related genomic information are needed to study how the structure affects function. As discussed in **Chapter 1**, the structure–function association is essential for studying all of the emerging knowledge domains, from pharmacogenomics to systems biology, from nutrigenomics to epigenetics.

In this diagram, the classes "Variance," "Genome," "NonDrugSubstrate," and their associated child classes such as "PopuInfo" and "GNInteraction" represent the features in pharmacogenomics and nutrigenomics. The classes "Disease," "PhysioRole," "Pathway," and "TissueDistribution" are critical for the study of systems biology. The class "Epigenetics" is designed specifically for the emerging field epigenetics and the studies of gene–environment interactions.

The characteristics of "Gene" are described from several aspects and levels including "GeneStructure," "Protein," and "Genome." As illustrated in the biomedical model of **Fig. 2.2**, detailed information of exons, introns, and characteristics of 5′ UTR and 3′ UTR in the sequence may be included in the "GeneStructure" class.

The function correlated with nucleotide structures (especially regulatory elements) is described in the class "RegulatoryElement." For example, promoter regions in a gene may influence the gene expression level. The subclasses of regulatory elements include "Enhancer," "Promoter," and "Silencer."

Genetic sequence variation may be crucial in functional variation. Variation is one of the most important features of pharmacogenomics and nutrigenomics, which studies different drug and nutrient responses in individuals. In **Fig. 2.5**, the class "Variance" is used to represent both sequence polymorphisms and mutations, because the definition of the concept "polymorphism" is somewhat narrow (36). The population information ("PopuInfo") of variations (especially polymorphisms) is used for analysis and selection of certain groups of patients. Sequence variations can alter the transporter function. The altered function and the pathological role of the variation are represented in the class "AlterFunction," which in turn can result in diseases ("Disease").

The position of the gene in the whole genome is identified in the "Genome" class. The classes in the protein structure domain include "Protein" and "Topology." "Topology" includes the domain type in the protein (such as transmembrane domains

(TMDs)). The abundance of transporter proteins in different tissues is described in the class "TissueDistribution."

The functions correlated with protein structures are described in the classes "PhysioRole," "NonDrugSubstrate," and "Inhibitor." "PhysioRole" describes the physiological functions of the transporter gene, such as their roles in the regulation of intracellular redox potential. The child class "Pathway" put the gene in the whole picture of its functioning processes and interactions. "NonDrugSubstrate" describes non-drug substrates that are known to interact with the transporter, including gene–nutrient interactions ("GNInteraction"). These classes are important for the study of nutrigenomics of transporters (*see* **Chapter 1**). The details of drug substrates will be described in the gene–drug correlation domain (*see* **Section 3.2.3.2**). The class "Inhibitor" describes those molecules that can inhibit the transporter function, such as those examples in **Fig. 2.2**.

3.2.3.2. Data Modeling of the Gene–Drug Interaction

The major concepts involved in the gene–drug interaction include gene, drug, and their correlation. Drug information such as drug structure and mechanisms is crucial for the understanding of drug-resistance and toxicity mechanisms. The information can be very helpful for predicting the response of new drugs with similar structure or action. Drug information is also important for designing strategies to reverse the resistance or toxicity associated with side effects and treatment failure. The model representing drug information is illustrated in **Fig. 2.6**.

The classes "Structure," "Mechanism," "Administration," and "DDInteraction" describe several aspects of the major class "Drug." "Mechanism" contains drug activities and target information. "DDInteraction" means drug–drug interaction. "Administration" describes detailed information of administration of the drug.

The interactions between gene/human and drug in **Figs. 2.2** and **2.3** incorporate two levels of response to drugs, both genotypic and phenotypic. The later level will be discussed in the next section. The class "GDInteraction" describes the overall characteristics and mechanisms of the gene–drug interaction.

Because the gene–drug interaction is mutual, both of the gene and drug can have responsive reactions influenced by each other. Considering the drug side, the actions of a drug may be changed by genetic alterations. For example, increased drug efflux or decreased drug influx may be caused by transporter variations, as illustrated in the biomedical model **Fig. 2.2**. The class "DrugEffect" in **Fig. 2.6** represents this kind of interaction result. The other side of the gene–drug interaction is "gene," which is represented in the class "GeneExpression." For example, overexpressed breast cancer-resistance protein (BCRP) was found to mediate resistance to mitoxantrone in breast cancer therapy (35).

Figures. 2.5 and **2.6** can be integrated to be one complete model. This can be done with extending and connecting the class "Gene" (in **Fig. 2.6**) to **Fig. 2.5**.

3.2.3.3. Data Modeling of the Genotype–Phenotype Correlation

The genotype–phenotype correlation includes normal, disease, and the drug-response phenotypes. Because the genotype–phenotype correlation is linked to structure–function and gene–drug correlations, the disease and drug-response aspects at this level of correlation are integrated in the two later correlations, as shown in **Figs. 2.5** and **2.6**, respectively. The disease aspect of phenotype is represented in class "Disease" in **Fig. 2.5**. This phenotypic response is correlated with genotypic factors through the correlation with the class "AlterFunction." The overall drug-response phenotypes are correlated with genotypic gene–drug interactions through class "DrugEffect." These phenotypes include "Resistance" and "Toxicity," as illustrated in **Fig. 2.6**. If we abstract the concepts in **Fig. 2.3**, drug activation, inactivation (such as clearance), absorption, distribution (include transportation) can be analyzed and described in the class "PharmacoRole," which describes the pharmacological role in the phenotypic response.

To evaluate this data model and see if it captures the most important aspects in the targeting knowledge domain, it can be checked back with the original biological facts in **Figs. 2.2** and **2.3**. This object model is consistent with the biological model and domain knowledge. For example, most of the objects in **Fig. 2.2**, such as the types of the molecules involved, are represented and included in **Figs. 2.5** and **2.6**. In the implementation of the system, if some modifications are found necessary, the model constructed here can still be changed and improved.

3.2.4. Sequence Diagrams

A sequence diagram uses dynamic views to describe a specific scenario or a real example of a use case (33). Sequence diagrams portray a more detailed view of the interaction between the objects of the main classes in the system. It shows how the actors interact directly with the system, and the system events the actors generate (*see* **Fig. 2.7**).

In a sequence diagram, the objects identified are listed along the top of the diagram, with a dotted line beneath each object (as shown in **Fig. 2.7**). These dotted lines are called lifelines. The objects are listed in the order they are employed in the scenario. The leftmost object is the one that makes the stimulus that starts the scenario. Events within a sequence, which happen later in the time order, are often represented lower on the chart. Objects are

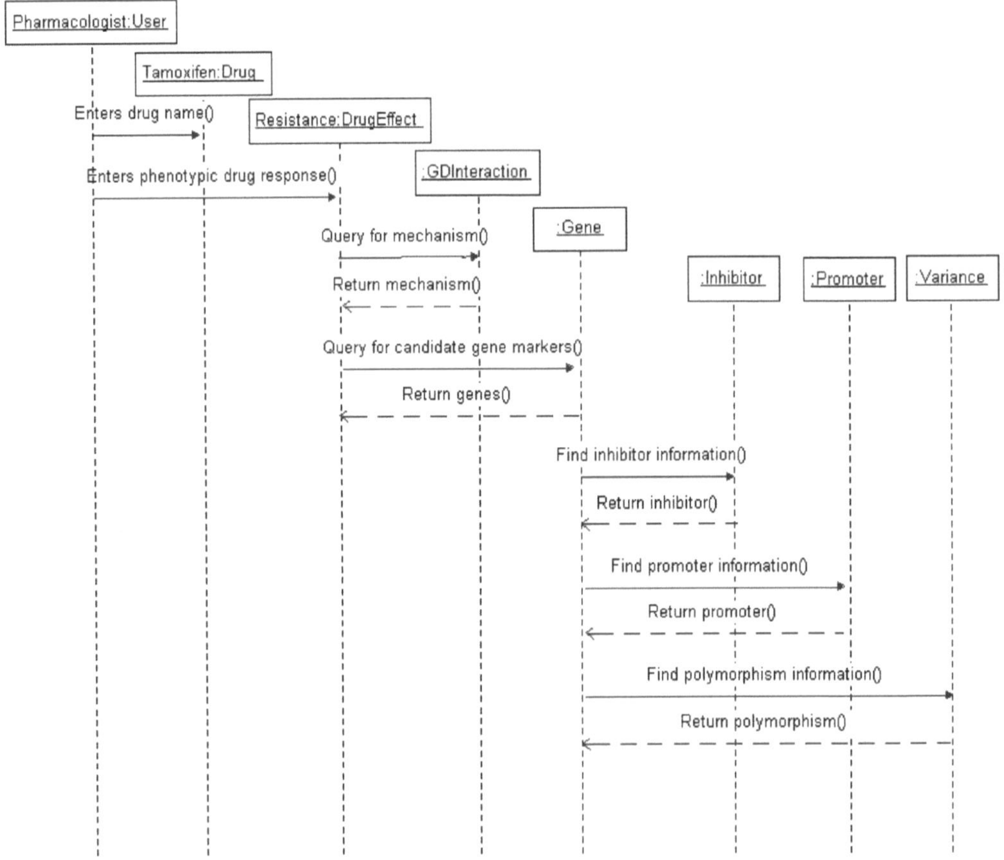

Fig. 2.7. The sequence diagram of the scenario "identifying candidate genetic markers in tamoxifen resistance."

linked by an event arrow, which suggests that a message is transferring between those two objects. The event ceases at an arrowhead.

Figure. 2.7 shows the sequence diagram of "identifying candidate genetic markers in tamoxifen resistance." In this scenario, a user, such as a pharmacologist or a drug developer, wants to find the mechanisms and design strategies (such as new drugs) to reverse the resistance to tamoxifen in breast cancer therapy. To do this, they need to know the candidate genes that may be responsible for tamoxifen resistance in breast cancer therapy. These genes can be potential targets for the reversal strategies.

This is a typical scenario with most of the important correlations of pharmacogenomics and systems biology involved (*see* **Chapter 1**). The goal in this decision-making process is to find *genotypes* that may be responsible for the resistance *phenotype*. With the drug name tamoxifen at the left side as input, what needs to be found out is the other side of the *drug–gene* interaction, the genes. Once the candidate genes and the interaction mechanisms are known, the information about the *structure–function* correlation is needed to identify the possible targets for finding reversal mechanisms. Such information may include the known inhibitors of the genes, the regulatory elements that affect the transcription activities, as well as variations including genetic polymorphisms for individualized strategies.

As shown in **Fig. 2.7**, the user first queries the system through entering the drug name "Tamoxifen," whose information is in the "Drug" class. The user then enters the possible phenotypic response "Resistance" to see the possible gene markers. The system looks for the information from the classes "DrugEffect," "GDInteraction," and "Gene," and extracts the genes that show altered expression as possible markers in the tamoxifen-resistance effect. The user may also want to ask how the genetic factors affect the drug actions, which can be retrieved from the class "GDInteraction." Now the user is ready to look for possible reversal targets such as some regulatory elements in the genes that can be used to inhibit the resistance effect. To do this, the information of "Inhibitor" and sequences of "Promoter" can be retrieved. To make the strategies specific for different individuals, the user wants to know the genetic variance that may occur in these genes, which can be obtained from the "Variance" class. With such information, the user can be ready to design reversal strategies.

This scenario describes how the information is extracted for the major correlations, and the information flow among the objects. This example shows how a complicated problem can be solved step by step with a decision support system (34).

4. Conclusion

The object-oriented UML approach provides data modeling capabilities and supports a systematic methodology for pharmacogenomics and systems biology studies in transporters. This methodology helps present data for multiple users including drug designers, pharmacologists, molecular biologists, clinicians, and microarray examination designers. A good methodology not only benefits software developers but also can improve and broaden the applications of the system by the users. It can give analysts the information necessary to make sound decisions about strategic issues for research, drug development, and treatment.

The methodology presented here is an attempt to provide the foundation for a comprehensive system that brings pharmacogenomics and systems biology into the clinic to benefit patients more directly (34). It is not limited for transporter studies but also can be used for other biomedical fields. The model here takes into account of the "variation" beyond the gene sequence, since "sequence variation is only one parameter, and certainly not the dominant parameter in human variation." (37) By OO modeling using UML, the problems of transporter pharmacogenomics and systems biology can be approached from different angles with a more complete view, which may greatly enhance the efforts in effective drug discovery and development.

The construction of the model demonstrates that UML, a modeling language that has been widely used in the business information technology (IT) industry, can be an appropriate common literacy in biomedicine if applied with appropriate methodologies. The main difficulty of applying UML in the biomedical domain is the barrier of domain knowledge. The "requirement analysis and biomedical models ⇒ concept abstraction and object design ⇒ UML OO models" methodology developed here is a useful measure for knowledge modeling in biomedicine. This methodology provides a generic way in how to build a computer OO model from the crude domain knowledge.

In biomedical science, knowledge modeling could play the role of mathematical modeling in physical sciences (18). Just as mathematicians, biomedical informaticians study models that are abstractions from the real biomedical world. As mathematical modeling encodes knowledge in a dynamic physical system, knowledge modeling (as the examples shown here) in biomedical systems also embraces both structural and dynamic behavioral aspects. This characteristic allows the dynamic representation of interactions and can be especially useful for the study of structure–function, gene–drug, and genotype–phenotype associations in transporter pharmacogenomics and systems biology (*see* **Chapter 1**).

References

1. Pipas, J.M. and McMahon, J.E. (1975) Method for predicting RNA secondary structure. *Proc. Natl. Acad. Sci. USA* **72**, 2017–2021.
2. Studnicka, G.M., Rahn, G.M., Cummings, I.W., Salser, W.A. (1978) Computer method for predicting the secondary structure of single-stranded RNA. *Nucleic Acids Res.* **5**, 3365–3387.
3. Erdmann, V.A. (1978) Collection of published 5S and 5.8S ribosomal RNA sequences. *Nucleic Acids Res.* **5**, r1–r13.
4. Dayhoff, M.O., Schwartz, R.M., Chen, H.R., Hunt, L.T., Barker, W.C., Orcutt, B.C. (1980) Nucleic acid sequence bank. *Science* **209**, 1182.
5. Korn, L.J., Queen, C.L., Wegman, M.N. (1977) Computer analysis of nucleic acid regulatory sequences. *Proc. Natl. Acad. Sci. USA* **74**, 4401–4405.
6. McCallum, D. and Smith, M. (1977) Computer processing of DNA sequence data. *J. Mol. Biol.* **116**, 29–30.
7. Fuchs,C., Rosenvold,E.C., Honigman,A., Szybalski, W. (1978) A simple method for identifying the palindromic sequences recognized by restriction endonucleases: the nucleotide sequence of the AvaII site. *Gene* **4**, 1–23.
8. Gingeras, T.R., Milazzo, J.P., Roberts, R.J. (1978) A computer assisted method for the determination of restriction enzyme recognition sites. *Nucleic Acids Res.* **5**, 4105–4127.
9. Paulsen IT, Nguyen L, Sliwinski MK, Rabus R, Saier MH Jr. (2000) Microbial genome analyses: comparative transport capabilities in eighteen prokaryotes. *J. Mol. Biol.* **4**, 75–100.
10. Chen, L., Ortiz-Lopez A., Jung A., Bush D.R. (2001) ANT1, an aromatic and neutral amino acid transporter in Arabidopsis. *Plant Physiol.* **125**, 1813–1820.
11. Dawson, P.A., Mychaleckyj, J.C., Fossey, S.C., Mihic, S.J., Craddock, A.L., Bowden, D.W. (2001) Sequence and functional

analysis of GLUT10: a glucose transporter in the Type 2 diabetes-linked region of chromosome 20q12-13.1. *Mol. Genet. Metab.* **74**, 186–199.

12. Kihara, D. and Kanehisa, M. (2000) Tandem clusters of membrane proteins in complete genome sequences. *Genome Res.* **10**, 731–743.
13. Yan, Q. (2001) *Informatics Support for Human Membrane Transporter Pharmacogenomics Studies.* ProQuest, Ann Arbor, MI, pp. 1–138.
14. http://tcdb.ucsd.edu/index.php (accessed in May 2009).
15. Nebert, D.W. (1999) Pharmacogenetics and pharmacogenomics: why is this relevant to the clinical geneticist? *Clin. Genet.* **56**, 247–258.
16. Sissung, T. M., Gardner, E. R., et al. (2008) Pharmacogenetics of membrane transporters: a review of current approaches. *Methods Mol. Biol.* **448**, 41–62.
17. Frishman, D., Heumann, K., Lesk, A., Mewes, H.W. (1998) Comprehensive, comprehensible, distributed and intelligent databases: current status. *Bioinformatics* **14**, 551–561.
18. Rechenmann, F. (2000) From data to knowledge. *Bioinformatics* **16**, 411.
19. Rumbaugh, J., Blaha, M., Premerlani, W., Eddy, F., Rumbaugh, J., Lorenson, W. (1991) *Object-Oriented Modeling and Design*. Prentice Hall, pp. 1–500.
20. Korson, T. and McGregor, J. (1990) Understanding Object-Oriented: A Unifying Paradigm. *CACM* **9**, 40–60.
21. Object Management Group. (1999) *OMG Unified Modeling Language Specification.* Object Management Group, Inc., pp. 1–808.
22. Martinez, R., Rozenblit, J., Cook, J.F., Chacko, A.K., and Timboe, H.L. (1999) Virtual management of radiology examinations in the virtual radiology environment using common object request broker architecture services. *J. Digit. Imaging* **12**, 181–185.
23. Quitadamo, L. R., Marciani, M. G., et al. (2008) Describing different brain computer interface systems through a unique model: a UML implementation. *Neuroinformatics* **6**, 81–96.
24. Fridsma, D. B., Evans, J., et al. (2008) The BRIDG project: a technical report. *J. Am. Med. Inform. Assoc.* **15**, 130–137.
25. Webb, K. and White, T. (2005) UML as a cell and biochemistry modeling language. *Biosystems* **80**, 283–302.
26. Taylor, C. F., Paton, N. W., et al. (2003) A systematic approach to modeling, capturing, and disseminating proteomics experimental data. *Nat. Biotechnol.* **21**, 247–254.
27. Xirasagar, S., Gustafson, S., et al. (2004) CEBS object model for systems biology data, SysBio-OM. *Bioinformatics* **20**, 2004–2015.
28. Jedlitschky, G., Leier, I., Buchholz, U., Barnouin, K., Kurz, G., and Keppler, D. (1996) Transport of glutathione, glucuronate, and sulfate conjugates by the MRP gene-encoded conjugate export pump. *Cancer Res.* **56**, 988–994.
29. Hipfner, D.R., Deeley, R.G., and Cole, S.P. (1999) Structural, mechanistic and clinical aspects of MRP1. *Biochim. Biophys. Acta.* **1461**, 359–376.
30. Hyde, S.C., Emsley, P., Hartshorn, M.J., et al. (1990) Structural model of ATP-binding proteins associated with cystic fibrosis, multidrug resistance and bacterial transport. *Nature* **346**, 362–365.
31. Cui, Y., Konig, J., Buchholz, J.K., et al. (1999) Drug resistance and ATP-dependent conjugate transport mediated by the apical multidrug resistance protein, MRP2, permanently expressed in human and canine cells. *Mol. Pharmacol.* **55**, 929–937.
32. Keppler, D., Leier, I., Jedlitschky, G., and Konig, J. (1998) ATP-dependent transport of glutathione S-conjugates by the multidrug resistance protein MRP1 and its apical isoform MRP2. *Chem. Biol. Interact.* **111–112**, 153–161.
33. Harmon, P. and Watson, M. (1997) *Understanding Uml: The Developer's Guide: With a Web-Based Application in Java.* Morgan Kaufmann Publishers, pp. 1–340.
34. http://pharmtao.com/transporter (accessed in May 2009).
35. Ross, D.D., Yang, W., Abruzzo, L.V., et al. (1999) Atypical multidrug resistance: breast cancer resistance protein messenger RNA expression in mitoxantrone-selected cell lines. *J. Natl. Cancer Inst.* **91**, 429–433.
36. Yan, Q., and Sadée, W. (2000) Human membrane transporter database: a Web-accessible relational database for drug transport studies and pharmacogenomics. *AAPS PharmSci* **2**, E20.
37. Marshall, A. (1997) Laying the foundations for personalized medicines. *Nat. Biotechnol.* 15, 954–957.

Chapter 3

Multidrug Resistance: Phylogenetic Characterization of Superfamilies of Secondary Carriers that Include Drug Exporters

Ming Ren Yen, Jonathan S. Chen, Jose L. Marquez, Eric I. Sun, and Milton H. Saier

Abstract

We here describe the application of novel programs that allow definition of phylogenetic relationships in transport protein superfamilies. These programs are used to provide information about the four major superfamilies of secondary carriers that include members that export hydrophobic and amphipathic compounds including drugs. These novel programs must be used when sequence divergence among superfamily members is too great to allow construction of reliable multiple alignments. We test the validity and demonstrate the reliability of these trees by conducting comparative analyses. We examine all of the largest superfamilies of secondary drug efflux pumps found in nature, the MOP, DMT, RND, and MFS superfamilies. Depending on the superfamily, phylogenetic clustering of the families and individual members of these families can occur according to organismal source, substrate type, polarity of transport, and/or mode of transport. In this chapter we define the phylogenetic relationships of sequence divergent drug exporters. The programs developed should be applicable to all classes of proteins and nucleic acids.

Key words: Phylogeny, protein family, superfamily, drug export, novel computer programs.

1. Introduction: The Problem In Perspective

For the past 20 years, our laboratory has been concerned with multidrug resistance (MDR) (1, 2, 3, 4, 5, 6, 7, 8) and the phylogenetic characterization of transport protein families (9, 10, 11). As our technology for establishing homology becomes more refined, our capability of identifying distant relationships improves (12). Thus, we have attempted to create superfamilies,

Q. Yan (ed.), *Membrane Transporters in Drug Discovery and Development*, Methods in Molecular Biology 637,
DOI 10.1007/978-1-60761-700-6_3, © Springer Science+Business Media, LLC 2010

identifying novel families and interconnecting previously recognized families (13). These endeavors are important because once a common origin is established, extrapolation of functional, structural, and mechanistic data for one protein is justifiable for all other members of the superfamily, with phylogenetic distance being inversely proportional to the confidence level of such an extrapolation. On the other hand, if common ancestry is not established, such extrapolations are not justified.

Our recent efforts have resulted in the identification and expansion of several large superfamilies of transport proteins [(12, 13); *see TCDB,* www.tcdb.org]. One problem that exists is that the more distantly related sequences, the less reliable phylogenetic tree construction becomes. This is true because sequences showing few identities and similarities cannot correctly be aligned with each other using any of the several programs that are currently available for this purpose. Any error in the multiple alignments creates error in the phylogenetic trees that are based on these alignments.

As a result of these considerations, there is a need for defining phylogenetic relationships without reliance on multiple alignments. In this report we present such programs. They rely on the bit scores obtained from BLAST searches. However, the procedure involves comparisons between hundreds of superfamily members, and then combining thousands of individual trees, to yield a single consensus tree. This approach provides probability values for each branch point, equivalent to bootstrapping. Increased confidence in comparisons with trees generated by classical multiple alignment approaches for less distantly related proteins results in concordance, increasing the confidence level of the methods we have developed. In this report, we apply this approach to some of the largest superfamilies of secondary carriers recorded in the Transporter Classification Database (TCDB) (14, 15). The results are presented here.

2. Methods: Phylogenetic Tree Construction

Using TCDB (www.tcdb.org), we generated a temporary database file including all members of the superfamily of interest. This file was used to define the criteria for that superfamily and how that superfamily would later be broken down into respective families and (when appropriate) subfamilies. The division of proteins into superfamilies, families, and subfamilies was conducted according to assignments in TCDB. In a few cases, assignments in TCDB proved incorrect as revealed by the use of the SuperfamilyTree1 (SFT1) program. Following confirmation, their phylogenetic classifications in TCDB were corrected.

The resultant database contained the selected protein sequences in FASTA file format so they could easily be used for rapid similarity searching. Using this database, we used PSI-BLAST (1) to search the NCBI "nonredundant" protein database and matched up potential members for each family. Searching brings up proteins of varying degrees of similarity as well as many redundant sequences. The BLAST hits were then classified. The large numbers of proteins gathered from NCBI were matched to their respective families based on sequence similarities and sorted by families. The sequences were either used to study the evolutionary pathway taken and/or to generate a phylogenetic tree.

To create a phylogenetic tree, the protein members of a large superfamily were inputted into the novel SuperfamilyTree1 (SFT1) program, which analyzes the proteins and generates comparative matrices of the superfamily through 100 repeat shuffles. Then the programs, Fitch and Consense, use this matrix information to generate a phylogenetic tree showing the relative phylogenetic positions of all members of the families within the superfamily. The information from SFT1 is then used to combine the sequences in each of the constituent families into a single file. The SFT1 program is then used with each file representing a distinct family. The Fitch and Consense programs (http://evolution.genetics.washington.edu/phylip.html) are again used to generate the SFT2 trees. They are viewed using the TreeView (TV) program (16). For optimal comprehension, this chapter should be read with continual reference to the Transporter Classification Database **[TCDB** (14, 15)**]**.

3. Overview

In this chapter we analyze the four largest superfamilies of electrochemically driven drug exporters present in TCDB as of June 2008–October 2008. All of these superfamilies are ubiquitous, being represented in all three domains of life: bacteria, archaea, and eukaryotes. These superfamilies include the MOP [TC# 2.A.66; (3)], DMT [2.A.7; (9)], RND [2.A.6; (17)], and MFS [2.A.1 plus 4 additional TC families (18, 10)]. The one remaining functional superfamily that includes MDR pumps, the ATP-binding cassette (ABC) superfamily (19, 20), exports drugs using primary active transport, dependent on ATP hydrolysis. It will be analyzed in a separate publication.

The phylogenetic analyses of each of the four superfamilies of secondary carriers reveal the relationship of the constituent proteins and families within each superfamily relative to each other. They provide a quantitative guide for extrapolation of structural, functional, and mechanistic data from a well-studied member of the superfamily to all others.

4. The Multidrug/Oligosaccharide/Polysaccharide (MOP) Superfamily

The current MOP superfamily, first characterized in 2003 (3), contains four well-defined families plus five poorly defined families. All members of the MOP superfamily appear to function in solute efflux. Phylogenetic trees were constructed using four different methods, and these are shown in **Fig. 3.1**. **Fig. 3.1a** shows the tree based on a ClustalX program-generated multiple alignment using the Tree View program [neighbor-joining (NJ)]. **Fig. 3.1b** shows a tree, also using a ClustalX-generated multiple alignment and Tree View, but based on the ProtPars program (Phylip package, parsimony (P)). **Fig. 3.1c** shows a tree based on the SuperfamilyTree1 (SFT1) program, and **Fig. 3.1d** shows the consensus tree based on the SuperfamilyTree2 (SFT2) program, essentially using the data from SFT1.

According to the ClustalX-NJ tree (**Fig. 3.1a**), all of the members of the ubiquitous MATE family of drug exporters (family 1) are found together in four distinct subclusters. The second family, the prokaryotic family of polysaccharide exporters (PST, family 2), includes members that are scattered throughout the tree. Thus the PST 1, 4, 10, and 13 proteins cluster loosely together, forming a subfamily. The PST 2, 6, and 7 proteins and the PST 3 and 11 proteins form two additional clusters. All other members of the PST family branch from points near the center of the tree without exhibiting appreciable clustering. Family 3, the eukaryotic oligosaccharide exporting OLF family, has three members that cluster together while the fourth member clusters separately. The two protein members of the mouse virulence factor (MVF) family included in TCDB (family 4; the MVF1 and MVF2 proteins) cluster loosely together. They and all other members of this family are of unknown specificity. The three Ank (family 9) proteins cluster together as do the two EPS-E (family 6) homologues. The members of each remaining family cluster together, but separate from members of other families. These results suggest that when the members of a family are sufficiently similar to generate a reliable multiple alignment, appropriate clustering is observed, but when their sequences are too divergent, the phylogenetic relationships are incorrectly predicted (see below).

The ProtPars (parsimony) tree (**Fig. 3.1b**) clusters all MATE family (family 1) proteins together. However, as for the

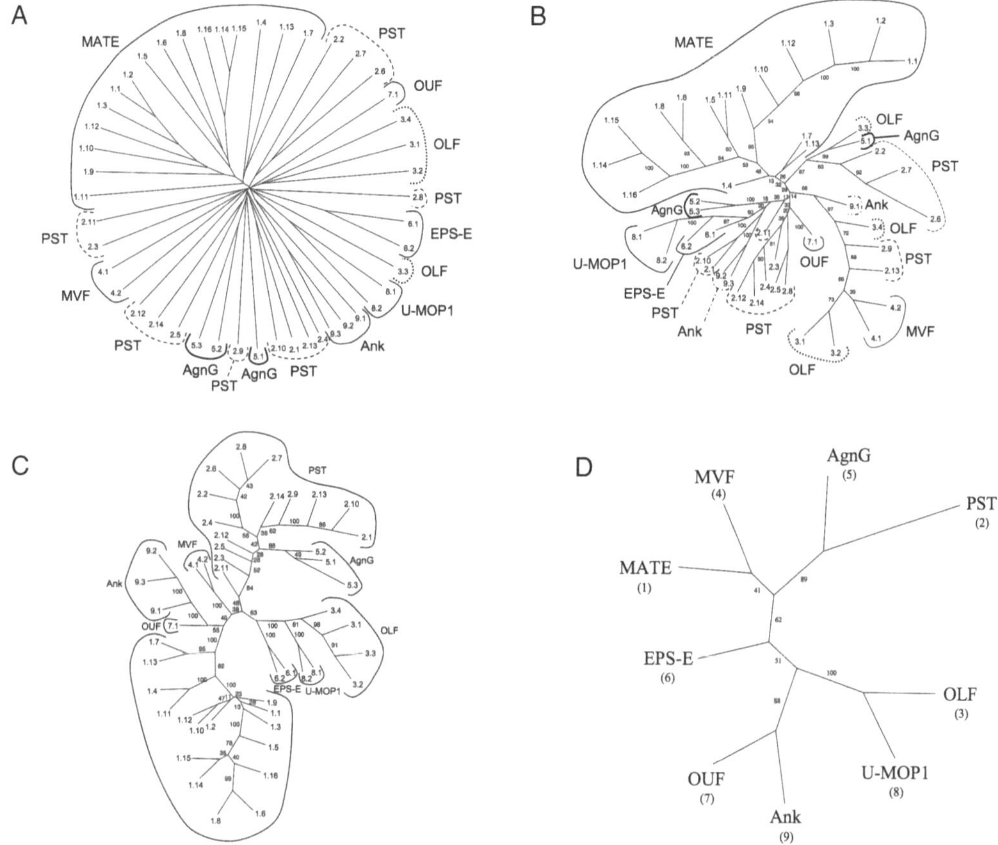

Fig. 3.1. Phylogenetic (Fitch) trees for the MOP superfamily using the proteins in TCDB as of October, 2008. Four different methods for tree construction were used: (**a**) Clustal X-based neighbor-joining (NJ), (**b**) ProtPars-based parsimony (P), (**c**) The Blast-derived SFT1-program showing all MOP superfamily protein members, and (**d**) The SFT2-program showing the families within the MOP superfamily [*see* **Section 2** and (12)]. In all trees, the abbreviations of the families are indicated. In (**a–c**) *numbers* indicate the protein TC#s as they are presented in TCDB. In (**d**), family numbers are indicated in *parentheses*. In (**b–d**), small numbers adjacent to the branches present the "bootstrap" values, indications of the reliability of the branching order.

neighbor-joining tree, members of the PST family (family 2) are scattered on several branches on the lower left- and right-hand sides of the tree. The U-MOP1 family proteins (all of unknown specificities) are loosely associated with a subgroup of the PST proteins as well as members of the AgnG, Ank, and EPS-E families. The OLF family proteins localize to three distinct branches on the right-hand sides of the tree, among branches of PST, AgnG, Ank, and MVF family members. The two MVF proteins cluster together as expected.

The SuperfamilyTree1 (SFT1) program-generated tree (**Fig. 3.1c**) differs from the neighbor-joining (NJ) and parsimony (P) trees in that the members of each family within the MOP superfamily cluster together. However, clustering patterns differ, depending on the program used. For example, the MATE 4 and 11 proteins cluster together in the SFT1 tree but not in the NJ or P tree. However, MATE 7 and 13 proteins cluster together in all three trees, as do MATE 14 and 15. The MATE 6, 8, and 16 proteins also cluster together in the SFT1 tree but only loosely in **Figs. 3.1a,b**. The SFT1 tree also clusters all PST proteins together (unlike the NJ and P trees), and the AgnG proteins cluster together within the PST family. This suggests a closer relationship between these two families than had been realized previously. The results suggest that AgnG may be a subfamily within the PST family and therefore may function for polysaccharide export. All four proteins of the OLF family cluster together. These proteins are at the ends of the branch that also bears the EPS-E and U-MOP1 proteins. Finally, the two MVF proteins cluster together as was true for the NJ and P trees.

The SFT2 tree (**Fig. 3.1d**) summarizes the relationships of the different families to each other. These are based on and in agreement with the results of the SFT1 tree. Thus, the MATE and MVF families prove to be more closely related to each other than to the other families, while the PST and AgnG families cluster together on a distinct branch (upper half of the tree). U-MOP1 and OLF cluster together, and OUF and Ank also cluster together on a distinct but nearby branch. It is therefore interesting to note that all prokaryotic proteins, except some of those in the ubiquitous MATE family, are localized at the top of the tree, while all eukaryotic proteins (including some bacterial proteins, e.g., in the ubiquitous OUF family, and one member of the Ank family) are localized at the bottom of the tree. It is important to note that MATE and OUF family members derive from both prokaryotes and eukaryotes. The results suggest that MVF (mouse virulence) family proteins, of unknown specificity, might be drug exporters like members of the MATE family; the AgnG family might be capable of exporting poly- or oligosaccharides as do PST family porters, and U-MOP1 proteins might export oligosaccharides as do OLF family transporters. The results of this analysis demonstrate the superiority of the SFT1 and SFT2 trees over those based on multiple alignments when sequence divergent proteins are examined. Further examples are provided below.

5. The Drug Metabolite Transporter (DMT) Superfamily

In 2001, we identified a superfamily of transporters that included the well-characterized four TMS small multidrug resistance (SMR) family (5, 7, 21). These secondary active transporters can function as nutrient uptake porters, drug/metabolite efflux pumps, or solute:solute exchangers (21). These porters are found in bacteria, archaea, and eukaryotes. We designated this ubiquitous superfamily the drug/metabolite transporter (DMT) superfamily (transporter classification number TC #2.A.7) and showed that it then consisted of 14 phylogenetic families, five of which included no functionally characterized members. The largest family in the DMT superfamily, the drug/metabolite exporter (DME) family, consists of over 500 sequenced members, several of which have been implicated in metabolite export, with some members transporting drugs (*see* TCDB).

Several families within the DMT superfamily consist of proteins with distinctive topologies: four, five, nine, or ten putative transmembrane α-helical spanners (TMSs) per polypeptide chain. The five TMS proteins include an N-terminal TMS lacking in the four TMS proteins. The full-length proteins of ten (or nine) putative TMSs apparently arose by intragenic duplication of an element encoding a primordial five TMS polypeptide (21). These five TMS repeat units are homologous throughout their lengths with the five TMS proteins of the DME family. Sequenced members of the 14 families were tabulated, and phylogenetic trees of the individual families were presented. Sequence and topological analyses allowed structural and functional predictions.

Since the completion of the work of Jack et al. (5), an additional ten families have been discovered and added to the DMT superfamily as tabulated in table 2.A.7 in TCDB. Prior to the work described here, no effort had been attempted to define the phylogenetic relationships of these 24 families.

Figures 3.2a,b show the SuperfamilyTree program-derived phylogenetic trees showing (**a**) all proteins of this superfamily listed in TCDB as of June 2008 (SFT1 program) and (**b**) all 24 families, where each family is shown on a distinct branch (SFT2 program). The latter tree was derived essentially by combining and averaging the results for individual proteins within any one family (*see* **Section 4** on the MOP superfamily above where the procedures are presented in greater detail and the reliability of the methods are examined).

In general, all members of a family cluster together, but there proved to be four exceptions, three in the DME family and one in the TPPT family (*see* **Fig. 3.2a**). These proteins group loosely together in the cluster that includes members of the DME family (family 3), LicB (family 18), RarD (family 7), TrpE (family 23), TPPT (family 24), and PE (family 17). Also included within this large cluster are BAT (family 2) and P-DME (family 4) members. As previously described (9), the P-DME family is actually a relatively small but coherent family within the DME family. DME family members derive exclusively from prokaryotes, while P-DME family members are from plants, possibly in chloroplasts and plastids, thus explaining the observed close relationships. We also discovered that the DME family is extremely diverse in sequence as are members of certain other families on the same phylogenetic branch. This, rather than a defect in the SuperfamilyTree program, appears to account for the apparent discrepancies noted for this cluster. All members of this cluster except members of the BAT family (with five TMSs) appear to have ten TMSs. All other clusters are coherent as predicted, based on TC family assignments. The results are in agreement with the conclusions cited above for the MOP superfamily.

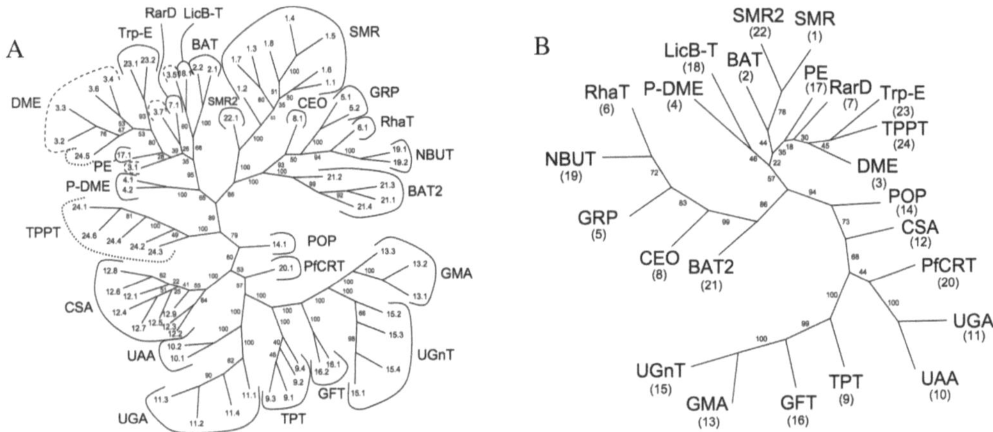

Fig. 3.2. Phylogenetic (Fitch) trees for the DMT superfamily. (**a**) The SFT1 tree showing all of the proteins included in TCDB as of August, 2008, and (**b**) the composite SFT2 tree showing the positions of the 24 families relative to each other. (**a**) The family names (*see* TCDB) are presented in **bold** type. Numbers at the ends of the branches refer to the protein TC numbers within each family. The smaller numbers next to the branches indicate the "bootstrap" values. (**b**) Family relationships based on the tree shown in (**a**). The conventions of presentation are the same as for (**a**) with bootstrap values presented next to the branches. Note that of the three clusters, only the one shown in the upper right-hand side of the tree exhibits low bootstrap values, indicative of uncertainty with respect to the relative branching orders. These numbers are roughly proportional to the confidence levels of the branch positions, expressed in percent.

Comparing **Fig. 3.2a**, which shows positions of all family members, with **Fig. 3.2b**, which shows only the positions of the families, we find excellent agreement. For example, the rhamnose uptake family, RhaT, the nucleobase uptake family, NBUT, and the glucose/ribose uptake family, GRP, cluster together. All three of these families consist of ten TMS nutrient uptake systems in prokaryotes. They thus have similar functions. The CEO and BAT2 families, branching from points closer to the base of the tree, are of unknown functions, but all current CEO family members are from a single animal, *Caenorhabditis elegans*,while all BAT2 family members are strictly from prokaryotes.

Closely related to this cluster is another consisting of the SMR and SMR2 families, both with four putative TMSs per polypeptide chain. It is interesting that they cluster together because of their substantial sequence divergence, causing them to be classified into two distinct families. Although no SMR2 family member has been characterized, these observations suggest that the two families may have similar specificities. In fact, these proteins share the conserved glutamate residue that plays crucial roles in the selection of cationic substrates and in proton antiport for members of the SMR family [(7); and unpublished results]. These observations provide the basis for the name of this family (SMR2) and the prediction that these porters function like members of the phylogenetically distinct SMR family (7).

The remainder of the tree (bottom half) exhibits no anomalies. In fact, this part of the tree coherently includes families of eukaryotic endoplasmic reticular nucleotide-sugar:nucleotide exchangers (UAA, UGA, GFT, UGnT, and GMA). Also present in this cluster are the triose phosphate:phosphate (TPPT) exchangers of eukaryotic chloroplasts and plastids. Finally, the CSA family of CMP-sialic acid:CMP antiporters also clusters loosely at the base of this cluster together with the chloroquine resistance porter, PfCRT, and the plant organocation permease (POP) family. The common functional attributes of these proteins thus correlate with their phylogenies. They form a coherent cluster suggesting a common ancestry after divergence from other DMT constituent families.

In summary, we find that family phylogeny in the DMT superfamily correlates reasonably well with 1. polarity and mechanism of transport, 2. substrate type, and 3. source organismal type. There is excellent agreement between the trees shown in **Figs. 3.2a,b**. These observations suggest that the families of porters with dissimilar mechanisms and specificities evolved early, while subdivision into families of similar functions evolved from their common ancestors much later. Segregation of porters of eukaryotes from those of prokaryotes suggests that genes encoding these proteins were not transferred between organisms of these two domains at least within the past two billion years.

6. The Resistance/Nodulation/Division (RND) Superfamily

All RND superfamily proteins (17) were analyzed using the SFT1 program (**Fig. 3.3a**) and the SFT2 program (**Fig. 3.3b**). The SFT1 tree reveals that the members of the two eukaryotic families (EST and Dispatched) cluster together. The many members of the eukaryotic sterol transporter (EST) family group together, while the Dispatched protein branches from a point near the center of the tree (center right). Closest to this branch is one bearing the HAE2 lipid exporters of Gram-positive bacteria, the HAE3 family of archaeal exporters of unknown specificity, and the bacterial

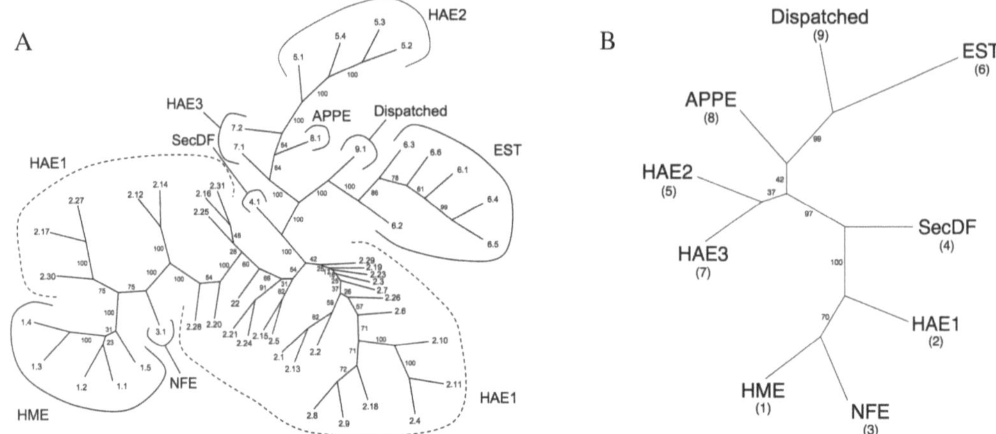

Fig. 3.3. Phylogenetic trees for the RND superfamily as determined with (**a**) the SFT1 and (**b**) the SFT2 programs. The conventions of presentation are the same as for the previous two figures.

APPE family with probable specificity for aryl-polyene pigments (21, 22).

The SecDF protein of *Escherichia coli* and its many prokaryotic homologues (not shown) branch from a central position in the tree, separating the two major branches, one (top) bearing families mentioned above and the other consisting of the Gram-negative bacterial HAE1 family proteins together with the heavy metal transporting HME porters and members of the NFE family, both represented primarily in Gram-negative bacteria. Thus, all of the Gram-negative bacterial proteins cluster together.

It is noteworthy that the HME and NFE families are sandwiched in between two clusters of the large HAE1 family. The tree indicates that these might represent two distinct subfamilies within the HAE1 family. Independent analyses (not presented) have confirmed this finding.

The tree shown in **Fig. 3.3b** was generated using the SFT2 program. As expected, the Gram-negative bacterial Heavy Metal Exporter (HME) and Lipooligosaccharide Nodulation Factor exporter (NFE) families cluster with the primarily Gram-negative bacterial HAE1 family. The prokaryotic SecDF family branches from a point nearer the center of the tree. Similarly, we find that the eukaryotic families, EST and Dispatched, cluster together, with the bacterial aryl-polyene pigment exporter (APPE) (23) family branching closer to the center of the tree with all prokaryotic families. The Gram-positive bacterial HAE2 family proteins and the archaeal HAE3 proteins cluster together nearer the center of the tree. The phylogenetic relationships of these 8 families, relative to each other, are thus defined, revealing clustering, in general, according to organismal type. HAE3 porters, of unknown function, may thus, like HAE2 family members, export hydrophobic/amphipathic compounds such as lipids, drugs, and antibiotics.

7. The Major Facilitator Superfamily (MFS)

The MFS is the largest superfamily of secondary active carriers found in living organisms on earth. The composite tree, obtained using SFT1 and SFT2, is shown in **Fig. 3.4**. All of the 67 currently recognized MFS families are included. In some cases, the protein families cluster according to function. Below, only families of known function will be discussed, except when phylogenetic clustering patterns allow novel functional predictions.

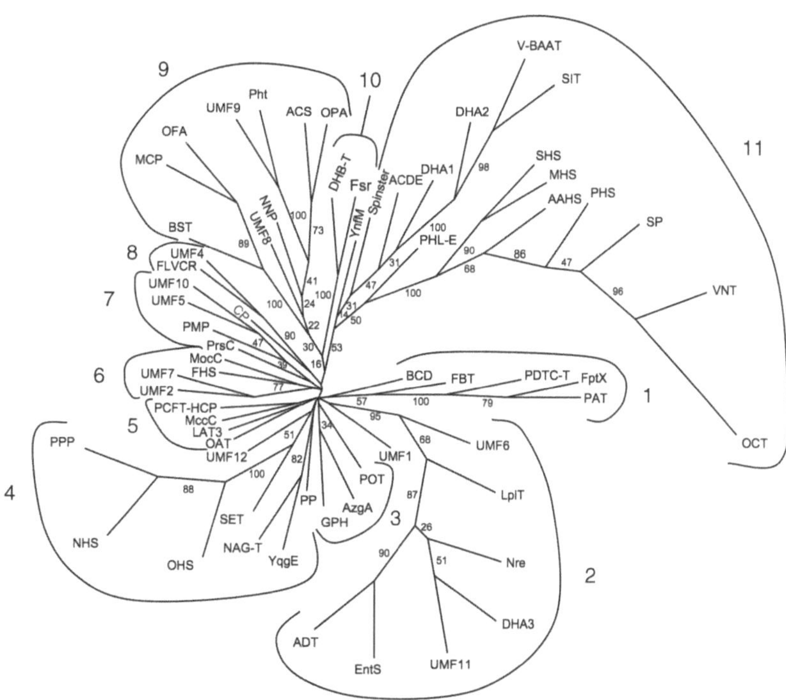

Fig. 3.4. Phylogenetic tree for 67 families within the Major Facilitator Superfamily (MFS) as derived with the SFT1 and SFT2 programs. The conventions of presentation are the same as for the trees shown in **Fig. 1d** and **2b**.

A branch (center right, cluster 1), displaying five families (BCD (bacteriochlorophyll), FBT (folate–biopterin), PDTC-T (iron-pyridine thiocarboxylate), FptX (ferripyochelin), and PAT (peptide/acetyl-CoA)) all take up aromatic anionic compounds. Just below this branch (cluster 2) is a large cluster of seven families, most of which transport hydrophobic or amphipathic compounds (except the Nre family which takes up nickel). However, some of these families catalyze substrate:H^+ symport (uptake), while others catalyze substrate:H^+ antiport (efflux).

Immediately below this branch is a small cluster (cluster 3) of uptake permeases specific for peptides (POT), purines (AzgA), and glycosides (GPH). These uptake transporters act on dissimilar types of substrates. Proceeding counterclockwise around the tree, the next cluster (cluster 4) includes uptake systems primarily for sugars: polyols (PP), *N*-acetyl glucosamine (Nag-T), oligosaccharides (OHS), nucleosides (NHS), and propionate (PPP). The only family in cluster 4 thought not to transport sugars is the PPP family. Only one family in this cluster, the SET family, catalyzes sugar export rather than import.

The next cluster of four families (cluster 5) includes only uptake symporters. They transport organic acids (OAT), amino acids (LAT3), a nucleotide–peptide microcin (MCC), and folate derivatives (PFCT-HCP). Thus, the proteins in this cluster prefer organic anionic substrates. Cluster 6 includes members that catalyze fucose uptake (FHS), uptake of opines (covalent condensation products of an aldehyde or ketone with a basic amino acid (MocC)), and efflux of siderophores (PrsC). These systems apparently can function with either inwardly or outwardly directed polarity and act on a variety of substrates.

Cluster 7 includes only four families, one that exports siderophores and the other three of unknown function. Cluster 8 consists of three families, one that takes up cyanide and the other two of unknown specificity. Cluster 9 is a coherent group of nine families which all share the characteristic that they function by cation symport, catalyzing the uptake of anions. Cluster 10 includes only two families, one (DHB-T) that catalyzes uptake of dihydroxybenzene and the other (Fsr) which catalyzes export of fosmidomycin, an anionic antibiotic.

Finally, cluster 11 is the largest cluster in the tree, including 15 MFS families. Within this cluster, we find both uptake and export families which transport a wide range of substrates. However, families tend to group according to both polarity of transport and the nature of the substrates. Thus, the substrates included in the ACDE, DHA1, and DHA2 families all catalyze drugs:H^+ antiport, while the two remaining families, V-BAAT and SIT, catalyze uptake of their substrates into eukaryotic vesicles (equivalent to export from the cell cytoplasm), also via a substrate:H^+ antiport mechanism. The last subcluster in cluster 11 includes eight families. Most take up anionic metabolites. These include the sugar anion, sialic acid (SHS), a range of anionic metabolites (MHS), aromatic anions (AAHS), and phosphate (PHS). However, the ubiquitous SP family takes up sugars, the VNT family members take up neurotransmitters into synaptic vesicles of animals, and the eukaryotic-specific OCT family members take up or export organic cations preferentially, but sometimes they transport organic anions as well. Several members of this last mentioned family transport a variety of drugs, in part due to their broad specificities.

8. Evolution of Drug Exporting Pumps

Active drug exporters are extremely important for medical applications (6). However, they did not develop with the advent of modern medicine, but instead became activated for drug export upon the use of these compounds by humans, particularly for use in farm animals. The amounts used for animals, raised for meat production, exceed those used for humans by over 1000-fold (5, 6). Our studies have led us to conclude that these transporters have been in bacteria, archaea, and eukaryotes for billions of years. They probably evolved for three distinct purposes: (a) to provide protection from toxic substances secreted by other organisms for the purpose of biological warfare, (b) to catalyze secretion of toxic substances made by the organism possessing these drug export pumps, thereby maintaining low, non-toxic levels of the substances in the cytoplasms of the producing organisms, and (c) to allow drug export via normal metabolite exporters when structural features of the drug and metabolite exhibit similarities (21). In this last regard, it should be noted that many normal cellular metabolites are toxic at high concentrations, and many toxic substances have been shown to serve valuable functions (e.g., communication, gene induction, metabolic coordination) not related to their toxic activities (24). Hence the distinction between "toxin" and "metabolite" blurs.

Examples of drug exporters that also catalyze extrusion of normal metabolites are numerous (25). For example (1) the DTX1 and the EDS5 pumps of *Arabidopsis thaliana* (TC#2.A.66.1.8 and 2.A.66.1.11) export plant alkaloids and salicylate, respectively. (2) The four TMS Small Multidrug Resistance (SMR) family within the DMT superfamily includes the MdtI protein of *E. coli* (TC#2.A.7.1.9) which is a spermidine exporter. (3) The NepAB homologue of *Arthrobacter nicotinovorans* (TC#2.A.7.1.8), within this same SMR family, may export methylamine, a normal product of nicotine degradation.

In the HAE1 family of drug resistance pumps in the RND superfamily, the AcrEF system of *E. coli* (2.A.6.2.1) has been shown to be required for chromosomal condensation and segregation as well as for cell division, while the AcrAB pump of *E. coli* (2.A.6.2.2) exports phospholipids. Other lipid exporters in the RND superfamily include the MtrCDE pump of *Neisseria gonorrhoeae* (2.A.6.2.5), and both MexAB of *Pseudomonas aerug-*

inosa (TC#2.A.6.2.6) and BpeAB of *Burkholderia pseudomallei* (TC#2.A.6.2.23). They may also export L-homoserine lactone autoinducers, essential for intercellular communication in these species. References for all of these systems can be found in TCDB.

Finally, in the MFS, many members of the drug:H^+ antiporter families (families DHA1, 2, and 3) export normal cellular metabolites. Just for the DHA1 family, these include (1) Bsu1 (Car1) of *Schizosaccharomyces pombe* (TC#2.A.1.2.1) which transports vitamins B1 and B6, (2) Blt of *Bacillus subtilis* (2.A.1.2.8) and TPO1 of *Saccharomyces cerevisiae* (TC#2.A.1.2.16) which export polyamines, (3) VMAT1, VMAT2, and VAT1 of mammals (TC#2.A.1.2.11, 12, and 29) which transport monoamines such as dopamine, norepinephrine, serotonin, and histamine, (4) Unc17 of *Caenorhabditis elegans* (TC#2.A.1.2.13) and VAChT of humans (TC#2.A.1.2.28) which secrete acetylcholine from the cytoplasm into synaptic vesicles, (5) bacterial exporters for sugars (TC#s 2.A.1.2.14, 15, 18, and 23), (6) The PbuE nucleobase exporters of *Bacillus subtilis* (TC#2.A.1.2.25), (7) siderophore exporters (TC#s 2.A.1.2.27 and 37), and (8) short-chain monocarboxylate exporters (TC#2.A.1.2.36).

The phylogenetic analyses reported here reveal that porters responsible for the extrusion of toxic hydrophobic and amphipathic substances generally fall into several families within superfamilies. These are sometimes (but not always) more closely related to each other than to families consisting of porters that transport hydrophilic substances. These observations suggest that these families have evolved independently only a few times during evolutionary history. In the four superfamilies examined here, those that function primarily as drug exporters may have evolved just once for the MATE family within the MOP superfamily (*see* **Fig. 3.1**) and for the closely related SMR and SMR2 families within the DMT superfamily (*see* **Fig. 3.2**). In the RND superfamily, we suggest that the HAE2 and HAE3 families, shown to be closely related (*see* **Fig. 3.3**), arose independently of the HAE1 family. Thus, two distinct hydrophobe/amphiphile exporter families, one in Gram-negative bacteria primarily and the other in Gram-positive bacteria and archaea, arose independently of each other. Finally, in the MFS, drug exporters apparently evolved independently at least three times (*see* **Fig. 3.4**; (a) DHA1 and DHA2, (b) DHA3, and (c) VNT and OCT).

In some cases we can reasonably predict the origins of these drug exporting families. Sometimes they probably arose by mutational change from complex carbohydrate efflux pumps (e.g., MOP1). In other cases, they may have come from lipid exporters (e.g., RND). In still other cases they most likely arose from metabolite exporters (e.g., DMT). In both the MFS (*see* **Fig. 3.4**) and the ABC superfamily (unpublished results) they almost certainly arose from multiple sources. In the ABC superfamily, a common source was probably peptide/protein secretion pumps. Our analyses thus not only allow us to predict the numbers of times drug resistance arose in each superfamily but also the nature of the transporters from which they derived.

While all members of a superfamily are likely to share certain structural and mechanistic features, this cannot be claimed for members of independently evolving superfamilies. Thus, while most DMT and MFS superfamily members probably function by drug:H^+ antiport, pumping their substrates from the cytoplasm, RND pumps of the HAE1 family may "suck" their hydrophobic substrates out of the cytoplasmic membranes or the periplasm (e.g., the vacuum cleaner model) (26). Further, most, but not all, MATE family members within the MOP superfamily prefer Na^+ as the antiported cation rather than H^+. It is therefore clear that distinct mechanistic features of these pumps may prove to be more advantageous to some organisms than to others. This fact may explain the dissimilar distributions of these transporters in different types of organisms (8).

9. Conclusions and Perspectives

For the past decade, our laboratory has established homology between distantly related transporters, thereby establishing protein superfamilies (13). We were, in fact, responsible for identifying, naming, and expanding all of the superfamilies of secondary active transporter described here, the MOP, DMT, RND, and MFS superfamilies (3, 5, 17, 18). These are the principal superfamilies of drug efflux porters, except for the ATP-binding cassette (ABC) superfamily of primary active transporters (8, 20). However, we had been frustrated by the inability of classical programs to construct reliable phylogenetic trees when sequence divergence between family members was extensive. Trees based on multiple alignments did not give consistent results and proved to be untrustworthy. This provided the incentive to design the SFT1 and SFT2 programs described and used here.

The SFT1 and SFT2 programs require sequential use because in essence the results of the SFT1 program are used to construct the SFT2 trees. Thus, SFT1 trees show the relationships of all inputted protein sequences, relative to each other, and they allow the user to evaluate the relationships of all proteins within a family or superfamily. By contrast, the SFT2 composite tree reveals the relationships of the represented families (not the proteins) to each other. Because it "averages" the results for all family members, the SFT2 tree is both more accurate and easier to read. Because it lacks the ability to reveal relative positions of the individual proteins, complete analyses of a superfamily require presentation of both SFT1 and SFT2 trees.

Comparisons with multiple alignment-based trees revealed excellent agreement with the SFT program-based trees as long as sequence similarities of the proteins analyzed are sufficient to allow construction of accurate multiple alignments (12, 27). However, with more divergent sequences, as occur in the large superfamilies analyzed here, multiple alignments become the results of a computerized "guessing game" and cannot be trusted. Given this situation, our results have shown that the SFT programs are superior to multiple alignment-based trees. These programs, which are as easy to use as the classical programs, should be applicable to all types of proteins (e.g., enzymes, structural proteins, regulatory protein) as well as to all types of stable nucleic acid molecules. It should be mentioned, however, that the establishment of homology is essential before their use is justified.

Why is it important to define the phylogenetic relationships of distantly related families? There are several reasons. Extrapolation of high resolution structural data is only justified when homology is established. Further, the degree of structural divergence is generally proportional to the phylogenetic distances between the two sequences being compared. The same is true for mechanistic and functional data as illustrated here. Moreover, we anticipate that the properties of proteins and nucleic acids (e.g., subcellular localization, biogenesis, protein associations) will prove to conform to the same principles. As genome sequence databases expand, the need for predictive algorithms based on phylogeny is likely to expand exponentially with database size.

For the first time, we can now estimate phylogenetic relationships for the largest macromolecular superfamilies found in living organisms on earth. The potential value of these programs is nearly unlimited, especially for predictive purposes. We hope the SFT programs will prove useful for designing systematic means for predicting and evaluating the evolutionary, functional, mechanistic, and structural relationships of proteins to each other and of nucleic acids or nucleic acid-based entities to each other. These programs should provide an indispensable aid for macromolecular classification, for drug design, for ligand binding predictions, and for bioinformatic extrapolations.

Acknowledgments

We thank Dorjee Tamang for useful discussions and technical assistance. This work was supported by NIH grant GMO77402.

References

1. Zhang, Z., Ma, C., Pornillos, O., Xiu, X., Chang, G., and Saier, M.H. Jr. (2007) Functional characterization of the heterooligomeric EbrAB multidrug efflux transporter of *Bacillus subtilis*. *Biochemistry* **46**, 5218–5225.
2. Kim, S.H., Chang, A.B., and Saier, M.H. Jr. (2004) Sequence similarity between multidrug resistance efflux pumps of the ABC and RND superfamilies. *Microbiology* **150**, 2493–2495.
3. Hvorup, R.N., Winnen, B., Chang, A.B., Jiang, Y., Zhou, X.F., and Saier, M.H. (2003) The multidrug/oligosaccharidyl-lipid/polysaccharide (MOP) exporter superfamily. *Eur. J. Biochem.* **270:** 799–813.
4. Chung, Y.J. and Saier, M.H. Jr. (2002) Overexpression of the *Escherichia coli sugE* gene confers resistance to a narrow range of quaternary ammonium compounds. *J. Bacteriol.* **184,** 2543–2545.
5. Jack, D.L., Yang, N.M., and Saier, M.H. Jr. (2001) The drug/metabolite transporter superfamily. *Eur. J. Biochem.* **268,** 3620–3639.
6. Paulsen, I.T., Chen, J., Nelson, K.E., and Saier, M.H. Jr. (2001) Comparative genomics of microbial drug efflux systems. *J. Mol. Microbiol. Biotechnol.* **3,** 145–150.
7. Chung, Y.-J. and Saier, M.H. Jr. (2001) SMR-type multidrug resistance pumps. *Curr. Opin. Drug Discov. Dev.* **4,** 237–245.
8. Saier, M.H. Jr. and Paulsen, I.T. (2001a) Phylogeny of multidrug transporters. *Semin. Cell Dev. Biol.* **12,** 205–213.
9. Saier, M.H. Jr. (1994) Computer-aided analyses of transport protein sequences: gleaning evidence concerning function, structure, biogenesis, and evolution. *Microbiol. Rev.* **58,** 71–93.
10. Saier, M.H. Jr. (2000). A functional-phylogenetic classification system for transmembrane solute transporters. *Microbiol. Mol. Biol. Rev.* **64,** 354–411.
11. Busch, W., and Saier, M.H. Jr. (2002) The transporter classification (TC) system. *CRC Crit. Rev. Biochem. Mol. Biol.* **37,** 287–337.
12. Yen, M.R., Choi, J., and Saier, M.H., Jr. (2009) Bioinformatic analyses of transmembrane transport: novel software for deducing protein phylogeny, topology, and evolution. Manuscript submitted for publication.
13. Chang, A.B., Lin, R., Studley, W.K., Tran, C.V., and Saier, M.H. Jr. (2004) Phylogeny as a guide to structure and function of membrane transport proteins. *Mol. Membrane Biol.* **21,** 171–181.
14. Saier, M.H. Jr., Tran, C.V., and Barabote, R.D. (2006) TCDB: the transporter classification database for membrane transport protein analyses and information. *Nucleic Acids Res.* **34,** D181–D186 (Database issue).
15. Saier, M.H. Jr., Yen, M.R., Noto, K., Tamang, D., and Elkan, C. (2009) The transporter classification database (TCDB): recent advances. *Nucleic Acids Res.* **37,** D274–D278 (Database issue).
16. Zhai, Y., Tchieu, J., and Saier, M.H. Jr. (2002) A web-based tree-view (TV) program for the visualization of phylogenetic trees. *J. Mol. Microbiol. Biotechnol.* **4,** 69–79.
17. Tseng, T.T., Gratwick, K.S., Kollman, J., Park, D., Nies, D.H., Goffeau, A., and Saier, M.H. Jr. (1999) The RND permease superfamily: an ancient, ubiquitous and diverse family that includes human disease and development proteins. *J. Mol. Microbiol. Biotechnol.* **1,** 107–125.
18. Pao, S.S., Paulsen, I.T., and Saier, M.H. Jr. (1998) Major facilitator superfamily. *Microbiol. Mol. Biol. Rev.* **62,** 1–34.
19. Higgins, C.F. (1992) ABC transporters: from microorganisms to man. *Annu. Rev. Cell Biol.* **8,** 67–113.
20. Davidson, A.L. and Maloney, P.C. (2007) ABC transporters: how small machines do a big job. *Trends Microbiol.* **15,** 448–455.
21. Saier, M.H. Jr., Paulsen, I.T., and Matin, A. (1997) A bacterial model system for understanding multi-drug resistance. *Microb. Drug Resist.* **3,** 289–295.
22. Goel, A.K., Rajagopal, L., Nagesh, N., and Sonti, R.V. (2002) Genetic locus encoding functions involved in biosynthesis and outer membrane localization of xanthomonadin in *Xanthomonas oryzae* pv. oryzae. *J. Bacteriol.* **184,** 3539–3548.
23. Poplawsky, A.R., Urban, S.C., and Chun, W. (2000) Biological role of xanthomonadin pigments in *Xanthomonas campestris* pv. campestris. *Appl. Environ. Microbiol.* **67,** 245–250.
24. Sahl, H.G. and Bierbaum, G. (2008) Multiple activities in natural antimicrobials. *Microbe* **3,** 467–473.
25. Saier, M.H. Jr. and Paulsen, I.T. (2001a) Phylogeny of multidrug transporters. *Semin. Cell Dev. Biol.* **12,** 205–213.
26. Aires J.R. and Nikaido H. (2005) Aminoglycosides are captured from both periplasm and cytoplasm by the AcrD multidrug efflux

transporter of Escherichia coli. *J. Bacteriol.* **187,** 1923–1929.
27. Lorca, G., Reddy, L., Yiu, C., Nguyen, A., Patel, S., Sun, E.I., Tamang, D., Wang, B., Wong, F.H., Yen, M.R., and Saier, M.H. Jr. (2010) Lactic acid bacteria: genomic analyses of transport systems. In *Biotechnology of Lactic Acid Bacteria: Novel Applications* (F. Mozzi, R.R. Raya, and G.M. Vignolo, eds.). Wiley-Blackwell, Oxford.

Chapter 4

Targeting Drug Transporters – Combining In Silico and In Vitro Approaches to Predict In Vivo

Praveen M. Bahadduri, James E. Polli, Peter W. Swaan, and Sean Ekins

Abstract

Transporter proteins are expressed throughout the human body in different vital organs. They play an important role to various extents in determining absorption, distribution, metabolism, excretion, and toxicity (ADME/Tox) properties of therapeutic molecules. Over the past decade, numerous drug transporters have been cloned and considerable progress has been made toward understanding the molecular characteristics of individual transporters. In this chapter several in vitro and in silico techniques are described with applications to understand transporter behavior. These include employing new techniques to rapidly identify novel ligands for transporters. Ultimately these methods should lead to a greater overall appreciation of the role of transporters in vivo.

Key words: ATPase assay, BBMV, BLMV, drug discovery, drug–transporter interaction, efflux assay, in vitro and in silico correlation, P-glycoprotein, transporters.

1. Introduction

The human genome contains 406 ion channel coding genes and 883 genes encoding transporters, of which 350 are intracellular transporters. Transporters are a group of transmembrane proteins responsible for carrying nutrient molecules and a range of therapeutic molecules of different size and shape mimicking nutrients across the membrane (1). Transporters can be classified into two distinct superfamilies, the solute carrier class (SLC) containing over 30 families and 200 members (http://www.bioparadigms.org/slc/menu.asp) (2) and the

Q. Yan (ed.), *Membrane Transporters in Drug Discovery and Development*, Methods in Molecular Biology 637,
DOI 10.1007/978-1-60761-700-6_4, © Springer Science+Business Media, LLC 2010

ATP-binding cassette (ABC) family containing 7 families and over 48 members, including the widely studied MDR1 gene (P-glycoprotein, P-gp) (3). These two superfamilies significantly contribute to xenobiotic disposition by influencing drug absorption (through uptake transporters), tissue distribution (through efflux and uptake transporters), metabolism (through hepatic transporters), and elimination (through renal and biliary transporters) (3, 4). Due to overlapping molecular pharmacophores of different therapeutic classes, successful oral delivery and adequate achievement of bioavailability make drug discovery challenging. This chapter will emphasize the importance of transporter proteins at various levels of ADME and as determinants of toxicological properties of therapeutic molecules. Transporter proteins as targets for drug delivery will also be discussed. We will provide a comprehensive account of experimental and computational methods adopted to study these significant proteins.

With only 10% of the human genome representing druggable protein targets identified and only half of those being relevant to diseases (5), it is important to be able to predict how druggable a novel target is in early drug discovery (6). On the basis of potential protein targets, transporter proteins are systematically explored to improve the overall efficacy of many discovery compounds and also a number of poorly absorbed drug molecules. Thus studying structure, function, and regulation of transporters should allow us to competently develop molecules with near-ideal pharmacokinetic properties. Consequently, drug discovery involving utilization of in vivo transporter mechanisms will require rapid high-throughput in vitro screening system for transporters. Methods allowing the rational prediction and extrapolation of in vivo drug disposition from in vitro data are urgently required (7).

Based on mechanism, drug uptake across epithelial barriers can be conveniently classified into two categories: passive absorption and active transport. Passive absorption involves carriers, channels, or direct diffusion through a membrane. This type of uptake always operates from regions of greater concentration to regions of lesser concentration. No cellular source of energy is required. Examples of passive absorption include simple diffusion, channel diffusion, and facilitated diffusion. Active transport can be categorized into primary active transport, secondary active transport, and tertiary active transport. Primary active transporters utilize ATP as their direct source of energy (ABC transporters). Secondary and tertiary active transporters take advantage of a previously existing concentration gradient or potential difference across cellular membranes as energy source (SLC proteins). **Tables 4.1** and **4.2** summarize the tissue expression and substrate specificities of some important members of the active transporter family.

Table 4.1
SLC transporter proteins and substrate specificities

SLC transporter	Acronym and gene	Tissue expression	Substrate specificities	References
Oligopeptide transporter 1	PEPT1 (*SLC15A1*)	Small intestine, kidney cortex	Dipeptides, tripeptides, β-lactam antibiotics, ACE inhibitors, bestatin, valacyclovir	(19, 42, 121)
Oligopeptide transporter 2	PEPT2 (*SLC15A2*)	Kidney medulla and cortex, brain, lung and kidney	Dipeptides, tripeptides, β-lactam antibiotics, ACE inhibitors, bestatin, valacyclovir	(199–202)
Organic anion transporter polypeptides	OATP (*SLC21A* family)	Intestine, liver, kidney, and brain	Bile salts, hormones, statins, cardiac glycosides, methotrexate, rifampicin	(203–207)
Organic anion transporters	OAT (*SLC22A* family)	Kidney proximal tubules	β-lactam antibiotics, probenecid, loop and thiazide diuretics, ACE inhibitors, NSAIDs, and methotrexate, cyclic nucleotides, prostaglandins, folate, neurotransmitter-metabolites, and hormone-conjugates	(208–212)
Organic cation transporters	OCT (*SLC22A* family)	Liver (OCT1) and kidney (OCT2)	Metformin, oxaliplatin and cisplatin, pindolol	(213–219)
Sodium taurocholate co-transporting polypeptide	NTCP (*SLC10A* Family)	Hepatocytes	Uptake of bile salt, bosentan	(220, 221)
Apical sodium-dependent bile acid transporter	ASBT (*SLC10A2*)	Cholangiocytes, distal ileum, renal proximal tubules	Bile salts, unconjugated bile acids	(46, 222–224)

Table 4.2
ABC transporter proteins and substrate specificities

ABC transporter	Acronym and gene	Tissue expression	Substrate specificities	References
P-glycoprotein	P-gp (MDR1)(*ABCB1*)	Tissues like colon, liver, kidney, lung, jejunum, brain, prostate, heart and kidney cortex; blood–tissue barriers like BBB, blood–testis barrier, and blood–placenta barrier, blood–ocular barrier, retinal endothelial cells, conjunctival epithelial cells, iris and ciliary epithelial cells	Typically neutral or positively charged hydrophobic compounds with MW >500, hydrogen bond donors >5, hydrogen bond acceptors >10 and clogP >5.0	(8, 40, 56, 225–228)
Breast cancer-resistance protein	BCRP (*ABCG2*)	Ovary germinal cells, placenta cytotrophoblast, small intestine, cervix epithelium, breast lobules, stomach epithelium, heart and myocardium	M itoxantrone, camptothecin-derived and indolocarbazole topoisomerase I inhibitors, methotrexate, flavopiridol, and quinazoline ErbB1 inhibitors, HIV protease inhibitors	(154, 158, 165, 229–232)
Multidrug-resistance protein superfamily	MRP 1 to 9 (*ABCC* superfamily)	MRP2 is expressed on apical membrane of hepatocytes, renal proximal tubular cells, enterocytes and syncytiotrophoblasts of the placenta	Phase II products of biotransformation (conjugates of lipophilic substances with glutathione, glucuronate, and sulfate)	(146, 233, 234)
Bile salt export pump	BSEP(*ABCB11*)	Apical side of canalicular membrane of liver	Bile salts, troglitazone and its metabolite troglitazone sulfate, pravastatin	(97, 235, 236)

Given our understanding of expression and tissue distribution of transporter proteins, it is important to decide early during the drug discovery process whether a drug–transporter interaction of the lead compound is desirable. Transporters therefore have significant clinical importance. Understanding and exploiting the transporter function either by targeting them to increase the bioavailability or by avoiding them completely to tackle the distribution issues, thereby preventing the toxic side effects of drugs are needed (8). With the overlapping substrate specificities of the transporters involved in pharmacokinetic processes, drug interactions involving transporters can often have a direct and adverse effect on therapeutic safety and efficacy of critical drugs (8).

2. Transporters of Therapeutic Relevance and Their In Vivo Expression

Membrane transporters have been utilized as targets for treatment of certain disorders and prodrug targeting. Due to their broad substrate specificities, some of the transporters are a major concern for potential drug–drug interactions (DDI). SGLT2 (SLC5A2) is a low-affinity sodium glucose transporter, expressed specifically in kidney and plays an important role in glucose reabsorption in the kidney. SGLT2 inhibitors enhance renal glucose excretion and consequently lower plasma glucose levels (9). Phlorizin and some thioglycosides were identified to be novel inhibitors of SGLT1 and SGLT2 but lacked specificity (10, 11). Sergliflozin-A, a prodrug of sergliflozin, was found to be a potent and highly selective inhibitor of human SGLT2. The drug was tested in rats and was found to exhibit a glucose-lowering effect independent of insulin secretion (12). An example of a transporter as target for drug delivery is gabapentin. Gabapentin is used to help control certain types of seizures in patients who have epilepsy. It is thought that gabapentin is absorbed from the intestine of humans and animals by a low-capacity solute transporter localized in the upper small intestine. XP13512 is a novel prodrug of gabapentin and was recently designed to be absorbed throughout the intestine by high-capacity nutrient transporters (13). It was shown that this prodrug inhibited transport of radiolabeled tracers across Caco-2 cell monolayers, in cells expressing human sodium-dependent multivitamin transporter (SMVT) and human embryonic kidney cells expressing the monocarboxylate transporter type-1 (MCT1). Thus, administration of this prodrug should lead to increased bioavailability of gabapentin through a transporter-dependent mechanism (13). Similar examples of transporter-targeted drug delivery include XP19986, a prodrug or *R*-baclofen, developed to treat gastroesophageal reflux disease. This prodrug is a substrate for the intestinal monocarboxylate transporter, MCT. XP21279 is a transporter-targeted prodrug of levodopa for the treatment of Parkinson's disease. XP20925 is a transporter-targeted prodrug of propofol for the treatment of postanesthesia, nausea, vomiting, and migraine (14).

The small peptide transporter, PEPT1, expressed in intestinal epithelium has been targeted for prodrug delivery of drugs with poor bioavailability. Valacyclovir and valganciclovir are ester prodrugs of acyclovir and ganciclovir targeted to peptide transporters, PEPT1 and PEPT2 (15–17). The bioavailability of acyclovir (20%) is significantly improved (55%) when administered as the prodrug valacyclovir, which is a substrate for the peptide transporter PEPT1, implying the key role of peptide transporters in determination of bioavailability (18). Recently, functional expression of the small peptide transporter PEPT2 was demonstrated in primary human lung epithelial cells. The human lung has a vast surface area and extensive vasculature and has been envisioned as a target for noninvasive systemic delivery of peptides and peptidomimetic drugs (19). PEPT2 expressed at the blood–brain barrier (BBB) is responsible for uptake of some opioid peptides. Opioid peptides play a critical role in a variety of biological processes, including analgesia, constipation, respiration, euphoria, and sedation (20). In addition PEPT2 expressed in the choroid plexus has been shown to limit the exposure of cefadroxil to the cerebral spinal fluid (CSF), an antibiotic used to treat bacterial meningitis. These findings demonstrate that PEPT2 has an important role in limiting the exposure of cefadroxil in the CSF. PEPT2, located at the apical membrane of the choroid plexus epithelium, acts in a unidirectional style in transporting cefadroxil from CSF into the cell (21). PEPT2 knockout mice studies have shown that cefadroxil content in the CSF was sixfold higher in PEPT2-null mice than in wild type (22).

Expression of certain solute transporters is also upregulated in certain disease states. Some cancerous cell lines like Caco-2, AsPc-1, and Capan-2 expressed oligopeptide transporter PEPT1 at high levels (23). Drug delivery methods utilizing the oligopeptide transporter expression activity in target tissues could be useful. The anticancer drug bestatin is one of the first examples of the transporter and tissue-targeted technique of drug delivery (24). Pravastatin, a HMG-CoA inhibitor, is another example of transporter-targeted drug delivery and treatment. The enterohepatic circulation mediated by OATP and MRP2 maintains a significant concentration of the drug in the liver. The mechanisms governing these transporters were identified later (25, 26). Similarly a novel cisplatin analog, cisplatin-ursodeoxycholic acid (Bamet-UD2), has been used for colon tumors by targeting it to the NTCP transporter expressed in sinusoidal membranes (27).

Drug transporters thus aid in improvement of the bioavailability of poorly absorbed molecules. On the other hand, undesirable and unintentional drug interactions with transporters have led to reduced oral bioavailability or drug-induced toxicity of many orally administered drug compounds. Currently, lead identification and optimization in the pharmaceutical industry have been confounded with the challenge of evaluating the candidates for their affinity toward key efflux transporters. The extensively studied efflux transporter, P-glycoprotein (P-gp), is a potential bottleneck for drug discovery projects because of reduced oral bioavailability and DDI resulting from its induction or inhibition by many clinically relevant drug candidates. Ubiquitous expression of P-gp in the human body and overlapping substrate specificities with drug metabolizing enzymes pose a significant challenge in drug affinity assessment. P-gp-related DDI can arise from either induction (mediated by the pregnane X receptor, PXR) or inhibition of P-gp (28, 29). One well-known DDI involving P-gp is the administration of paclitaxel with cyclosporin-A (CsA) (30). But as both paclitaxel and CsA are inducers of CYP3A4, alternate strategies to administer paclitaxel with second and third generation inhibitors like PSC-833 (valspodar), GF120918 (elacridar), and LY335979 (zosuquidar) have been proposed (31–33). Co-administration of topotecan, a topoisomerase-1 inhibitor, with elacridar, a breast cancer resistance protein (BCRP)/P-gp inhibitor, improves the bioavailability of topotecan, an anticancer drug used to treat ovarian cancer and lung cancer (30, 34). Anti-HIV drugs used in combination antiretroviral therapy have been shown to inhibit BCRP to differing extents which may contribute to their efficacy and at the same time could contribute to other drug–drug interactions (35). St. John's Wort (SJW) is available over the counter and is used to treat mild depression and anxiety but is also known to induce CYP3A4 and P-gp expression through PXR receptor activation. SJW-induced P-gp and CYP3A4 are a major cause of clinically significant DDI with CsA (36), tacrolimus (37), talinolol (38), and several other clinically important drug candidates (38).

Though it is understood that drug transporters play a major role in the ADME/Tox of drugs, the mechanisms and their function are complex while the rate of understanding drug–transporter interactions has been slow. The possibility of transporter-based DDI increases with co-administration of drugs. As we understand the importance of such transporter-related DDI and possible involvement of transporters in drug-related toxicity, there is a need to rapidly identify the range of substrates as well as transporter targets with established roles in drug disposition, DDI, and other essential physiological processes.

3. Approaches to Identifying Ligands for Drug Transporters

With the advent of technology and new emerging tools like virtual screening and high-throughput screening, it is now possible to predict the drug properties with more accuracy (39). This has significantly increased the number of compounds with early data on ADME/Tox before proceeding to advanced stages of drug discovery. A variety of in silico and in vitro approaches aid in accurate prediction of these physicochemical and pharmacological characteristics and can maximize the output in drug discovery (40). With the lack of available crystal structures for therapeutically important transporter membrane proteins, the complimentary methods of virtual screening and in vitro approaches should significantly reduce the attrition rates of bringing an ideal candidate molecule to market (21, 40, 41). Ligand identification for drug transporters could be divided into two broad categories: in vitro and in silico approaches.

3.1. In Vitro Transporter Assays for Drug Discovery

Over the past decade explosion of molecular biology techniques has enabled a number of drug transporters to be cloned and expressed into different mammalian cell lines. Transporter cell systems or isolated cell membrane fractions expressing transporter proteins have been a valuable in vitro tool in identifying therapeutic molecules as either substrates, inhibitors, or inducers of transporters. Single transporters can be overexpressed in mammalian cell lines by stable or transient transfection techniques to study definitive effects and eliminate false results through contribution from other transporters. The selection of an in vitro tool for activity assessment depends on the overall objective of the study.

3.1.1. Mammalian Cell-Based Transporter Assays

The common cell-based assays include uptake inhibition assays, bidirectional and unidirectional transport assays, and cytotoxicity assays. These assays utilize fully differentiated monolayers of mammalian cell lines to conduct a range of studies. The cells are seeded onto cell culture-treated plates for a simple uptake or cytotoxicity assay. For transport assays and membrane permeability functions, cells are seeded onto semipermeable membrane support that develops into a monolayer resembling the physiological epithelial barrier in vivo.

The uptake assay enables the determination of Michaelis–Menten (MM)-type kinetic parameters (V_{max} and K_m) and K_i or IC_{50} values in a simple cell accumulation assay (42–45). Uptake assays are performed by measurement of the uptake concentration through direct treatment of cell monolayer with putative compounds followed by whole cell lysis and compound analysis using sensitive instrumental techniques like HPLC or LC-MS/MS or by radiolabeling the putative compounds which provide for direct measurement of compound activity. A popular technique is to use a prototypical radiolabeled tracer which is substrate for the transporter. Generally, MM type parameters are determined using a fixed concentration of radiolabeled tracer and saturating log concentrations of putative compounds under study. Some examples of radiolabeled tracers include ^{3}H-Gly-Sar for PEPT1 and PEPT2 (19, 42), ^{3}H-taurocholate for studies with ASBT (46), ^{3}H-digoxin and ^{3}H-amprenavir for P-gp studies (47, 48), and ^{3}H-uridine for nucleoside transporters. Alternately, fluorescently labeled probes offer sensitivity and speed for in vitro assessment of compound activity for transporters. Calcein–AM is a non-fluorescent, highly membrane permeable dye and a substrate for P-gp. Inhibition of P-gp causes increased cellular accumulation of calcein–AM, which is hydrolyzed to the membrane impermeant and fluorescent calcein by intracellular esterases. The amount of fluorescence is proportional to the activity of the inhibitor of P-gp being tested (48, 49). The general rule is that all P-gp substrates inhibit calcein–AM uptake (50). Hoechst 33342 was recently identified as fluorescent marker for studying functional expression of BCRP in hematopoietic cells (51–53). Rhodamine 123 for P-gp (51), LysoTracker for BCRP, and BODIPY-prazosin for BSEP (54, 55) are other examples of fluorescent substrates used to perform functional studies on drug transporters. Cell accumulation and transport assays can also aid in prediction of possible DDI (19, 42).

Transepithelial transport assays use polarized mammalian cells cultured on a semipermeable membrane supports. Polarized cell lines cultured on a semipermeable membrane differentiate into a monolayer with functional tight junctions that resemble the physiological epithelial barrier in human tissues. Measurement of transepithelial electrical resistance (TEER) or flux of membrane integrity markers like radiolabeled mannitol, polyethylene glycol, or the fluorescent dye Lucifer yellow allows one to ensure the quality of monolayers. Transepithelial transport models are useful to determine passive permeability as well as the affinity of a drug in a transport assay setting (19, 56). These assays can be developed in a high-throughput setting to screen large databases of compounds for transporter activity (57, 58). The human intestinal cell lines Caco-2, HT29, and T84 have been used to study the intestinal transport processes (59). Bidirectional transport assays are intended to study mainly efflux transporters to determine if the selected drug candidates are substrates, non-substrates, or inhibitors for an efflux transporter (40, 48). Apparent permeability (P_{app}) is calculated by normalizing flux (J) with area of membrane (A) and the concentration (C_0) of the compound. For efflux transporters, the efflux ratio (ER) is calculated by dividing the P_{app} basolateral to apical by P_{app} apical-to-basolateral directions. Typically an ER greater than two signifies limited absorption and increased secretion of a molecule. Currently widely used cell lines for efflux transport assays are the human colon adenocarcinoma cell line Caco-2 (48), Madin–Darby canine kidney cell line transfected with the*MDR1*gene, MDCKII-MDR1 [(60), Chang, 2006, #1], pig kidney cell line transfected with the MDR1 gene, L-MDR1 (61), and MDCKII transfected with MRP2 (62).

Caco-2 cells have been extensively used over the last 20 years as a model of the intestine (63). In culture Caco-2 cells undergo differentiation forming brush border apical membrane and functional tight junctions expressing several morphological and functional characteristics of the mature enterocytes (64). Caco-2 cells express several ABC transporters like P-gp (65), BCRP (66), MRP1 (67), and MRP2 (66) as well as SLC transporters like PEPT1 (68), OCT, ASBT, OAT, and MCT (69, 70). However, expression of specific transporters in Caco-2 cells can be variable. Alternatively, single transporters can be overexpressed in mammalian cell lines by stable or transient transfection to study and eliminate false results through the contribution of other transporters, as well as to provide potentially more reliable high expression. MDCKII, LLC-PK1, HeLa, and CHO-K1 represent model cell lines that are used to transfect different transporter genes because of their low background transporter expression. The MDCKII-MDR1 cell model is extensively used to conduct definitive P-glycoprotein studies. This model demonstrates a differential affinity toward the same substrates when compared with Caco-2 cells (62).

The limitations of using transfected cell lines as well as immortalized cell lines like Caco-2 are that they lose their capacity to express proteins over a period of time due to repeated culturing and variation in culture conditions. They also lose their characteristic to resemble the physiological epithelial barrier (48, 71, 72). This limitation can be overcome by using positive control substrates and maintaining consistency among cell culture conditions. Primary cell lines offer the advantage of overcoming this limitation to a certain extent. Primary cells isolated directly from human subjects resemble native tissue in terms of phenotype and genotype. It has been demonstrated that primary cell cultures offer a robust model to study drug transporters and evaluate the functional presence of transporters (19). Some examples of human primary cell models include the primary human lung epithelial cells (hLEC) to study human peptide transporter PEPT2 (19) and primary cultures of human proximal tubular cells (hPT) used as a model to study several key transporters for organic anions (OAT1 and OAT3) and cations (OCTN1 and OCTN2), peptides (PEPT2), MRP2, MRP5, P-glycoprotein, and neutral amino acids (73). Other notable examples include primary cultured bovine brain microvessel endothelial cells to study the blood–brain barrier (74) and rabbit primary corneal epithelial cell culture (75). Limited availability, the laborious isolation procedure of cells, short lifespan, and restricted growth potential are some of the disadvantages of using primary cell lines in studying transporters.

Since the objective of these in vitro transepithelial transport models is to predict the in vivo absorption of drug molecules, it is important to establish an in vitro–in vivo correlation. Some of the in vitro models developed to date have been shown to have a good correlation with in vivo activity of drug molecules (60–62, 76).

Cytotoxicity assays are performed to determine the potentiation or attenuation of toxicity of the putative compounds in drug transporter expressing cells while comparing them to wild-type cells that do not express these transporters. Generally the estimation parameters are the determination of IC_{50} values for the compound under study and are expressed as fold reversion or MDR ratio. Fold reversion is the ratio of IC_{50} of cytotoxic drug alone to the IC_{50} of the cytotoxic drug in the presence of the putative inhibitor (77). For example, the BCRP inhibitors Ko143, Ko132, and Ko134 were identified by this technique and were successful in reversing the BCRP-mediated resistance to mitoxantrone and topotecan in MEF3.8 cells with a murine *Bcrp1* expression (78).

3.1.2. Non-cell-Based Transporter Assays

Certain non-cell-based assays use membrane fractions isolated from cells expressing the transporter protein. Three main types of non-cell-based approaches are *ATPase* activation assays, membrane vesicular transport assays, and ligand binding studies. The advantages of using non-cell-based systems include their high-throughput capabilities, elimination of the effects from other transporters, relatively easier handling of the membrane preparations, and faster procedure.

Non-cell-based approaches use membrane fractions that express ABC transporters like MDR1, BCRP, and MRP which depend on ATP as the energy source for their function. The energy released as a result of ATP hydrolysis postbinding to nucleotide binding domain (NBD) is utilized by the transporter as the "driving force." This cytoplasmic activity is vanadate sensitive and can be stimulated or inhibited by substrates for these transporters. At the molecular level, the complex generated with ADP, inorganic vanadate, and divalent metal ions like Mg^{+2}, Mn^{+2}, and Co^{+2} is trapped in the transition state. The conformation of the trapped state resembles that of the catalytic transition state and the complex trapped at one of the NBD completely inhibits the *ATPase* activity at both NBD sites. The substrates or inhibitors of the transporter can stimulate or inhibit the formation of the transition state complex formation (79, 80). The extent of stimulation or inhibition of the *ATPase* activity is a measure of the affinity toward the ABC transporter. ABC transporter expressing membranes are prepared either from insect cell membranes or mammalian cell membranes. P-gp, MRP, and BCRP cDNA is cloned using baculovirus-infected insect cells (High Five, BTI-TN5B1-4) (81–83). A membrane preparation that is prepared after infection with wild-type virus is used as a control for native activities. The membrane preparations from insect cell or mammalian cell membranes also contain non-ABC transporter-related ion translocating *ATPases*. Inhibitors such as sodium azide (mitochondrial *ATPase* inhibitor), oligomycin and ouabain (Na^+/K^+ *ATPase* inhibitor), and EGTA (Ca^{+2} *ATPase* inhibitor) are often used to eliminate potential false and interfering results (84, 85). These inhibitors are non-interfering with the *ATPase* activity related to P-gp, MRP, and BCRP. The inorganic phosphate released in the process can be quantitated by colorimetric assay using visualizing reagents under acidic conditions (81) or by using luciferase generated luminescence signal (86). The drug-stimulated *ATPase* activity is determined (nmol/min/mg of protein) as the difference between the amounts of inorganic phosphate released from ATP in the absence and presence of vanadate. Drug-stimulated ABC transporter activity is usually reported as fold stimulation relative to the basal ABC transporter activity in the drug (vehicle control). Typically, test compounds that stimulate the *ATPase* activity by twofold are considered P-gp ligands.

Based on the fold activation pattern, the compounds that stimulate can be divided into three classes. Class I compounds like vinblastine, verapamil, and paclitaxel stimulate *ATPase* activity at low concentrations but inhibit at higher concentrations. Class II compounds like bisantrene, valinomycin, and tetraphenylphosphonium activate the *ATPase* in a concentration-dependent manner. Class III compounds such as cyclosporin A, rapamycin, and gramicidin D inhibit both basal and verapamil stimulated *ATPase*activity (50, 77). The *ATPase* activation assay is easy to perform in a high-throughput setting as membrane preparations are commercially available and the method does not require radiolabeled compounds. However, this assay is not a functional assay and cannot distinguish between inhibitors and substrates. The data generated from the *ATPase* assay should be used with caution. Polli et al. have studied an exhaustive list of compounds with different in vitro P-gp assay techniques (48). Compounds that activate *ATPase* are not necessarily P-gp substrates or inhibitors. Well-known P-gp ligands like diltiazem, cyclosporine A, testosterone, mitomycin C, digoxin, GF120918, and Hoechst 33342 did not change *ATPase* activity, whereas compounds like ketoconazole, mebendazole, nifidipine, nicardipine, and verapamil are not apparently effluxed well but are *ATPase* activators (48, 77). Such apparent false-negative results could be due to the low affinity of the compounds for P-gp or the high concentration required to activate the *ATPase* reaction. Another reason is that moderate passive permeability is necessary for P-gp to reduce the apical-to-basolateral drug permeability (87). In particular, high passive permeability compounds that are efflux transporter substrates are less able to exhibit reduced apical-to-basolateral permeability due to efflux (88).

Membrane vesicles are another simplified assay system to study transporter behavior. Despite their disadvantages and limitations, membrane vesicles offer another easy system to work with. Brush border membrane vesicles (BBMV) and basolateral membrane vesicles (BLMV) are membrane closed unilamellar shells formed from membranes either in physiological transport processes or else when membranes are mechanically disrupted. They form spontaneously when the membrane is broken because the free ends of a lipid bilayer are highly unstable. They allow a closer look at the substrate interaction with only the brush border or basolateral membrane of the enterocyte that is devoid of underlying musculature and submucosa (89). Membrane vesicles can be prepared from either intestinal scrapings, isolated enterocytes, or mammalian cells that are stably or transiently transfected to overexpress the desired transporter (90). Membrane vesicles are also prepared from various parts of mammalian tissue like brush border membranes of intestine, kidney and choroid plexus (91), hepatic sinusoidal and canalicular membranes (92), and placenta (77, 93, 94). The advantages of using vesicles are that they allow one to isolate carrier systems from either apical or basolateral membranes of intestinal cells and they also allow the effect of a substance on the enzymatic and lipid composition of the intestine to be studied. The disadvantage of using the tissue-based membrane vesicles is the potential for interference due to endogenous transporters. BLMV may also contain membrane and organelle fragments. The small volume of vesicles also limits their use for studying concentrative transport, i.e., active carrier-mediated processes (89). Additionally, vesicle preparations are nonconsistent and vary from animal to animal or with small changes in the experimental conditions. Membrane vesicular transport assays can be performed in high-throughput mode and are mainly used to evaluate the ABC transporters like P-gp (95), BCRP (96), MRPs, and BSEP (97) and some uptake transporters like OCT3 (94) and NTCP (98). Alternatively, inside-out membrane vesicles are also used to evaluate affinity of new ligands for P-gp (99). One of the common drawbacks of using membrane vesicles is that preparation and purification are a time-consuming and laborious process; however, they are also easy to store at –80°C for months for subsequent later use.

Photolabeling studies are employed to study the substrate binding sites and affinities of various compounds in ABC transporters. These types of studies typically employ photoaffinity radiolabels like ^{3}H, ^{125}I, ^{32}P, and ^{35}S (100–102). A specific compound under investigation is radiolabeled by chemical processes and incubated with membranes prepared as membrane vesicles or protein in solution. The solution is then UV irradiated for set amount of time and proteins are then solubilized with Laemmli's buffer before running them on SDS-PAGE gel. The gel is then dried onto blotting paper and exposed to X-ray film at room temperature or at –70°C for 1–3 days. The extent of affinity is reflected by the intensity of the band after exposure. Commonly used photoaffinity labels are ^{3}H-azidopine (for P-gp), ^{125}I-iodoaryl azidoprazosin (P-gp and BCRP), ^{125}I-11-azidophenyl agosterol A (for P-gp and MRP), ^{3}H-LTC_4 (for MRP), ^{125}I-iodoaryl azido-rhodamine123 (IAARh123) (for P-gp, MRP, and BCRP), and 8-azido-(α-^{32}P)ATP (for ABC proteins) (77). **Table 4.3** summarizes the in vitro assay techniques available for studying drug transporters.

Table 4.3
Summary of some in vitro models used in drug discovery

In vitro transporter models

Cell-based models	Non-cell-based models
Uptake assays	*ATPase* activation assay
Transepithelial transport assays	Membrane vesicles
Cytotoxicity assays	Ligand binding assays

3.2. In Silico Transporter Models for Drug Discovery

Membrane proteins are known for their promiscuous nature for transporting molecules with structural resemblance to nutrients, across the epithelial barrier. Drug–drug and drug–nutrient structural similarity and overlapping specificity can be seen as a confounding yet advantageous factor in drug discovery and design. Correctly designing molecules with affinity for specific transporter proteins and accurately identifying molecules from large databases that have optimum interactions with different transporters are useful in early discovery. Molecular modeling techniques encompass various techniques to rapidly identify molecule structures that will interact or are likely to have an interaction with a transporter (103). The computational approaches to transporter-targeted drug discovery can be generally classified into protein modeling-based approaches and ligand-based approaches. Comprehensive reviews of these two methods have been published (56, 104).

3.2.1. Ligand-Based Transporter Modeling

Ligand-based molecular modeling using three-dimensional quantitative structure activity relationships (3D-QSAR) and pharmacophore methods are widely used methods for assessing the drug–transporter interactions (104). These include statistical tools to search correlations between a given properties and a set of molecular and structural descriptors (properties like log P, molecular mass, polar surface area) (103). The pharmacophore or QSAR model is generally trained using a known set of compounds with demonstrated experimental affinity against a transporter protein in question. The model is then used to predict whether molecules not in the training set are likely to have affinity based on the similarity to the molecules in the training sets. Methods such as Comparative Molecular Field Analysis (CoMFA) which correlate electrostatic and steric interactions with the biological activity of the molecules through partial least square (PLS) analysis (105) have also been widely used for transporter modeling. The pharmacophore modeling approach is commonly used when there are data sets with structurally diverse and conformationally flexible compounds (47). Three popular pharmacophore modeling tools are DIStance COmparisons (DISCO) (106), Genetic Algorithm Similarity Program (GASP) (107), and Catalyst/HIPHOP (47). The latter Catalyst program (Accelrys, Inc., San Diego, CA) has two distinct modules using the common chemical features of a few active drug molecules (108) or based on a series of molecules with varying structural activity and features (109). This method has been very widely used to model transporters by our group (40, 42, 110–114) (**Fig. 4.1**) due to its ease for database searching.

3.2.2. Transporter Protein Modeling

The dearth of membrane protein structures makes computational prediction a potentially important means of obtaining novel virtual protein structures. Computational techniques have been adopted to use the existing protein sequence data to model sequence alignment, motif search, functional residue identification, transmembrane segment and protein topology predictions, homology and ab initio modeling (115). For successful transporter protein modeling the availability of a correct protein template is necessary. Generally, for homology modeling >70% sequence similarity is essential to build a high-quality transporter homology model (116). It has been recognized that homology among membrane proteins can be extremely low, but the overall folding pattern is highly similar. In protein homology modeling, the protein mutual sequence similarity is converted into a 3D protein structure. The process usually starts with template identification followed by sequence alignment, model generation, model optimization, and model validation. Several automated algorithms like MODELLER, Sybyl COMPOSER, Insight

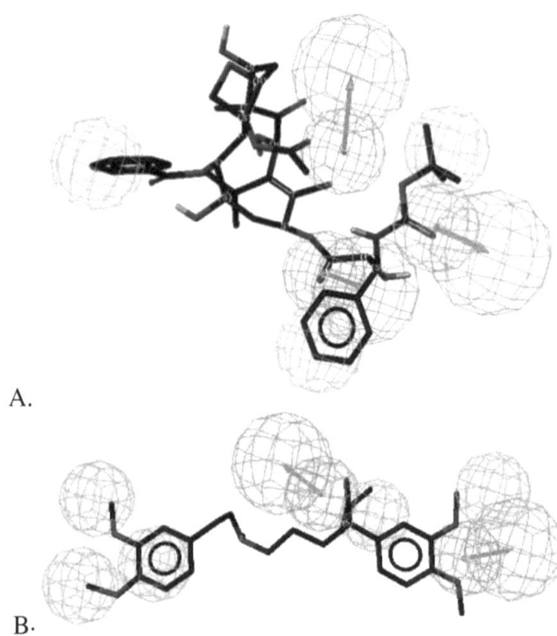

Fig. 4.1. (**a**) Docetaxol mapping to a BCRP inhibitor pharmacophore (56) and (**b**) verapamil mapping to a P-gp substrate pharmacophore (110).

II Homology, and ICM are available (116). The ab initio technique of protein modeling is independent of template identification. This technique has seen some success with transporter proteins because of their large structure and modeling of the transmembrane domains is generally difficult with this technique (117, 118). **Table 4.4** summarizes the computational techniques available for studying in silico modeling of drug transporters.

Table 4.4
Summary of some computational techniques available for transporter modeling

Transporter modeling		
Data modeling		**Protein modeling**
Pharmacophore modeling	3D-QSAR modeling	1. Homology modeling
		2. Ab initio modeling
1. DISCO	1. CoMFA	
2. GASP	2. CoMSIA	
3. Catalyst HIPHOP	3. GOLPE	
4. Catalyst Hypogen		

4. Combined In Silico and In Vitro Approaches to Drug Discovery

Combined in vitro and in silico approaches may improve the accurate profiling of ADME/Tox properties and assist in faster discovery of ligands for transporter-targeted drug delivery. In silico pharmacophore-based modeling has been used increasingly to develop models for the rapid screening of libraries of molecules for prioritization and in vitro testing, predicting the probability that compounds will have affinity for transporters.

The human PEPT1 is a clinically relevant and widely studied transporter involved in the transport of a broad range of substrates. In addition to facilitating transport of its natural substrates di- and tripeptides occurring in food products (119), peptidomimetics such as β-lactam antibiotics (120), ACE inhibitors (121), and bestatin (24). The PEPT1 transporter makes an attractive prodrug target because of its high capacity and relatively broad substrate specificity. Prodrugs like valacyclovir, valganciclovir (122), and LY544344 (123) were designed to target PEPT1. In order to rapidly identify compounds that have affinity for PEPT1, we developed a HIPHOP pharmacophore model for human peptide transporter PEPT1 using well-known substrates for PEPT1, each representing a different therapeutic class and being structurally different. The compounds were Gly-Sar (dipeptide), bestatin (peptidomimetic), and enalapril (ACE inhibitor) (113). The key features of the pharmacophore consisted of two hydrophobic features, a hydrogen bond donor, hydrogen bond acceptor, and a negative ionizable group. The pharmacophore was used to search a database of over 8,000 "drug-like" molecules in an attempt to identify other hPEPT1 ligands. One hundred and forty-five virtual hits mapped to the pharmacophore features. Seven of the best scoring molecules with drug-like properties (i.e., MW < 500) were selected and purchased for in vitro testing to ascertain the predictability of the pharmacophore model. Two commonly prescribed drug molecules, fluvastatin (antihyperlipidemic) and repaglinide (antidiabetic), and one component of the sugar substitute and pharmaceutical component, aspartame, were mapped to the pharmacophore features and were verified experimentally in vitro to be hPEPT1 inhibitors. This pharmacophore has also been used to assess the potential affinity of selected bacterial dipeptides. We observed that γ-iE-DAP was scored highest (124). A second pharmacophore developed with three high-affinity PEPT2 molecules (125) contained two hydrogen bond acceptors, two hydrogen bond donors, and one hydrophobic feature and γ-iE-DAP fit to all these features. These two transporters are quite similar, sharing steric, electrostatic, hydrophobic, and hydrogen bonding features, and their magnitude and localization are spatially different and distinguished by subtle differences in substrate specificity.

Other recent pharmacophore and QSAR models for hPEPT1 include those derived from conformational analyses of dipeptides which suggested the intramolecular π–π stacking, a distance of 5.6 Å from carboxyl to the basic amine, and a hydrogen bond acceptor was important (126). A set of 25 tripeptides and tripeptide mimetics with hPEPT1 K_i data were used with VolSurf descriptors and PLS to describe the activity. Molecules with more hydrophobic surface areas were shown to be preferred, while hydrogen bonding, negative charge were also suggested as important (127). Biegel and colleagues (125) generated CoMFA and CoMSIA models using 83 molecules for PEPT1 and PEPT2 and observed that higher hydrophobicity and electron density increase affinity for PEPT2 relative to PEPT1.

Small molecule substrates for P-gp results in reduced oral drug absorption and enhanced renal and biliary excretion. Limiting the exposure of xenobiotics to P-gp at the BBB level and placental barrier may also be important. As a consequence, P-gp plays a key role in determining drug distribution of many important drugs candidates. It is of great interest whether a drug candidate is a P-gp substrate or inhibitor as attempts to co-administer P-gp modulators, inhibitors, or inducers to increase cellular availability of other drugs by blocking the actions of P-gp have been met with only limited success. Thus, the rapid identification of P-gp substrates or inhibitors would be advantageous but this comes at increasing cost of identifying clinical drug candidates and circumvention of P-gp activity. This has resulted in a large number of experimental data sets charactering P-gp function, which has enabled extensive computational modeling (110, 111, 128, 129) that may assist in the rational design of potential inhibitors, substrates, or in some cases molecules that circumvent it (129).

For example, Cianchetta and colleagues used GRID alignment-independent descriptors (GRIND) and Almond and Volsurf descriptors with 129 substrates with a range of calcein–AM assay (130). The best PLS model was derived with Volsurf and Almond descriptors ($r^2 = 0.82$, $q^2 = 0.73$). A pharmacophore developed by the same group had some overlap in the features and distances with previously published models. A second group used the same types of descriptors for a set of 53 diverse drugs classifying them as substrates or non-substrates using PLS discriminant analysis (128). The model was tested with 272 proprietary molecules and attained a 72% prediction accuracy. A second model by the same group used 30 of the 53 molecules with calcein–AM inhibitor data and created a model for discriminating between substrates and inhibitors with >82% accuracy for a test set of 125 molecules. A GRIND 3D-pharmacophore containing multiple hydrophobic areas and two hydrogen bond acceptors was also proposed to be similar to other published models (128).

Globisch et al. evaluated a series of 32 anthranilamide Tariquidar analogs as P-gp inhibitors using 3D-QSAR methods: CoMFA and CoMSIA (131). Hydrogen bond acceptor, steric and hydrophobic fields were found to be most important. A second group looked at 49 anthranilamide analogs using the same methods and had similar findings (132). CoMFA and CoMSIA analysis has also been used with a series of 32 natural and synthetic coumarins to indicate the importance of the phenyl at position C4 as well as the α-(hydroisopropyl)dihydrofuran which possess a favorable electrostatic and steric volume (133). Linear discriminant analysis was used with topological (TOPS-MODE) descriptors for 163 molecules in the training set and 40 external molecules producing ~77% prediction accuracy (134). A recursive partitioning classification model using 125 molecules and 5 Cerius2 descriptors was generated for molecules that were P-gp substrates/non-substrates as determined in Caco-2 cells. When tested with 46 molecules the model performed better at predicting substrates (89%) than non-substrates (72%) (135).

Two studies independently described using multiple different descriptor types and algorithms with P-gp substrates in an effort to discover the most predictive model. One group termed this a Competitive Workflow comparing neural networks and classification trees using descriptors from a pool of over 2,000 (136). These models performed poorly on non-substrates due to the uneven balance of classes in the training data. A second group called their approach combinatorial QSAR comparing kNN, decision trees, binary QSAR, and support vector machines and using MolconnZ, atom pair, VolSurf, and MOE descriptors (137). The training set of 144 molecules was used with each algorithm and descriptor type, and a test set 51 molecules was classified. The best models were generated with SVM with Volsurf descriptors, outperforming a study that used the same training and test set for pharmacophore models (138). Interestingly, the group did not look at the combination of different descriptor types.

There has also been recent extensive protein-based modeling of small molecule–P-gp interactions. One group used rigid body molecular dynamics simulations to produce a structure used in docking six molecules with Autodock that resulted in a reasonable correlation with actual K_i data (139). A second group used the same software to dock 24 flavonoids into a model of the nucleotide binding domain, obtaining a correlation of 0.67 for observed versus predicted K_d(140). One study used a combination of QSAR for wild type and variants of P-gp as well as molecular dynamics simulation to demonstrate the importance and structural function of various polymorphisms (141). The convergence of different experimental (photoaffinity labeling and site-directed mutagenesis) and in silico (residue importance and homology modeling) methods has also be exploited to show the importance of the transmembrane domains in P-gp for substrate binding and transport, predicting the formation of two pseudosymmetric pockets (142).

Recently several P-gp pharmacophores have been used for database screening and identification of potential P-gp substrates and inhibitors. Rebitzer and colleagues used a propafenone derivative MDR modulator-based pharmacophore model to screen the Derwent World Drug Index which retrieved among the returned 28 hits, 9 that were previously described MDR-modulators (143). A more detailed discussion of the P-gp pharmacophore-based database screening study was later published (144). A more recent publication by Kaiser et al. used a set of 131 propafenone-type P-gp inhibitors with 2D topological autocorrelation vectors to generate Kohonen self-organizing maps (145). The library of over 130,000 molecules from the SPECS database was mined using the training set map to identify molecules for testing. Two compounds with EC_{50} less than 1 μM were discovered that were also used with Tanimoto similarity searching of the same database to discover further active analogs (145). Another study has undertaken a comparison of different P-gp pharmacophore models for substrates and inhibitors and applied them to search structurally diverse molecule databases to identify new molecules and verify their P-gp inhibitor and substrate status (40). Two inhibitor pharmacophore models and one substrate pharmacophore model were used after first evaluating their capabilities with a database of 189 known P-gp substrates and non-substrates (138). After this quantitative validation, all three models were applied to screen a database of over 500 commonly prescribed drugs retrieving seven drugs with previously undocumented P-gp affinities that were selected and purchased for in vitro testing. Both (^{3}H)-digoxin transport assay and ATPase activation assay were carried out to characterize each candidate. All seven drugs were either μM inhibitors or substrates of P-gp (40).

An important member of the ABC transporter family is the multidrug resistance protein 1 (MRP1) which transports a broad range of substrates, ranging from anticancer drugs like vincristine, mitoxantrone, daunorubicin to organic anionic substrates like the conjugates of glutathione, glucuronide, and sulfates (146, 147). Due to its increasing significance in MDR, there is a major interest in the discovery of MRP1 inhibitors as MDR reversal agents (148). We selected five diverse highly potent MRP1 inhibitors that have been developed as MDR reversal agents (LY329146, LY402913, dehydrosilybin, indolopyrimidine, and phenoxymethyl quinoxalinone II) from a recent review (148) and generated a catalyst HIPHOP model in the same manner as those previously described for P-gp and hPEPT1. The model contained three ring aromatic and three HBA features and was applied to screen the database of over 500 clinically used drugs. Eight hits were retrieved including candesartan, eprosartan, fexofenadine, losartan, sulfasalazine, telmisartan, vancomycin, and zafirlukast, among these three drugs have been previously documented to have MRP1 affinity [losartan (149), sulfasalazine (150), and zafirlukast (151)]. In addition, recent work (149) indicates that tetrazole compounds are susceptible to P-gp and MRP1-mediated efflux. This suggests that the tetrazole-containing candesartan could also inhibit MRP1. Our model was generated based on synthesized high-affinity MRP1 inhibitors, yet it successfully discovered more diverse commonly used drugs as MRP1 inhibitors. The literature data are relatively recent indicating the ongoing discovery of MRP1 ligands is a new occurrence and deserving of further study. More experimental verification is required to test the selected compounds. Interestingly, two groups have derived pharmacophores and QSAR models for the rat Mrp-2 transporter. One of the groups used a narrow series of 25 methotrexate analogs with MOE descriptors and simulated annealing-PLS as well as catalyst for pharmacophore generation (152). A long pharmacophore with three hydrophobic, a negative ionisable and ring aromatic feature was derived. An earlier study used 16 more diverse molecules with SUPERPOSE and CoMFA and suggested two hydrophobic and two negative charged or hydrogen bond acceptor features in the compact pharmacophore (153). Neither group's pharmacophores were evaluated with large test sets.

BCRP is expressed on the apical side at the subcellular level in placenta, breast, liver hepatocytes, and endothelium (154, 155). BCRP expression confers resistance to several anticancer drugs like mitoxantrone and anthracyclines (156–158), camptothecin-derived topoisomerase I inhibitors (159), methotrexate (160), and flavoperidol (54). This half transporter, unlike P-gp and MRP1, has one ATP-binding site and consumes one ATP molecule per substrate molecule transferred (161). The diverse range of drugs that are exported by BCRP has prompted the design of many high-affinity inhibitors (162). The data accumulated were helpful in generating a model for BCRP. We recently derived a pharmacophore model to describe BCRP inhibition requirements. Four BCRP inhibitors (GF120918, Ko143, nelfinavir, nicardipine) used as the training set and a HIPHOP pharmacophore containing three HBA and three hydrophobic features were generated. The model was again utilized to search the database of over 500 commercially available drugs. Among the 37 retrieved molecules, 6 were previously identified BCRP ligands, digoxin (163), docetaxel (164) (**Fig. 4.1**), indinavir, lopinavir, ritonavir, saquinavir (165), and the training set compound nicardipine. A preliminary QSAR model for BCRP was also developed with 44 molecules with percent inhibition data of [^{3}H]methotrexate transport in Sf9 cells expressing the transporter, which was recently published (166). The TOPFRAG topological descriptors were used with multiple linear regression and one molecule, gefitinib was used to test the model.

The dopamine transporter (DAT) has been implicated as the monoamine transporter that is most closely associated with the reinforcing effects of cocaine (167, 168). Identification of DAT inhibitors as a potential treatment for cocaine abuse has met with many of these cocaine analogs demonstrating a cocaine-like behavioural profile in animal models of drug abuse (169, 170). Pharmacophore-based database screening is helpful in picking leads with different structural features than cocaine and has been used in DAT inhibitor searching studies (171). Wang and colleagues generated a pharmacophore model based on accumulated data on cocaine and its analogs and screened the NCI 3D-database (172). The crucial binding features identified by their model had a nitrogen atom, carbonyl group, and the aromatic ring centre and retrieved 4094 hits from the NCI database. The total number of compounds was filtered to 385 by using molecular weight and structural diversity criteria. In vitro assay data were generated using the [^{3}H]mazindol reuptake assay which verified the hit rate of 63% out of 70 tested molecules as DAT inhibitors. Among the 44 positive hits, 13 novel DAT inhibitors were further characterized in the [^{3}H]dopamine reuptake assay. The most promising compound was selected and subject to modifications to further improve its activity and selectivity.

The methods above all illustrate how computational models can be used to discover new inhibitors or substrates for transporters using a combination of database searching and in vitro data generation. These may represent a method to computationally predict whether new molecules in a pharmaceutical company are also likely to interact with particular transporters and therefore prioritize molecules for in vitro testing as well as potentially rationalize which areas on a molecule should be adjusted to maximize or minimize interaction with one or more transporter. As described above, available computational models have a very high hit rate at identifying new molecules with interesting levels of affinity to transporters and this may also suggest they could be a useful method for potentially screening known drugs, orphan drug databases, and other FDA-approved chemicals for likely therapeutic transporter inhibitors. Such drug repurposing would represent a mechanism to rapidly identify clinically useful molecules, for example, discovery of more potent BCRP inhibitors that could be dosed with anticancer compounds.

Similarly, the FDA's Critical Path Initiative highlights the need to speed up the process of bringing a pharmacokinetically sound dosage forms to market at lower cost (173). One of the challenges of the biopharmaceutical research is to correlate the in vitro drug release information of different drug formulations to the in vivo drug release profiles (IVIVC). The BCS guidance by the FDA explains when a waiver for in vivo bioavailability and bioequivalence studies may be requested. For an immediate release orally active dosage form, the rate and extent of its absorption are determined by its aqueous solubility and permeability in the gastrointestinal tract. Accordingly, certain drug classes can be considered for a biowaiver, i.e., approval of products based on their in vitro drug dissolution tests instead of their human bioequivalence data, which is a costly task for drug manufacturers. Such waivers significantly improve the speed and decrease the cost of bringing orally administered therapeutics to market. Currently, the BCS system allows a waiver of in vivo bioequivalence testing of immediate-release solid dosage forms for Class I drugs (174), whereas waivers for Class III drugs are recommended only based on scientific justifications (175, 176). Recently, Wu and Benet (177) extensively examined about 167 BCS classified drugs. They noticed that pharmacokinetic considerations like effects of food, absorptive transporters, efflux transporters, and routes of elimination (renal/biliary) were important determinants of overall drug absorption and bioavailability for immediate-release oral dosage forms. Thus, they suggested that classifying molecules based on the extent of metabolism is less ambiguous as compared to permeability or extent of absorption. This classification may also increase the number of Class I drugs that would become eligible for biowaivers (178). A challenge for both BCS and BDDCS is the actual classification of drugs based on the required in vitro data for metabolism, solubility, or permeability. The BDDCS represents an area where in silico approaches could aid the pharmaceutical industry in speeding drugs to the patient and reducing costs. More recently, we have shown that the BDDCS class can be predicted for new compounds from molecular structure alone, using readily available molecular descriptors and QSAR modeling methods (179). This could also have applications in research to identify molecules with future developability issues or assist drug companies to rapidly select candidate molecules for development under the FDA biowaiver system.

5. Protein Disorder in Transporters

Proteins may be functional in any one of the three states: ordered, random coil, and molten globule which accommodates the many proteins which possess intrinsic disorder (ID) in some part of their sequence (180). A disordered region of >30 AA is classed as important, predictions for smaller regions may be less relevant. Protein disorder has also been associated with alternative splicing in human genes that may enable regulatory and functional diversity, facilitating more rapid evolution (181). Protein sequences for several transporters were downloaded from the UniProt database (182) followed by ID prediction using the PONDR VL3H algorithm (183) (http://www.ist.temple.edu/disprot/predictor.php). The three proteins P-gp, BCRP, and hPEPT1 demonstrate differences in their predicted disorder (**Fig. 4.2**). For example, P-gp possesses a region of >50 amino acids that have a disorder probability >50% and are therefore classified as disordered. In contrast BCRP and

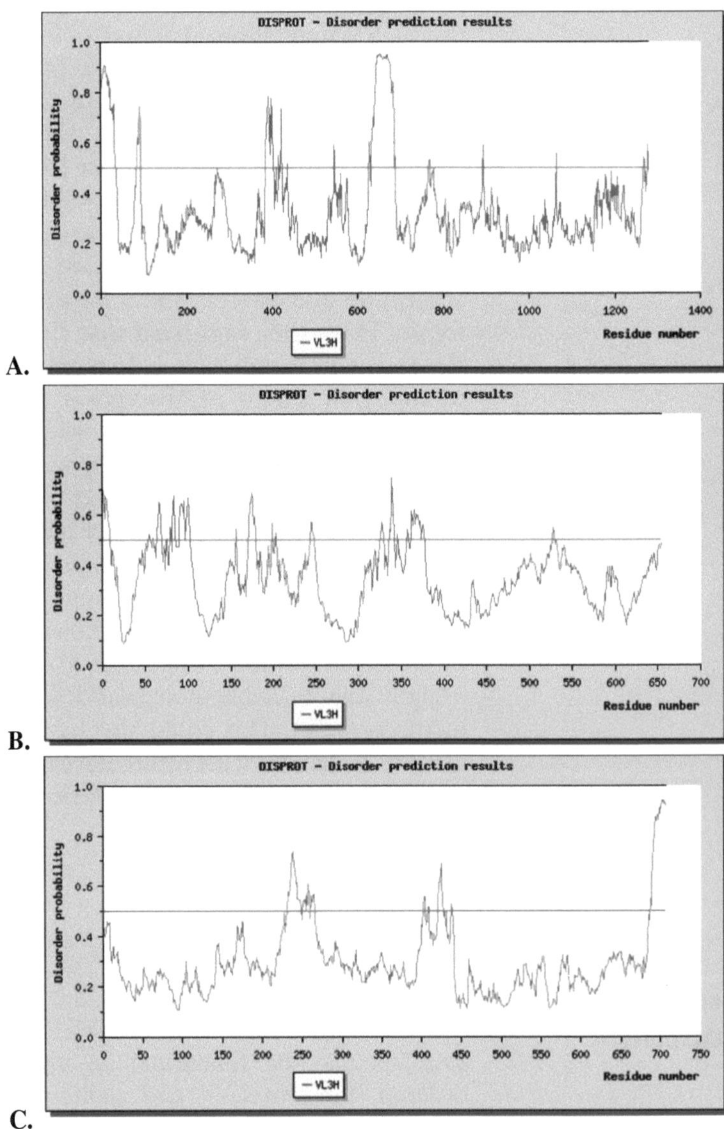

Fig. 4.2. Predicted intrinsic disorder for transporters. Values greater than 50% are considered to be disordered. (**a**) P-glycoprotein 1, (**b**) BCRP, (**c**). PEPT1.

hPEPT1 have only very short regions that are disordered. This may indicate that the intracellular region in P-gp (amino acids 633–687) could be disordered, while likely not involved in substrate binding it could have some functional importance. A more exhaustive analysis of ID in other transporters may be warranted as has recently been pursued with nuclear receptors (184).

6. Regulation and Network Analysis for Transporters

The nuclear hormone receptors play a key role in the regulation of transporters. For example, PXR is a transcriptional regulator of the enzyme human MDR1 (P-gp), MRPs, and OATP (55) (**Fig. 4.3**) as well as many other genes involved in the transport, metabolism (185–188), and biosynthesis of bile acids (189). Additional receptors such as the constitutive androstane receptor (CAR), farnesoid X receptor (FXR), liver X receptor (LXR), and other nuclear receptors and transcriptional factors take part in a complex network of interactions. Elucidation of the regulatory networks, which control the expression of efflux transporters and uptake transporters such as OATP (93), is of considerable interest to researchers in this area. The gene expression of 50 transporters across human tissues and between monkey, dog, rat, and mouse has been recently described (190). This indicated differences in expression of some genes between species and across tissues generally in line with studies that had previously been published, as well as noting that variability of transporter expression was generally not as wide as for the cytochrome P450s (190).

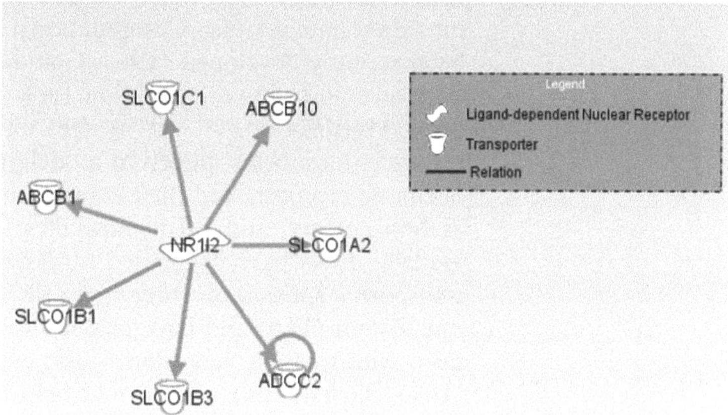

Fig. 4.3. Ingenuity pathways analysis version 7.1 was used to create a network of direct interacting transporter genes around the nuclear receptor PXR (NR1I2) in human.

Other smaller scale experimental studies have shown that the bile salt export pump in human hepatocytes is regulated by FXR and not LXR, as it was inducible by 22(*R*)-hydroxycholesterol and appeared to have different ligand binding determinants in the receptor than chenodeoxycholic acid (191). FXR has also been suggested to regulate the organic solute transporters α and β in human adrenal gland, kidney, and intestine (192). Although the exact physiological function of these transporters has not been defined, there may be a role in bile acid resorption.

In species commonly used for in vivo toxicology studies such as the rat, orthologs of the transporters such as oatp2 are expressed and are inducible with PXR ligands such as PCN (193). This is a useful knowledge because the advent of high-content and high-throughput genomics, proteomics/microarray technologies enables one to dose a rat with a xenobiotic and assess thousands of genes/proteins simultaneously in a particular tissue. These then allow one to look at the effects of a compound on the regulation of enzymes and transporters that could in turn influence clearance, excretion, or uptake. For instance, animals dosed with known nephrotoxins demonstrate upregulation of the Na–K–Cl transporter (194). Some transporters may be differentially targeted by drugs in different tissues (e.g., the CNS) but these may also be expressed elsewhere, representing a site for off-target toxic effects. Well-known examples are P-gp, expressed at the blood–brain barrier and intestine, impacting the efficacy and bioavailability of drugs and the serotonin transporter, expressed in the lungs and brain, where substrates such as fenfluramine can result in primary pulmonary hypertension as they accumulate in lung cells (195).

Biological knowledge has been traditionally captured on static maps and pathways. Due to our ever expanding biological knowledge we need methods to rapidly and dynamically find connections between molecules and proteins, providing insight into these interactions. Computational biological networks have been recently developed (196). Commercially available pathway databases and network building tools enable network comparisons, visualization, as well as data mining. Pathway tools and resources have been applied to modeling the networks of nuclear hormone receptors and their connections with other genes (such as transporters) and small molecules (42, 197). Commercially available tools have been used to demonstrate that the ABCA1 transporter appears on three manually curated pathway maps in one system (116) and have enabled the visualization of associations with toxicity, xenobiotics, and endobiotics in another tool (104). Such visualizations could help identify pathways around a transporter or molecules of interest for further study that may have off-target effects. Several groups have taken different approaches to studying the gene–drug relationship (198) which will be useful for identifying which molecules may be substrates for different transporters (1) and this in turn will assist data mining efforts (56).

7. Conclusions

This review has described how considerable attention has been given to the rapid and accurate identification of substrates and inhibitors for transporter proteins in the hope of improving the drug development process. Significant efforts have been made to understand the mechanisms underlying the overall transport process and increase our knowledge and generate databases of substrate and inhibitor affinity for various drug transporters. The wealth of molecular biology techniques to identify and clone therapeutically relevant as well as unidentified transporters potentially useful for drug delivery is being paralleled by the application of computational approaches in drug discovery. The combined approach of prospective in silico and in vitro methods for transporter substrate and inhibitor discovery suggests considerable promise to prioritize testing in early discovery to minimize time, cost, and labor intensiveness of screening. Our own studies with clinically relevant drug transporters suggest that the application of in vitro and in silico transporter models will likely provide information that will apply in vivo.

Acknowledgments

SE gratefully acknowledges the considerable efforts of Dr. Shikha Varma (Accelrys, San Diego, CA) for making Discovery Studio Catalyst available and Ingenuity for kindly providing IPA. Dr. David Lawson is thanked for assistance with intrinsic disorder predictions, and we acknowledge the contributions of our colleagues and collaborators that contributed to some of the studies referenced above.

References

1. Anderle, P., Huang, Y., and Sadee, W. (2004) Intestinal membrane transport of drugs and nutrients: genomics of membrane transporters using expression microarrays. *Eur. J. Pharm. Sci.* **21**, 17–24.
2. Hediger, M. A., Romero, M. F., Peng, J. B., Rolfs, A., Takanaga, H., and Bruford, E. A. (2004) The ABCs of solute carriers: physiological, pathological and therapeutic implications of human membrane transport

proteins Introduction. *Pflugers Arch.* **447**, 465—468.

3. Zhang, E. Y., Phelps, M. A., Cheng, C., Ekins, S., and Swaan, P. W. (2002) Modeling of active transport systems. *Adv. Drug. Deliv. Rev.* **54**, 329–354.
4. Ware, J. A. (2006) Membrane transporters in drug discovery and development: A new mechanistic ADME era. *Mol. Pharm.* **3**, 1–2.
5. Venter, J. C., Adams, M. D., and Others. (2001) The sequence of the human genome. *Science* **291**, 1304–1351.
6. Cheng, A. C., Coleman, R. G., Smyth, K. T., Cao, Q., Soulard, P., Caffrey, D. R., Salzberg, A. C., and Huang, E. S. (2007) Structure-based maximal affinity model predicts small-molecule druggability. *Nat. Biotechnol.* **25**, 71–75.
7. Sugiyama, Y. (2005) Druggability: selecting optimized drug candidates. *Drug Discov. Today* **10**, 1577–1579.
8. Mizuno, N., Niwa, T., Yotsumoto, Y., and Sugiyama, Y. (2003) Impact of drug transporter studies on drug discovery and development. *Pharmacol. Rev.* **55**, 425–461.
9. Isaji, M. (2007) Sodium-glucose cotransporter inhibitors for diabetes. *Curr. Opin. Investig. Drugs* **8**, 285–292.
10. Handlon, A. (2005) Sodium glucose co-transporter 2 (SGLT2) inhibitors as potential antidiabetic agents. *Expert. Opin. Ther. Patients* **15**, 1531–1540.
11. Castaneda, F., Burse, A., Boland, W., and Kinne, R. K. (2007) Thioglycosides as inhibitors of hSGLT1 and hSGLT2: potential therapeutic agents for the control of hyperglycemia in diabetes. *Int. J. Med. Sci.* **4**, 131–139.
12. Katsuno, K., Fujimori, Y., Takemura, Y., Hiratochi, M., Itoh, F., Komatsu, Y., Fujikura, H., and Isaji, M. (2007) Sergliflozin, a novel selective inhibitor of low-affinity sodium glucose cotransporter (SGLT2), validates the critical role of SGLT2 in renal glucose reabsorption and modulates plasma glucose level. *J. Pharmacol. Exp. Ther.* **320**, 323–330.
13. Cundy, K. C., Branch, R., Chernov-Rogan, T., Dias, T., Estrada, T., Hold, K., Koller, K., Liu, X., Mann, A., Panuwat, M., Raillard, S. P., Upadhyay, S., Wu, Q. Q., Xiang, J. N., Yan, H., Zerangue, N., Zhou, C. X., Barrett, R. W., and Gallop, M. A. (2004) XP13512 [(+/-)-1-([(alpha-isobutanoyloxyethoxy)carbonyl] aminometh yl)-1-cyclohexane acetic acid], a novel gabapentin prodrug: I. Design, synthesis, enzymatic conversion to gabapentin, and transport by intestinal solute transporters. *J. Pharmacol. Exp. Ther.* **311**, 315–323.
14. Xenoport. (2006) Xenoport, Inc. www.xenoport.com.
15. Amidon, G. L., and Walgreen, C. R., Jr. (1999) "5′-Amino acid esters of antiviral nucleosides, acyclovir, and AZT are absorbed by the intestinal PEPT1 peptide transporter," *Pharm. Res.* **16**, 175.
16. Thomsen, A. E., Christensen, M. S., Bagger, M. A., and Steffansen, B. (2004) Acyclovir prodrug for the intestinal di/tri-peptide transporter PEPT1: comparison of in vivo bioavailability in rats and transport in Caco-2 cells. *Eur. J. Pharm. Sci.* **23**, 319–325.
17. Sugawara, M., Huang, W., Fei, Y. J., Leibach, F. H., Ganapathy, V., and Ganapathy, M. E. (2000) Transport of valganciclovir, a ganciclovir prodrug, via peptide transporters PEPT1 and PEPT2. *J. Pharm. Sci.* **89**, 781–789.
18. Granero, G. E., and Amidon, G. L. (2006) Stability of valacyclovir: implications for its oral bioavailability. *Int. J. Pharm* **317**, 14–18.
19. Bahadduri, P. M., D′Souza, V. M., Pinsonneault, J. K., Sadee, W., Bao, S., Knoell, D. L., and Swaan, P. W. (2005) Functional characterization of the peptide transporter PEPT2 in primary cultures of human upper airway epithelium. *Am. J. Respir. Cell Mol. Biol.* **32**, 319–325.
20. Ganapathy, V., and Miyauchi, S. (2005) Transport systems for opioid peptides in mammalian tissues. *Aaps. J.* **7**, E852–856.
21. Shen, H., Keep, R. F., Hu, Y., and Smith, D. E. (2005) PEPT2 (Slc15a2)-mediated unidirectional transport of cefadroxil from cerebrospinal fluid into choroid plexus. *J. Pharmacol. Exp. Ther.* **315**, 1101–1108.
22. Shen, H., Ocheltree, S. M., Hu, Y., Keep, R. F., and Smith, D. E. (2007) Impact of genetic knockout of PEPT2 on cefadroxil pharmacokinetics, renal tubular reabsorption, and brain penetration in mice. *Drug. Metab. Dispos.* **35**, 1209–1216.
23. Gonzalez, D. E., Covitz, K. M., Sadee, W., and Mrsny, R. J. (1998) An oligopeptide transporter is expressed at high levels in the pancreatic carcinoma cell lines AsPc-1 and Capan-2. *Cancer Res.* **58**, 519–525.
24. Nakanishi, T., Tamai, I., Takaki, A., and Tsuji, A. (2000) Cancer cell-targeted drug delivery utilizing oligopeptide transport activity. *Int. J. Cancer* **88**, 274–280.
25. Sasaki, M., Suzuki, H., Ito, K., Abe, T., and Sugiyama, Y. (2002) Transcellular transport of organic anions across a double-transfected Madin-Darby canine

kidney II cell monolayer expressing both human organic anion-transporting polypeptide (OATP2/SLC21A6) and Multidrug resistance-associated protein 2 (MRP2/ABCC2). *J. Biol. Chem.* **277**, 6497–6503.
26. Nakai, D., Nakagomi, R., Furuta, Y., Tokui, T., Abe, T., Ikeda, T., and Nishimura, K. (2001) Human liver-specific organic anion transporter, LST-1, mediates uptake of pravastatin by human hepatocytes. *J. Pharmacol. Exp. Ther.* **297**, 861–867.
27. Ballestero, M. R., Monte, M. J., Briz, O., Jimenez, F., Gonzalez-San Martin, F., and Marin, J. J. (2006) Expression of transporters potentially involved in the targeting of cytostatic bile acid derivatives to colon cancer and polyps. *Biochem. Pharmacol.* **72**, 729–738.
28. Lin, J. H. (2003) Drug-drug interaction mediated by inhibition and induction of P-glycoprotein. *Adv. Drug Deliv. Rev.* **55**, 53–81.
29. Chen, J., and Raymond, K. (2006) Roles of rifampicin in drug-drug interactions: underlying molecular mechanisms involving the nuclear pregnane X receptor. *Ann. Clin. Microbiol. Antimicrob.* **5**, 3.
30. Kuppens, I. E., Breedveld, P., Beijnen, J. H., and Schellens, J. H. (2005) Modulation of oral drug bioavailability: from preclinical mechanism to therapeutic application. *Cancer Invest.* **23**, 443–464.
31. Kemper, E. M., van Zandbergen, A. E., Cleypool, C., Mos, H. A., Boogerd, W., Beijnen, J. H., and van Tellingen, O. (2003) Increased penetration of paclitaxel into the brain by inhibition of P-Glycoprotein. *Clin. Cancer Res.* **9**, 2849–2855.
32. Nekhayeva, I. A., Nanovskaya, T. N., Hankins, G. D., and Ahmed, M. S. (2006) Role of human placental efflux transporter P-glycoprotein in the transfer of buprenorphine, levo-alpha-acetylmethadol, and paclitaxel. *Am. J. Perinatol.* **23**, 423–430.
33. Bardelmeijer, H. A., Ouwehand, M., Beijnen, J. H., Schellens, J. H., and van Tellingen, O. (2004) Efficacy of novel P-glycoprotein inhibitors to increase the oral uptake of paclitaxel in mice. *Invest. New Drugs* **22**, 219–229.
34. Kuppens, I. E., Witteveen, E. O., Jewell, R. C., Radema, S. A., Paul, E. M., Mangum, S. G., Beijnen, J. H., Voest, E. E., and Schellens, J. H. (2007) A phase I, randomized, open-label, parallel-cohort, dose-finding study of elacridar (GF120918) and oral topotecan in cancer patients. *Clin. Cancer Res.* **13**, 3276–3285.
35. Weiss, J., Rose, J., Storch, C. H., Ketabi-Kiyanvash, N., Sauer, A., Haefeli, W. E., and Efferth, T. (2007) Modulation of human BCRP (ABCG2) activity by anti-HIV drugs. *J. Antimicrob. Chemother.* **59**, 238–245.
36. Bauer, S., Stormer, E., Johne, A., Kruger, H., Budde, K., Neumayer, H. H., Roots, I., and Mai, I. (2003) Alterations in cyclosporin A pharmacokinetics and metabolism during treatment with St John's wort in renal transplant patients. *Br. J. Clin. Pharmacol.* **55**, 203–211.
37. Pawarode, A., Shukla, S., Minderman, H., Fricke, S. M., Pinder, E. M., O'Loughlin, K. L., Ambudkar, S. V., and Baer, M. R. (2007) Differential effects of the immunosuppressive agents cyclosporin A, tacrolimus and sirolimus on drug transport by multidrug resistance proteins. *Cancer Chemother. Pharmacol.* **60**, 179–188.
38. Schwarz, U. I., Hanso, H., Oertel, R., Miehlke, S., Kuhlisch, E., Glaeser, H., Hitzl, M., Dresser, G. K., Kim, R. B., and Kirch, W. (2007) Induction of intestinal P-glycoprotein by St John's wort reduces the oral bioavailability of talinolol. *Clin. Pharmacol. Ther.* **81**, 669–678.
39. Bajorath, J. (2002) Integration of virtual and high-throughput screening. *Nat. Rev. Drug Discov.* **1**, 882–894.
40. Chang, C., Bahadduri, P. M., Polli, J. E., Swaan, P. W., and Ekins, S. (2006) Rapid identification of P-glycoprotein substrates and inhibitors. *Drug Metab. Dispos.* **34**, 1976–1984.
41. Kerns, E. H., and Di, L. (2003) Pharmaceutical profiling in drug discovery. *Drug Discov. Today* **8**, 316–323.
42. Ekins, S., Johnston, J. S., Bahadduri, P., D'Souza, V. M., Ray, A., Chang, C., and Swaan, P. W. (2005) In vitro and pharmacophore-based discovery of novel hPEPT1 inhibitors. *Pharm. Res.* **22**, 512–517.
43. Han, H. K., Oh, D. M., and Amidon, G. L. (1998) Cellular uptake mechanism of amino acid ester prodrugs in Caco-2/hPEPT1 cells overexpressing a human peptide transporter. *Pharm. Res.* **15**, 1382–1386.
44. Han, H. K., Rhie, J. K., Oh, D. M., Saito, G., Hsu, C. P., Stewart, B. H., and Amidon, G. L. (1999) CHO/hPEPT1 cells overexpressing the human peptide transporter (hPEPT1) as an alternative in vitro model for peptidomimetic drugs. *J. Pharm. Sci.* **88**, 347–350.
45. Surendran, N., Covitz, K. M., Han, H., Sadee, W., Oh, D. M., Amidon, G. L., Williamson, R. M., Bigge, C. F., and Stewart, B. H.

(1999) Evidence for overlapping substrate specificity between large neutral amino acid (LNAA) and dipeptide (hPEPT1) transporters for PD 158473, an NMDA antagonist. *Pharm. Res.* **16**, 391–395.

46. Banerjee, A., and Swaan, P. W. (2006) Membrane topology of human ASBT (SLC10A2) determined by dual label epitope insertion scanning mutagenesis. New evidence for seven transmembrane domains. *Biochemistry* **45**, 943–953.
47. Chang, C., and Swaan, P. W. (2006) Computational approaches to modeling drug transporters. *Eur. J. Pharm. Sci.* **27**, 411–424.
48. Polli, J. W., Wring, S. A., Humphreys, J. E., Huang, L., Morgan, J. B., Webster, L. O., and Serabjit-Singh, C. S. (2001) Rational use of in vitro P-glycoprotein assays in drug discovery. *J. Pharmacol. Exp. Ther.* **299**, 620–628.
49. Tiberghien, F., and Loor, F. (1996) Ranking of P-glycoprotein substrates and inhibitors by a calcein-AM fluorometry screening assay. *Anticancer Drugs* 7, 568–578.
50. Ambudkar, S. V., Dey, S., Hrycyna, C. A., Ramachandra, M., Pastan, I., and Gottesman, M. M. (1999) Biochemical, cellular, and pharmacological aspects of the multidrug transporter. *Annu. Rev. Pharmacol. Toxicol.* **39**, 361–398.
51. Tang, F., Ouyang, H., Yang, J. Z., and Borchardt, R. T. (2004) Bidirectional transport of rhodamine 123 and Hoechst 33342, fluorescence probes of the binding sites on P-glycoprotein, across MDCK-MDR1 cell monolayers. *J. Pharm. Sci.* **93**, 1185–1194.
52. Kim, M., Turnquist, H., Jackson, J., Sgagias, M., Yan, Y., Gong, M., Dean, M., Sharp, J. G., and Cowan, K. (2002) The multidrug resistance transporter ABCG2 (breast cancer resistance protein 1) effluxes Hoechst 33342 and is overexpressed in hematopoietic stem cells. *Clin. Cancer Res.* **8**, 22–28.
53. Scharenberg, C. W., Harkey, M. A., and Torok-Storb, B. (2002) The ABCG2 transporter is an efficient Hoechst 33342 efflux pump and is preferentially expressed by immature human hematopoietic progenitors. *Blood* **99**, 507–512.
54. Robey, R. W., Honjo, Y., van de Laar, A., Miyake, K., Regis, J. T., Litman, T., and Bates, S. E. (2001) A functional assay for detection of the mitoxantrone resistance protein, MXR (ABCG2). *Biochim. Biophys. Acta* **1512**, 171–182.
55. Wang, E. J., Casciano, C. N., Clement, R. P., and Johnson, W. W. (2003) Fluorescent substrates of sister-P-glycoprotein (BSEP) evaluated as markers of active transport and inhibition: evidence for contingent unequal binding sites. *Pharm. Res.***20**, 537–544.
56. Chang, C., Ekins, S., Bahadduri, P., and Swaan, P. W. (2006) Pharmacophore-based discovery of ligands for drug transporters. *Adv. Drug Deliv. Rev.* **58**, 1431–1450.
57. Balimane, P. V., Patel, K., Marino, A., and Chong, S. (2004) Utility of 96 well Caco-2 cell system for increased throughput of P-gp screening in drug discovery. *Eur. J. Pharm. Biopharm.* **58**, 99–105.
58. Marino, A. M., Yarde, M., Patel, H., Chong, S., and Balimane, P. V. (2005) Validation of the 96 well Caco-2 cell culture model for high throughput permeability assessment of discovery compounds. *Int. J. Pharm.* **297**, 235–241.
59. Hidalgo, I. J. (1996) *Cultured Intestinal Epithelial Cell Models*, Plenum Press, New York.
60. Irvine, J. D., Takahashi, L., Lockhart, K., Cheong, J., Tolan, J. W., Selick, H. E., and Grove, J. R. (1999) MDCK (Madin-Darby canine kidney) cells: A tool for membrane permeability screening. *J. Pharm. Sci.* **88**, 28–33.
61. Yamazaki, M., Neway, W. E., Ohe, T., Chen, I., Rowe, J. F., Hochman, J. H., Chiba, M., and Lin, J. H. (2001) In vitro substrate identification studies for p-glycoprotein-mediated transport: species difference and predictability of in vivo results. *J. Pharmacol. Exp. Ther.* **296**, 723–735.
62. Tang, F., Horie, K., and Borchardt, R. T. (2002) Are MDCK cells transfected with the human MRP2 gene a good model of the human intestinal mucosa? *Pharm. Res.* **19**, 773–779.
63. Hidalgo, I. J., Raub, T. J. and Borchardt, R. T. (1989) Characterization of the Human Colon Carcinoma Cell Line (Caco-2) as a model system for Intestinal Epithelial Permeability. *Gastroenterology* **96**, 736–749.
64. Sambuy, Y., De Angelis, I., Ranaldi, G., Scarino, M. L., Stammati, A., and Zucco, F. (2005) The Caco-2 cell line as a model of the intestinal barrier: influence of cell and culture-related factors on Caco-2 cell functional characteristics. *Cell Biol. Toxicol.* **21**, 1–26.
65. Hunter, J., Jepson, M. A., Tsuruo, T., Simmons, N. L., and Hirst, B. H. (1993) Functional expression of P-glycoprotein in apical membranes of human intestinal Caco-2 cells. Kinetics of vinblastine secretion and interaction with modulators. *J. Biol. Chem.* **268**, 14991–14997.
66. Videmann, B., Tep, J., Cavret, S., and Lecoeur, S. (2007) Epithelial transport

of deoxynivalenol: Involvement of human P-glycoprotein (ABCB1) and multidrug resistance-associated protein 2 (ABCC2). *Food Chem. Toxicol.*
67. Nakamura, T., Sakaeda, T., Ohmoto, N., Tamura, T., Aoyama, N., Shirakawa, T., Kamigaki, T., Nakamura, T., Kim, K. I., Kim, S. R., Kuroda, Y., Matsuo, M., Kasuga, M., and Okumura, K. (2002) Real-time quantitative polymerase chain reaction for MDR1, MRP1, MRP2, and CYP3A-mRNA levels in Caco-2 cell lines, human duodenal enterocytes, normal colorectal tissues, and colorectal adenocarcinomas. *Drug Metab. Dispos.* **30**, 4–6.
68. Irie, M., Terada, T., Tsuda, M., Katsura, T., and Inui, K. (2006) Prediction of glycylsarcosine transport in Caco-2 cell lines expressing PEPT1 at different levels. *Pflugers Arch.* **452**, 64–70.
69. Calcagno, A. M., Ludwig, J. A., Fostel, J. M., Gottesman, M. M., and Ambudkar, S. V. (2006) Comparison of drug transporter levels in normal colon, colon cancer, and Caco-2 cells: impact on drug disposition and discovery. *Mol. Pharm.* **3**, 87–93.
70. Artursson, P., Palm, K., and Luthman, K. (2001) Caco-2 monolayers in experimental and theoretical predictions of drug transport. *Adv. Drug Deliv. Rev.* **46**, 27–43.
71. Behrens, I., and Kissel, T. (2003) Do cell culture conditions influence the carrier-mediated transport of peptides in Caco-2 cell monolayers? *Eur. J. Pharm. Sci.* **19**, 433–442.
72. Tang, F., Horie, K., and Borchardt, R. T. (2002) Are MDCK cells transfected with the human MDR1 gene a good model of the human intestinal mucosa? *Pharm. Res.* **19**, 765–772.
73. Lash, L. H., Putt, D. A., and Cai, H. (2006) Membrane transport function in primary cultures of human proximal tubular cells. *Toxicology* **228**, 200–218.
74. Bachmeier, C. J., Trickler, W. J., and Miller, D. W. (2006) Comparison of drug efflux transport kinetics in various blood-brain barrier models. *Drug Metab. Dispos.* **34**, 998–1003.
75. Talluri, R. S., Katragadda, S., Pal, D., and Mitra, A. K. (2006) Mechanism of L-ascorbic acid uptake by rabbit corneal epithelial cells: evidence for the involvement of sodium-dependent vitamin C transporter 2. *Curr. Eye Res.* **31**, 481–489.
76. Artursson, P. (1991) Cell cultures as models for drug absorption across the intestinal mucosa. *Crit. Rev. Ther. Drug Carrier Syst.* **8**, 305–330.
77. Xia, C. Q., Milton, M. N., and Gan, L. S. (2007) Evaluation of drug-transporter interactions using in vitro and in vivo models. *Curr. Drug Metab.* **8**, 341–363.
78. Allen, J. D., van Loevezijn, A., Lakhai, J. M., van der Valk, M., van Tellingen, O., Reid, G., Schellens, J. H., Koomen, G. J., and Schinkel, A. H. (2002) Potent and specific inhibition of the breast cancer resistance protein multidrug transporter in vitro and in mouse intestine by a novel analogue of fumitremorgin C. *Mol. Cancer Ther.* **1**, 417–425.
79. Ramachandra, M., Ambudkar, S. V., Gottesman, M. M., Pastan, I., and Hrycyna, C. A. (1996) Functional characterization of a glycine 185-to-valine substitution in human P-glycoprotein by using a vaccinia-based transient expression system. *Mol. Biol. Cell.* **7**, 1485–1498.
80. Taguchi, Y., Yoshida, A., Takada, Y., Komano, T., and Ueda, K. (1997) Anti-cancer drugs and glutathione stimulate vanadate-induced trapping of nucleotide in multidrug resistance-associated protein (MRP). *FEBS Lett.* **401**, 11–14.
81. Drueckes, P., Schinzel, R., and Palm, D. (1995) Photometric microtiter assay of inorganic phosphate in the presence of acid-labile organic phosphates. *Anal. Biochem.* **230**, 173–177.
82. Sarkadi, B., Price, E. M., Boucher, R. C., Germann, U. A., and Scarborough, G. A. (1992) Expression of the human multidrug resistance cDNA in insect cells generates a high activity drug-stimulated membrane ATPase. *J. Biol. Chem.* **267**, 4854–4858.
83. Chang, K. H., Lee, J. M., Jeon, H. K., and Chung, I. S. (2004) Improved production of recombinant tumstatin in stably transformed Trichoplusia ni BTI Tn 5B1-4 cells. *Protein Expr. Purif.* **35**, 69–75.
84. Senior, A. E., al-Shawi, M. K., and Urbatsch, I. L. (1995) ATP hydrolysis by multidrug-resistance protein from Chinese hamster ovary cells. *J. Bioenerg. Biomembr.* **27**, 31–36.
85. Ozvegy, C., Varadi, A., and Sarkadi, B. (2002) Characterization of drug transport, ATP hydrolysis, and nucleotide trapping by the human ABCG2 multidrug transporter. Modulation of substrate specificity by a point mutation. *J. Biol. Chem.* **277**, 47980–47990.
86. Promega. (http://www.promega.com/tbs/tb341/tb341.pdf).
87. Lentz, K. A., Polli, J. W., Wring, S. A., Humphreys, J. E., and Polli, J. E. (2000) Influence of passive permeability on apparent P-glycoprotein kinetics. *Pharm. Res.* **17**, 1456–1460.

88. Tolle-Sander, S., Rautio, J., Wring, S., Polli, J. W., and Polli, J. E. (2003) Midazolam exhibits characteristics of a highly permeable P-glycoprotein substrate. *Pharm. Res.* **20**, 757–764.
89. Hillgren, K. M., Kato, A., and Borchardt, R.T. (1995) In vitro Systems for Studying Intestinal Drug Absorption. *Med. Res. Revs.* **15**, 83–109.
90. Zhang, L., Lin, G., Kovacs, B., Jani, M., Krajcsi, P., and Zuo, Z. (2007) Mechanistic study on the intestinal absorption and disposition of baicalein. *Eur. J. Pharm. Sci.* **31**, 221–231.
91. Whittico, M. T., Hui, A. C., and Giacomini, K. M. (1991) Preparation of brush border membrane vesicles from bovine choroid plexus. *J. Pharmacol. Methods* **25**, 215–227.
92. Meier, P. J., and Boyer, J. L. (1990) Preparation of basolateral (sinusoidal) and canalicular plasma membrane vesicles for the study of hepatic transport processes. *Methods Enzymol.* **192**, 534–545.
93. Ushigome, F., Koyabu, N., Satoh, S., Tsukimori, K., Nakano, H., Nakamura, T., Uchiumi, T., Kuwano, M., Ohtani, H., and Sawada, Y. (2003) Kinetic analysis of P-glycoprotein-mediated transport by using normal human placental brush-border membrane vesicles. *Pharm. Res.* **20**, 38–44.
94. Sata, R., Ohtani, H., Tsujimoto, M., Murakami, H., Koyabu, N., Nakamura, T., Uchiumi, T., Kuwano, M., Nagata, H., Tsukimori, K., Nakano, H., and Sawada, Y. (2005) Functional analysis of organic cation transporter 3 expressed in human placenta. *J. Pharmacol. Exp. Ther.* **315**, 888–895.
95. Aanismaa, P., and Seelig, A. (2007) P-Glycoprotein kinetics measured in plasma membrane vesicles and living cells. *Biochemistry* **46**, 3394–3404.
96. Jin, J., Shahi, S., Kang, H. K., van Veen, H. W., and Fan, T. P. (2006) Metabolites of ginsenosides as novel BCRP inhibitors. *Biochem. Biophys. Res. Commun.* **345**, 1308–1314.
97. Hirano, M., Maeda, K., Hayashi, H., Kusuhara, H., and Sugiyama, Y. (2005) Bile salt export pump (BSEP/ABCB11) can transport a nonbile acid substrate, pravastatin. *J. Pharmacol. Exp. Ther.* **314**, 876–882.
98. McRae, M. P., Lowe, C. M., Tian, X., Bourdet, D. L., Ho, R. H., Leake, B. F., Kim, R. B., Brouwer, K. L., and Kashuba, A. D. (2006) Ritonavir, saquinavir, and efavirenz, but not nevirapine, inhibit bile acid transport in human and rat hepatocytes. *J. Pharmacol. Exp. Ther.* **318**, 1068–1075.
99. Pak, Y., Emerick, R., Perry III, W., and KM, H. (2005) Use of Inside-out Membrane Vesicles to Characterize Substrates and Inhibitors of P-glycoprotein, in *AAPS Workshop on Drug Transporters in ADME: From the Bench to the Bedside*, AAPS, Parsippany, NJ.
100. Pouliot, J. F., L'Heureux, F., Liu, Z., Prichard, R. K., and Georges, E. (1997) Reversal of P-glycoprotein-associated multidrug resistance by ivermectin. *Biochem. Pharmacol.* **53**, 17–25.
101. Qian, Y. M., Grant, C. E., Westlake, C. J., Zhang, D. W., Lander, P. A., Shepard, R. L., Dantzig, A. H., Cole, S. P., and Deeley, R. G. (2002) Photolabeling of human and murine multidrug resistance protein 1 with the high affinity inhibitor [125I]LY475776 and azidophenacyl-[35S]glutathione. *J. Biol. Chem.* **277**, 35225–35231.
102. Safa, A. R., Glover, C. J., Meyers, M. B., Biedler, J. L., and Felsted, R. L. (1986) Vinblastine photoaffinity labeling of a high molecular weight surface membrane glycoprotein specific for multidrug-resistant cells. *J. Biol. Chem.* **261**, 6137–6140.
103. van de Waterbeemd, H., and Gifford, E. (2003) ADMET in silico modelling: towards prediction paradise? *Nat. Rev. Drug Discov.* **2**, 192–204.
104. Ekins, S., Ecker, G.F., Chiba P. and Swaan, P.W. (2007) Future Directions for Drug Transporter Modeling. *Xenobiotica.* **37**, 1152–1170.
105. Cramer, R. D., 3rd, Patterson, D. E., and Bunce, J. D. (1989) Recent advances in comparative molecular field analysis (CoMFA). *Prog. Clin. Biol. Res.* **291**, 161–165.
106. Martin, Y. C., Bures, M. G., Danaher, E. A., DeLazzer, J., Lico, I., and Pavlik, P. A. (1993) A fast new approach to pharmacophore mapping and its application to dopaminergic and benzodiazepine agonists, *J. Comput. Aided Mol. Des.* **7**, 83–102.
107. Jones, G., Willett, P., and Glen, R. C. (1995) A genetic algorithm for flexible molecular overlay and pharmacophore elucidation, *J. Comput. Aided. Mol. Des.* **9**, 532–549.
108. Clement, O. O. a. M., A.T. (2000) *HipHop: Pharmacophore Based on Multiple common-feature alignments.*, IUL, San Diego.
109. Evans, D. A., Doman, T. N., Thorner, D. A., and Bodkin, M. J. (2007) 3D QSAR methods: Phase and Catalyst compared. *J. Chem. Inf. Model.* **47**, 1248–1257.
110. Ekins, S., Kim, R. B., Leake, B. F., Dantzig, A. H., Schuetz, E. G., Lan, L. B., Yasuda, K., Shepard, R. L., Winter, M. A., Schuetz, J. D., Wikel, J. H., and Wrighton, S. A. (2002) Application of three-dimensional quantitative structure-activity relationships of

P-glycoprotein inhibitors and substrates. *Mol. Pharmacol.* **61**, 974–981.

111. Ekins, S., Kim, R. B., Leake, B. F., Dantzig, A. H., Schuetz, E. G., Lan, L. B., Yasuda, K., Shepard, R. L., Winter, M. A., Schuetz, J. D., Wikel, J. H., and Wrighton, S. A. (2002) Three-dimensional quantitative structure-activity relationships of inhibitors of P-glycoprotein. *Mol. Pharmacol.* **61**, 964–973.
112. Bednarczyk, D., Ekins, S., Wikel, J. H., and Wright, S. H. (2003) Influence of molecular structure on substrate binding to the human organic cation transporter, hOCT1. *Mol. Pharmacol.* **63**, 489–498.
113. Chang, C., Pang, K. S., Swaan, P. W., and Ekins, S. (2005) Comparative pharmacophore modeling of organic anion transporting polypeptides: a meta-analysis of rat Oatp1a1 and human OATP1B1. *J. Pharmacol. Exp. Ther.* **314**, 533–541.
114. Suhre, W. M., Ekins, S., Chang, C., Swaan, P. W., and Wright, S. H. (2005) Molecular determinants of substrate/inhibitor binding to the human and rabbit renal organic cation transporters hOCT2 and rbOCT2. *Mol. Pharmacol.* **67**, 1067–1077.
115. Punta, M., Forrest, L. R., Bigelow, H., Kernytsky, A., Liu, J., and Rost, B. (2007) Membrane protein prediction methods. *Methods* **41**, 460–474.
116. Chang, C. S., PW (2006) *Computational Modeling of Drug Disposition*, John Wiley & Sons, Inc., Hoboken, New Jersey.
117. Adamian, L., and Liang, J. (2006) Prediction of transmembrane helix orientation in polytopic membrane proteins, *BMC Struct. Biol.* **6**, 13.
118. Campagna-Slater, V., and Weaver, D. F. (2007) Molecular modelling of the GABAA ion channel protein. *J. Mol. Graph. Model.* **25**, 721–730.
119. Daniel, H. (2004) Molecular and integrative physiology of intestinal peptide transport. *Annu. Rev. Physiol.* **66**, 361–384.
120. Covitz, K. M., Amidon, G. L., and Sadee, W. (1996) Human dipeptide transporter, hPEPT1, stably transfected into Chinese hamster ovary cells. *Pharm. Res.* **13**, 1631–1634.
121. Fei, Y. J., Kanai, Y., Nussberger, S., Ganapathy, V., Leibach, F. H., Romero, M. F., Singh, S. K., Boron, W. F., and Hediger, M. A. (1994) Expression cloning of a mammalian proton-coupled oligopeptide transporter. *Nature* **368**, 563–566.
122. Ganapathy, M. E., Huang, W., Wang, H., Ganapathy, V., and Leibach, F. H. (1998) Valacyclovir: a substrate for the intestinal and renal peptide transporters PEPT1 and PEPT2. *Biochem. Biophys. Res. Commun.* **246**, 470–475.
123. Perkins, E. J., and Abraham, T. (2007) Pharmacokinetics, Metabolism, and Excretion of the PepT1-Targeted Prodrug (1S,2S,5R,6S)-2-[(2′S)-(2-Amino) propionyl]aminobicyclo[3.1.0.]hexen-2,6-di carboxylic acid (LY544344) in Rats and Dogs: Assessment of First-pass Bioactivation and Dose-Linearity. *Drug Metab. Dispos.*
124. Swaan, P. W., Bensman, T., Bahadduri, P. M., Hall, M. W., Sarkar, A., Bao, S., Khantwal, C. M., Ekins, S., and Knoell, D. L. (2008) Bacterial peptide recognition and immune activation facilitated by human peptide transporter PEPT2. *Am. J. Respir. Cell Mol. Biol.* **39**, 536–542.
125. Biegel, A., Gebauer, S., Brandsch, M., Neubert, K., and Thondorf, I. (2006) Structural requirements for the substrates of the H+/peptide cotransporter PEPT2 determined by three-dimensional quantitative structure-activity relationship analysis. *J. Med. Chem.* **49**, 4286–4296.
126. Vig, B. S., Stouch, T. R., Timoszyk, J. K., Quan, Y., Wall, D. A., Smith, R. L., and Faria, T. N. (2006) Human PEPT1 pharmacophore distinguishes between dipeptide transport and binding. *J. Med. Chem.* **49**, 3636–3644.
127. Andersen, R., Jorgensen, F. S., Olsen, L., Vabeno, J., Thorn, K., Nielsen, C. U., and Steffansen, B. (2006) Development of a QSAR model for binding of tripeptides and tripeptidomimetics to the human intestinal di-/tripeptide transporter hPEPT1. *Pharm. Res.* **23**, 483–492.
128. Crivori, P., Reinach, B., Pezzetta, D., and Poggesi, I. (2006) Computational models for identifying potential P-glycoprotein substrates and inhibitors. *Mol. Pharm.* **3**, 33–44.
129. Raub, T. J. (2006) P-glycoprotein recognition of substrates and circumvention through rational drug design. *Mol. Pharm.* **3**, 3–25.
130. Cianchetta, G., Singleton, R. W., Zhang, M., Wildgoose, M., Giesing, D., Fravolini, A., Cruciani, G., and Vaz, R. J. (2005) A pharmacophore hypothesis for P-glycoprotein substrate recognition using GRIND-based 3D-QSAR. *J. Med. Chem.* **48**, 2927–2935.
131. Globisch, C., Pajeva, I. K., and Wiese, M. (2006) Structure-activity relationships of a series of tariquidar analogs as multidrug resistance modulators. *Bioorg. Med. Chem.* **14**, 1588–1598.

132. Labrie, P., Maddaford, S. P., Fortin, S., Rakhit, S., Kotra, L. P., and Gaudreault, R. C. (2006) A comparative molecular field analysis (CoMFA) and comparative molecular similarity indices analysis (CoMSIA) of anthranilamide derivatives that are multidrug resistance modulators. *J. Med. Chem.* **49**, 7646–7660.
133. Raad, I., Terreux, R., Richomme, P., Matera, E. L., Dumontet, C., Raynaud, J., and Guilet, D. (2006) Structure-activity relationship of natural and synthetic coumarins inhibiting the multidrug transporter P-glycoprotein. *Bioorg. Med. Chem.* **14**, 6979–6987.
134. Cabrera, M. A., Gonzalez, I., Fernandez, C., Navarro, C., and Bermejo, M. (2006) A topological substructural approach for the prediction of P-glycoprotein substrates. *J. Pharm. Sci.* **95**, 589–606.
135. Zhang, L., Balimane, P. V., Johnson, S. R., and Chong, S. (2007) Development of an in silico model for predicting efflux substrates in Caco-2 cells. *Int. J. Pharm.* **343**, 98–105.
136. Cartmell, J., Enoch, S., Krstajic, D., and Leahy, D. E. (2005) Automated QSPR through Competitive Workflow. *J. Comput. Aided. Mol. Des.* **19**, 821–833.
137. de Cerqueira Lima, P., Golbraikh, A., Oloff, S., Xiao, Y., and Tropsha, A. (2006) Combinatorial QSAR modeling of P-glycoprotein substrates. *J. Chem. Inf. Model.* **46**, 1245–1254.
138. Penzotti, J. E., Lamb, M. L., Evensen, E., and Grootenhuis, P. D. (2002) A computational ensemble pharmacophore model for identifying substrates of P-glycoprotein. *J. Med. Chem.* **45**, 1737–1740.
139. Vandevuer, S., Van Bambeke, F., Tulkens, P. M., and Prevost, M. (2006) Predicting the three-dimensional structure of human P-glycoprotein in absence of ATP by computational techniques embodying crosslinking data: insight into the mechanism of ligand migration and binding sites. *Proteins* **63**, 466–478.
140. Badhan, R., and Penny, J. (2006) In silico modelling of the interaction of flavonoids with human P-glycoprotein nucleotide-binding domain. *Eur. J. Med. Chem.* **41**, 285–295.
141. Sakurai, A., Onishi, Y., Hirano, H., Seigneuret, M., Obanayama, K., Kim, G., Liew, E. L., Sakaeda, T., Yoshiura, K., Niikawa, N., Sakurai, M., and Ishikawa, T. (2007) Quantitative structure–activity relationship analysis and molecular dynamics simulation to functionally validate nonsynonymous polymorphisms of human ABC transporter ABCB1 (P-glycoprotein/MDR1). *Biochemistry* **46**, 7678–7693.
142. Chiba, P., Mihalek, I., Ecker, G. F., Kopp, S., and Lichtarge, O. (2006) Role of transmembrane domain/transmembrane domain interfaces of P-glycoprotein (ABCB1) in solute transport. Convergent information from photoaffinity labeling, site directed mutagenesis and in silico importance prediction. *Curr. Med. Chem.* **13**, 793–805.
143. Rebitzer, S., Annibali, D., Kopp, S., Eder, M., Langer, T., Chiba, P., Ecker, G. F., and Noe, C. R. (2003) In silico screening with benzofurane- and benzopyrane-type MDR-modulators. *Farmaco.* **58**, 185–191.
144. Langer, T., Eder, M., Hoffmann, R. D., Chiba, P., and Ecker, G. F. (2004) Lead identification for modulators of multidrug resistance based on in silico screening with a pharmacophoric feature model. *Arch. Pharm. (Weinheim)* **337**, 317–327.
145. Kaiser, D., Terfloth, L., Kopp, S., Schulz, J., de Laet, R., Chiba, P., Ecker, G. F., and Gasteiger, J. (2007) Self-organizing maps for identification of new inhibitors of P-glycoprotein. *J. Med. Chem.* **50**, 1698–1702.
146. Jedlitschky, G., Hoffmann, U., and Kroemer, H. K. (2006) Structure and function of the MRP2 (ABCC2) protein and its role in drug disposition. *Expert. Opin. Drug Metab. Toxicol.* **2**, 351–366.
147. Morrow, C. S., Peklak-Scott, C., Bishwokarma, B., Kute, T. E., Smitherman, P. K., and Townsend, A. J. (2006) Multidrug resistance protein 1 (MRP1, ABCC1) mediates resistance to mitoxantrone via glutathione-dependent drug efflux. *Mol. Pharmacol.* **69**, 1499–1505.
148. Boumendjel, A., Baubichon-Cortay, H., Trompier, D., Perrotton, T., and Di Pietro, A. (2005) Anticancer multidrug resistance mediated by MRP1: recent advances in the discovery of reversal agents. *Med. Res. Rev.* **25**, 453–472.
149. Young, A. M., Audus, K. L., Proudfoot, J., and Yazdanian, M. (2006) Tetrazole compounds: the effect of structure and pH on Caco-2 cell permeability. *J. Pharm. Sci.* **95**, 717–725.
150. Mols, R., Deferme, S., and Augustijns, P. (2005) Sulfasalazine transport in in-vitro, ex-vivo and in-vivo absorption models: contribution of efflux carriers and their modulation by co-administration of synthetic nature-identical fruit extracts. *J. Pharm. Pharmacol.* **57**, 1565–1573.

151. van Brussel, J. P., Oomen, M. A., Vossebeld, P. J., Wiemer, E. A., Sonneveld, P., and Mickisch, G. H. (2004) Identification of multidrug resistance-associated protein 1 and glutathione as multidrug resistance mechanisms in human prostate cancer cells: chemosensitization with leukotriene D4 antagonists and buthionine sulfoximine. *BJU Int.* **93**, 1333–1338.

152. Ng, C., Xiao, Y. D., Lum, B. L., and Han, Y. H. (2005) Quantitative structure-activity relationships of methotrexate and methotrexate analogues transported by the rat multispecific resistance-associated protein 2 (rMrp2). *Eur. J. Pharm. Sci.* **26**, 405–413.

153. Hirono, S., Nakagome, I., Imai, R., Maeda, K., Kusuhara, H., and Sugiyama, Y. (2005) Estimation of the three-dimensional pharmacophore of ligands for rat multidrug-resistance-associated protein 2 using ligand-based drug design techniques. *Pharm. Res.* **22**, 260–269.

154. Yeboah, D., Sun, M., Kingdom, J., Baczyk, D., Lye, S. J., Matthews, S. G., and Gibb, W. (2006) Expression of breast cancer resistance protein (BCRP/ABCG2) in human placenta throughout gestation and at term before and after labor. *Can. J. Physiol. Pharmacol.* **84**, 1251–1258.

155. Choudhuri, S., and Klaassen, C. D. (2006) Structure, function, expression, genomic organization, and single nucleotide polymorphisms of human ABCB1 (MDR1), ABCC (MRP), and ABCG2 (BCRP) efflux transporters. *Int. J. Toxicol.* **25**, 231–259.

156. Breuzard, G., Piot, O., Angiboust, J. F., Manfait, M., Candeil, L., Del Rio, M., and Millot, J. M. (2005) Changes in adsorption and permeability of mitoxantrone on plasma membrane of BCRP/MXR resistant cells. *Biochem. Biophys. Res. Commun.* **329**, 64–70.

157. Honjo, Y., Hrycyna, C. A., Yan, Q. W., Medina-Perez, W. Y., Robey, R. W., van de Laar, A., Litman, T., Dean, M., and Bates, S. E. (2001) Acquired mutations in the MXR/BCRP/ABCP gene alter substrate specificity in MXR/BCRP/ABCP-overexpressing cells. *Cancer Res.* **61**, 6635–6639.

158. Doyle, L. A., and Ross, D. D. (2003) Multidrug resistance mediated by the breast cancer resistance protein BCRP (ABCG2). *Oncogene.* **22**, 7340–7358.

159. Nagashima, S., Soda, H., Oka, M., Kitazaki, T., Shiozawa, K., Nakamura, Y., Takemura, M., Yabuuchi, H., Fukuda, M., Tsukamoto, K., and Kohno, S. (2006) BCRP/ABCG2 levels account for the resistance to topoisomerase I inhibitors and reversal effects by gefitinib in non-small cell lung cancer. *Cancer Chemother. Pharmacol.* **58**, 594–600.

160. Volk, E. L., and Schneider, E. (2003) Wild-type breast cancer resistance protein (BCRP/ABCG2) is a methotrexate polyglutamate transporter. *Cancer Res.* **63**, 5538–5543.

161. Rocchi, E., Khodjakov, A., Volk, E. L., Yang, C. H., Litman, T., Bates, S. E., and Schneider, E. (2000) The product of the ABC half-transporter gene ABCG2 (BCRP/MXR/ABCP) is expressed in the plasma membrane. *Biochem. Biophys. Res. Commun.* **271**, 42–46.

162. Maliepaard, M., van Gastelen, M. A., Tohgo, A., Hausheer, F. H., van Waardenburg, R. C., de Jong, L. A., Pluim, D., Beijnen, J. H., and Schellens, J. H. (2001) Circumvention of breast cancer resistance protein (BCRP)-mediated resistance to camptothecins in vitro using non-substrate drugs or the BCRP inhibitor GF120918. *Clin. Cancer Res.* **7**, 935–941.

163. Pavek, P., Merino, G., Wagenaar, E., Bolscher, E., Novotna, M., Jonker, J. W., and Schinkel, A. H. (2005) Human breast cancer resistance protein: interactions with steroid drugs, hormones, the dietary carcinogen 2-amino-1-methyl-6-phenylimidazo(4,5-b)pyridine, and transport of cimetidine. *J. Pharmacol. Exp. Ther.* **312**, 144–152.

164. Huisman, M. T., Chhatta, A. A., van Tellingen, O., Beijnen, J. H., and Schinkel, A. H. (2005) MRP2 (ABCC2) transports taxanes and confers paclitaxel resistance and both processes are stimulated by probenecid. *Int. J. Cancer* **116**, 824–829.

165. Gupta, A., Zhang, Y., Unadkat, J. D., and Mao, Q. (2004) HIV protease inhibitors are inhibitors but not substrates of the human breast cancer resistance protein (BCRP/ABCG2). *J. Pharmacol. Exp. Ther.* **310**, 334–341.

166. Saito, H., Hirano, H., Nakagawa, H., Fukami, T., Oosumi, K., Murakami, K., Kimura, H., Kouchi, T., Konomi, M., Tao, E., Tsujikawa, N., Tarui, S., Nagakura, M., Osumi, M., and Ishikawa, T. (2006) A new strategy of high-speed screening and quantitative structure-activity relationship analysis to evaluate human ATP-binding cassette transporter ABCG2-drug interactions. *J. Pharmacol. Exp. Ther.* **317**, 1114–1124.

167. Kuhar, M. J., Ritz, M. C., and Boja, J. W. (1991) The dopamine hypothesis of the reinforcing properties of cocaine. *Trends Neu-*

rosci. **14**, 299–302.
168. Self, D. W., and Nestler, E. J. (1995) Molecular mechanisms of drug reinforcement and addiction. *Annu. Rev. Neurosci.* **18**, 463–495.
169. Tomlinson, I. D., Mason, J. N., Blakely, R. D., and Rosenthal, S. J. (2006) High affinity inhibitors of the dopamine transporter (DAT): novel biotinylated ligands for conjugation to quantum dots. *Bioorg. Med. Chem. Lett.* **16**, 4664–4667.
170. Cline, E. J., Terry, P., Carroll, F. I., Kuhar, M. J., and Katz, J. L. (1992) Stimulus generalization from cocaine to analogs with high in vitro affinity for dopamine uptake sites. *Behav. Pharmacol.* **3**, 113–116.
171. Huang, X., and Zhan, C. G. (2007) How Dopamine Transporter Interacts with Dopamine: Insights from Molecular Modeling and Simulation. *Biophys. J.*
172. Wang, S., Sakamuri, S., Enyedy, I. J., Kozikowski, A. P., Deschaux, O., Bandyopadhyay, B. C., Tella, S. R., Zaman, W. A., and Johnson, K. M. (2000) Discovery of a novel dopamine transporter inhibitor, 4-hydroxy-1-methyl-4-(4-methylphenyl)-3-piperidyl 4-methylphenyl ketone, as a potential cocaine antagonist through 3D-database pharmacophore searching. Molecular modeling, structure-activity relationships, and behavioral pharmacological studies. *J. Med. Chem.* **43**, 351–360.
173. FDA. (http://www.fda.gov/oc/initiatives/criticalpath/initiative.html) FDA's Critical Path Initiative—Science Enhancing the Health and Well-Being of All Americans (2004).
174. WHO. (2002) Waiver of In vivo Bioavailability and Bioequivalence Studies for Immediate-Release Solid Oral Dosage Forms Based on a Biopharmaceutics Classification System, FDA Guidance for Industry, Federal Drug and Food Administration, Rockville, MD.
175. Blume, H. H., and Schug, B. S. (1999) The biopharmaceutics classification system (BCS): class III drugs - better candidates for BA/BE waiver? *Eur. J. Pharm. Sci.* **9**, 117–121.
176. Polli, J. E., Yu, L. X., Cook, J. A., Amidon, G. L., Borchardt, R. T., Burnside, B. A., Burton, P. S., Chen, M. L., Conner, D. P., Faustino, P. J., Hawi, A. A., Hussain, A. S., Joshi, H. N., Kwei, G., Lee, V. H., Lesko, L. J., Lipper, R. A., Loper, A. E., Nerurkar, S. G., Polli, J. W., Sanvordeker, D. R., Taneja, R., Uppoor, R. S., Vattikonda, C. S., Wilding, I., and Zhang, G. (2004) Summary workshop report: biopharmaceutics classification system–implementation challenges and extension opportunities. *J. Pharm. Sci.* **93**, 1375–1381.
177. Wu, C. Y., and Benet, L. Z. (2005) Predicting drug disposition via application of BCS: transport/absorption/ elimination interplay and development of a biopharmaceutics drug disposition classification system. *Pharm. Res.* **22**, 11–23.
178. Takagi, T., Ramachandran, C., Bermejo, M., Yamashita, S., Yu, L. X., and Amidon, G. L. (2006) A provisional biopharmaceutical classification of the top 200 oral drug products in the United States, Great Britain, Spain, and Japan. *Mol. Pharm.* **3**, 631–643.
179. Khandelwal, A., Bahadduri, P. M., Chang, C., Polli, J. E., Swaan, P. W., and Ekins, S. (2007) Computational Models to Assign Biopharmaceutics Drug Disposition Classification from Molecular Structure. *Pharm. Res.*
180. Dunker, A. K., Lawson, J. D., Brown, C. J., Williams, R. M., Romero, P., Oh, J. S., Oldfield, C. J., Campen, A. M., Ratliff, C. M., Hipps, K. W., Ausio, J., Nissen, M. S., Reeves, R., Kang, C., Kissinger, C. R., Bailey, R. W., Griswold, M. D., Chiu, W., Garner, E. C., and Obradovic, Z. (2001) Intrinsically disordered protein. *J. Mol. Graph. Model.* **19**, 26–59.
181. Romero, P. R., Zaidi, S., Fang, Y. Y., Uversky, V. N., Radivojac, P., Oldfield, C. J., Cortese, M. S., Sickmeier, M., LeGall, T., Obradovic, Z., and Dunker, A. K. (2006) Alternative splicing in concert with protein intrinsic disorder enables increased functional diversity in multicellular organisms. *Proc. Natl. Acad. Sci. U S A* **103**, 8390–8395.
182. Bairoch, A., Apweiler, R., Wu, C. H., Barker, W. C., Boeckmann, B., Ferro, S., Gasteiger, E., Huang, H., Lopez, R., Magrane, M., Martin, M. J., Natale, D. A., O′Donovan, C., Redaschi, N., and Yeh, L. S. (2005) The Universal Protein Resource (UniProt). *Nucleic. Acids Res.* **33**, D154–159.
183. Peng, K., Vucetic, S., Radivojac, P., Brown, C. J., Dunker, A. K., and Obradovic, Z. (2005) Optimizing long intrinsic disorder predictors with protein evolutionary information, *J. Bioinform. Comput. Biol.* **3**, 35–60.
184. Krasowski, M. D., Reschly, E. J., and Ekins, S. (2008) Intrinsic disorder in nuclear hormone receptors, *J. Proteome. Res.* 7, 4359–4372.
185. Bertilsson, G., Heidrich, J., Svensson, K., Asman, M., Jendeberg, L., Sydow-Backman,

M., Ohlsson, R., Postlind, H., Blomquist, P., and Berkenstam, A. (1998) Identification of a human nuclear receptor defines a new signaling pathway for CYP3A induction. *Proc. Natl. Acad. Sci. U S A* **95**, 12208–12213.

186. Blumberg, B., and Evans, R. M. (1998) Orphan nuclear receptors–new ligands and new possibilities. *Genes Dev.* **12**, 3149–3155.

187. Kliewer, S. A., Moore, J. T., Wade, L., Staudinger, J. L., Watson, M. A., Jones, S. A., McKee, D. D., Oliver, B. B., Willson, T. M., Zetterstrom, R. H., Perlmann, T., and Lehmann, J. M. (1998) An orphan nuclear receptor activated by pregnanes defines a novel steroid signaling pathway. *Cell* **92**, 73–82.

188. Synold, T. W., Dussault, I., and Forman, B. M. (2001) The orphan nuclear receptor SXR coordinately regulates drug metabolism and efflux. *Nat. Med.* **7**, 584–590.

189. Staudinger, J., Liu, Y., Madan, A., Habeebu, S., and Klaassen, C. D. (2001) Coordinate regulation of xenobiotic and bile acid homeostasis by pregnane X receptor. *Drug Metab. Dispos.* **29**, 1467–1472.

190. Bleasby, K., Castle, J. C., Roberts, C. J., Cheng, C., Bailey, W. J., Sina, J. F., Kulkarni, A. V., Hafey, M. J., Evers, R., Johnson, J. M., Ulrich, R. G., and Slatter, J. G. (2006) Expression profiles of 50 xenobiotic transporter genes in humans and preclinical species: a resource for investigations into drug disposition. *Xenobiotica* **36**, 963–988.

191. Deng, R., Yang, D., Yang, J., and Yan, B. (2006) Oxysterol 22(R)-hydroxycholesterol induces the expression of the bile salt export pump through nuclear receptor farsenoid X receptor but not liver X receptor. *J. Pharmacol. Exp. Ther.* **317**, 317–325.

192. Lee, H., Zhang, Y., Lee, F. Y., Nelson, S. F., Gonzalez, F. J., and Edwards, P. A. (2006) FXR regulates organic solute transporters alpha and beta in the adrenal gland, kidney, and intestine. *J. Lipid Res.* **47**, 201–214.

193. Guo, G. L., Staudinger, J., Ogura, K., and Klaassen, C. D. (2002) Induction of rat organic anion transporting polypeptide 2 by pregnenolone-16alpha-carbonitrile is via interaction with pregnane X receptor. *Mol. Pharmacol.* **61**, 832–839.

194. Fleck, C., Schwertfeger, M., and Taylor, P. M. (2003) Regulation of renal amino acid (AA) transport by hormones, drugs and xenobiotics - a review. *Amino Acids* **24**, 347–374.

195. Rothman, R. B., Ayestas, M. A., Dersch, C. M., and Baumann, M. H. (1999) Aminorex, fenfluramine, and chlorphentermine are serotonin transporter substrates. Implications for primary pulmonary hypertension. *Circulation* **100**, 869–875.

196. Barabasi, A. L., and Oltvai, Z. N. (2004) Network biology: understanding the cell's functional organization. *Nat. Rev. Genet.* **5**, 101–113.

197. Apic, G., Ignjatovic, T., Boyer, S., and Russell, R. B. (2005) Illuminating drug discovery with biological pathways. *FEBS Lett.* **579**, 1872–1877.

198. Huang, J. C., Sakata, T., Pfleger, L. L., Bencsik, M., Halloran, B. P., Bikle, D. D., and Nissenson, R. A. (2004) PTH differentially regulates expression of RANKL and OPG. *J. Bone Miner. Res.* **19**, 235–244.

199. Shu, C., Shen, H., Hopfer, U., and Smith, D. E. (2001) Mechanism of intestinal absorption and renal reabsorption of an orally active ace inhibitor: uptake and transport of fosinopril in cell cultures. *Drug Metab. Dispos.* **29**, 1307–1315.

200. Mitsuoka, K., Kato, Y., Kubo, Y., and Tsuji, A. (2007) Functional expression of stereoselective metabolism of cephalexin by exogenous transfection of oligopeptide transporter PEPT1. *Drug Metab. Dispos.* **35**, 356–362.

201. Launay-Vacher, V., Izzedine, H., Karie, S., Hulot, J. S., Baumelou, A., and Deray, G. (2006) Renal tubular drug transporters. *Nephron. Physiol.* **103**, p97–106.

202. MacDougall, C., and Guglielmo, B. J. (2004) Pharmacokinetics of valaciclovir. *J. Antimicrob. Chemother.* **53**, 899–901.

203. Shitara, Y., Hirano, M., Sato, H., and Sugiyama, Y. (2004) Gemfibrozil and its glucuronide inhibit the organic anion transporting polypeptide 2 (OATP2/OATP1B1:SLC21A6)-mediated hepatic uptake and CYP2C8-mediated metabolism of cerivastatin: analysis of the mechanism of the clinically relevant drug-drug interaction between cerivastatin and gemfibrozil. *J. Pharmacol. Exp. Ther.* **311**, 228–236.

204. Kameyama, Y., Yamashita, K., Kobayashi, K., Hosokawa, M., and Chiba, K. (2005) Functional characterization of SLCO1B1 (OATP-C) variants, SLCO1B1*5, SLCO1B1*15 and SLCO1B1*15+C1007G, by using transient expression systems of HeLa and HEK293 cells. *Pharmacogenet Genomics* **15**, 513–522.

205. Abe, T., Unno, M., Onogawa, T., Tokui, T., Kondo, T. N., Nakagomi, R., Adachi, H.,

Fujiwara, K., Okabe, M., Suzuki, T., Nunoki, K., Sato, E., Kakyo, M., Nishio, T., Sugita, J., Asano, N., Tanemoto, M., Seki, M., Date, F., Ono, K., Kondo, Y., Shiiba, K., Suzuki, M., Ohtani, H., Shimosegawa, T., Iinuma, K., Nagura, H., Ito, S., and Matsuno, S. (2001) LST-2, a human liver-specific organic anion transporter, determines methotrexate sensitivity in gastrointestinal cancers. *Gastroenterology* **120**, 1689–1699.

206. Konig, J., Seithel, A., Gradhand, U., and Fromm, M. F. (2006) Pharmacogenomics of human OATP transporters. *Naunyn. Schmiedebergs Arch. Pharmacol.* **372**, 432–443.

207. Mikkaichi, T., Suzuki, T., Tanemoto, M., Ito, S., and Abe, T. (2004) The organic anion transporter (OATP) family. *Drug Metab. Pharmacokinet.* **19**, 171–179.

208. Masuda, S. (2003) Functional characteristics and pharmacokinetic significance of kidney-specific organic anion transporters, OAT-K1 and OAT-K2, in the urinary excretion of anionic drugs. *Drug Metab. Pharmacokinet.* **18**, 91–103.

209. Kaler, G., Truong, D. M., Khandelwal, A., Nagle, M., Eraly, S. A., Swaan, P. W., and Nigam, S. K. (2007) Structural variation governs substrate specificity for organic anion transporters (oat) homologs: potential remote sensing by oat family members, *J. Biol. Chem.*

210. Eraly, S. A., Bush, K. T., Sampogna, R. V., Bhatnagar, V., and Nigam, S. K. (2004) The molecular pharmacology of organic anion transporters: from DNA to FDA? *Mol. Pharmacol.* **65**, 479–487.

211. Sweet, D. H., Bush, K. T., and Nigam, S. K. (2001) The organic anion transporter family: from physiology to ontogeny and the clinic. *Am. J. Physiol. Renal. Physiol.* **281**, F197–205.

212. Russel, F. G., Masereeuw, R., and van Aubel, R. A. (2002) Molecular aspects of renal anionic drug transport. *Annu. Rev. Physiol.* **64**, 563–594.

213. Gorboulev, V., Ulzheimer, J. C., Akhoundova, A., Ulzheimer-Teuber, I., Karbach, U., Quester, S., Baumann, C., Lang, F., Busch, A. E., and Koepsell, H. (1997) Cloning and characterization of two human polyspecific organic cation transporters. *DNA Cell Biol.* **16**, 871–881.

214. Muller, J., Lips, K. S., Metzner, L., Neubert, R. H., Koepsell, H., and Brandsch, M. (2005) Drug specificity and intestinal membrane localization of human organic cation transporters (OCT). *Biochem. Pharmacol.* **70**, 1851–1860.

215. Shu, Y., Brown, C., Castro, R. A., Shi, R. J., Lin, E. T., Owen, R. P., Sheardown, S. A., Yue, L., Burchard, E. G., Brett, C. M., and Giacomini, K. M. (2007) Effect of Genetic Variation in the Organic Cation Transporter 1, OCT1, on Metformin Pharmacokinetics. *Clin. Pharmacol. Ther.*.

216. Zhang, S., Lovejoy, K. S., Shima, J. E., Lagpacan, L. L., Shu, Y., Lapuk, A., Chen, Y., Komori, T., Gray, J. W., Chen, X., Lippard, S. J., and Giacomini, K. M. (2006) Organic Cation Transporters Are Determinants of Oxaliplatin Cytotoxicity. *Cancer Res.* **66**, 8847–8857.

217. Yonezawa, A., Masuda, S., Yokoo, S., Katsura, T., and Inui, K. (2006) Cisplatin and oxaliplatin, but not carboplatin and nedaplatin, are substrates for human organic cation transporters (SLC22A1-3 and multidrug and toxin extrusion family). *J. Pharmacol. Exp. Ther.* **319**, 879–886.

218. Kaewmokul, S., Chatsudthipong, V., Evans, K. K., Dantzler, W. H., and Wright, S. H. (2003) Functional mapping of rbOCT1 and rbOCT2 activity in the S2 segment of rabbit proximal tubule. *Am. J. Physiol. Renal. Physiol.* **285**, F1149–1159.

219. Dresser, M. J., Xiao, G., Leabman, M. K., Gray, A. T., and Giacomini, K. M. (2002) Interactions of n-tetraalkylammonium compounds and biguanides with a human renal organic cation transporter (hOCT2). *Pharm. Res.* **19**, 1244–1247.

220. Trauner, M., and Boyer, J. L. (2003) Bile salt transporters: molecular characterization, function, and regulation. *Physiol. Rev.* **83**, 633–671.

221. Leslie, E. M., Watkins, P. B., Kim, R. B., and Brouwer, K. L. (2007) Differential Inhibition of Rat and Human Na+-Dependent Taurocholate Cotransporting Polypeptide (NTCP/SLC10A1)by Bosentan: A Mechanism for Species Differences in Hepatotoxicity. *J. Pharmacol. Exp. Ther.* **321**, 1170–1178.

222. Alpini, G., Glaser, S., Robertson, W., Phinizy, J. L., Rodgers, R. E., Caligiuri, A., and LeSage, G. (1997) Bile acids stimulate proliferative and secretory events in large but not small cholangiocytes. *Am. J. Physiol.***273**, G518–529.

223. Alpini, G., Glaser, S. S., Rodgers, R., Phinizy, J. L., Robertson, W. E., Lasater, J., Caligiuri, A., Tretjak, Z., and LeSage, G. D. (1997) Functional expression of the apical Na+-dependent bile acid transporter in large but not small rat cholangiocytes. *Gastroenterology* **113**, 1734–1740.

224. Wong, M. H., Oelkers, P., Craddock, A. L., and Dawson, P. A. (1994) Expression cloning and characterization of the hamster ileal sodium-dependent bile acid transporter. *J. Biol. Chem.* **269**, 1340–1347.
225. Ho, G. T., Moodie, F. M., and Satsangi, J. (2003) Multidrug resistance 1 gene (P-glycoprotein 170): an important determinant in gastrointestinal disease? *Gut.* **52**, 759–766.
226. Fojo, A. T., Ueda, K., Slamon, D. J., Poplack, D. G., Gottesman, M. M., and Pastan, I. (1987) Expression of a multidrug-resistance gene in human tumors and tissues. *Proc. Natl. Acad. Sci. U S A* **84**, 265–269.
227. Dey, S., Patel, J., Anand, B. S., Jain-Vakkalagadda, B., Kaliki, P., Pal, D., Ganapathy, V., and Mitra, A. K. (2003) Molecular evidence and functional expression of P-glycoprotein (MDR1) in human and rabbit cornea and corneal epithelial cell lines. *Invest. Ophthalmol. Vis Sci.* **44**, 2909–2918.
228. Chan, L. M., Lowes, S., and Hirst, B. H. (2004) The ABCs of drug transport in intestine and liver: efflux proteins limiting drug absorption and bioavailability. *Eur. J. Pharm. Sci.* **21**, 25–51.
229. Maliepaard, M., van Gastelen, M. A., de Jong, L. A., Pluim, D., van Waardenburg, R. C., Ruevekamp-Helmers, M. C., Floot, B. G., and Schellens, J. H. (1999) Overexpression of the BCRP/MXR/ABCP gene in a topotecan-selected ovarian tumor cell line. *Cancer Res.* **59**, 4559–4563.
230. Yang, C. H., Schneider, E., Kuo, M. L., Volk, E. L., Rocchi, E., and Chen, Y. C. (2000) BCRP/MXR/ABCP expression in topotecan-resistant human breast carcinoma cells. *Biochem. Pharmacol.* **60**, 831–837.
231. Taipalensuu, J., Tornblom, H., Lindberg, G., Einarsson, C., Sjoqvist, F., Melhus, H., Garberg, P., Sjostrom, B., Lundgren, B., and Artursson, P. (2001) Correlation of gene expression of ten drug efflux proteins of the ATP-binding cassette transporter family in normal human jejunum and in human intestinal epithelial Caco-2 cell monolayers. *J. Pharmacol. Exp. Ther.* **299**, 164–170.
232. Kobayashi, D., Ieiri, I., Hirota, T., Takane, H., Maegawa, S., Kigawa, J., Suzuki, H., Nanba, E., Oshimura, M., Terakawa, N., Otsubo, K., Mine, K., and Sugiyama, Y. (2005) Functional assessment of ABCG2 (BCRP) gene polymorphisms to protein expression in human placenta. *Drug Metab. Dispos.* **33**, 94–101.
233. Aye, I. L., Paxton, J. W., Evseenko, D. A., and Keelan, J. A. (2007) Expression, Localisation and Activity of ATP Binding Cassette (ABC) Family of Drug Transporters in Human Amnion Membranes. *Placenta* **28**, 868–877.
234. Nies, A. T., and Keppler, D. (2007) The apical conjugate efflux pump ABCC2 (MRP2). *Pflugers Arch.* **453**, 643–659.
235. Thompson, R., and Strautnieks, S. (2001) BSEP: function and role in progressive familial intrahepatic cholestasis. *Semin Liver Dis.* **21**, 545–550.
236. Funk, C., Pantze, M., Jehle, L., Ponelle, C., Scheuermann, G., Lazendic, M., and Gasser, R. (2001) Troglitazone-induced intrahepatic cholestasis by an interference with the hepatobiliary export of bile acids in male and female rats. Correlation with the gender difference in troglitazone sulfate formation and the inhibition of the canalicular bile salt export pump (Bsep) by troglitazone and troglitazone sulfate. *Toxicology* **167**, 83–98.

Chapter 5

Methods to Evaluate Transporter Activity in Cancer

Takeo Nakanishi, Douglas D. Ross, and Keisuke Mitsuoka

Abstract

Plasma membrane transporter proteins play an important role in taking up nutrients into and effluxing xenobiotics out of cells to sustain cell survival. In the last decade, a number of studies have shown that these physiologically important transporters affect absorption, distribution, and excretion of major anticancer agents in clinical use. More importantly, many transporters have been reported to be differentially upregulated in cancer cells compared to normal tissues, suggesting that the differential expression of transporters in cancer cells may become good targets for enhancing drug delivery as well as diagnostic markers for cancer therapy. Hence, utilizing the knowledge of transporter functions likely provides us with the possibility of delivering a drug to the target tissues, avoiding distribution to other tissues, and improving oral bioavailability. This chapter focuses on methodology to analyze the activity of transporters that are involved in drug transport.

Key words: SLC transporter, ABC transporter, cancer, transport study, vesicle preparation.

1. Introduction

Plasma membrane transporter proteins are classified into two superfamilies, the solute carrier (SLC) and the ATP binding cassette (ABC) transporters. SLC transporters import nutrient substances essential for cell survival in a facilitated manner or by utilizing the downhill flow of an ion to transport a substrate against its concentration gradient across the plasma membrane.

Neoplastic transformation is usually accompanied with an adaptive increase in biosynthesis of nucleotide and protein to maintain a high proliferation rate in malignant tumors. In fact, several classes of SLC transporters for amino acids and nucleosides are well known to be functionally upregulated in some types

Q. Yan (ed.), *Membrane Transporters in Drug Discovery and Development*, Methods in Molecular Biology 637, DOI 10.1007/978-1-60761-700-6_5,

of human cancer cells. Since we discovered that an oligopeptide transport activity was elevated in human fibrosarcoma cell line, HT1080 (1), expression of the human peptide transporter PEPT1 has been investigated in a variety of human cancer cells (2–4). More recently, a PEPT1-targeted positron emission tomography (PET) probe has been shown to be a useful tool to detect early presence of malignant tissues in in vivo experimental animal models (4). In addition to PEPT1, other SLC transporters including amino acid transporters (e.g., LAT1, ASCT2, $ATB^{0,+}$), organic cation transporters (e.g., OCT1), and nucleoside transporters (e.g., CNT1 and ENT1) are also known molecular targets for anticancer chemotherapy (5).

On the other hand, most of the ABC transporters function as efflux pump that play a role in exporting xenobiotics out of cells, while some form specific membrane channels. In terms of anticancer chemotherapy, three major ABC transporters, P-glycoprotein (a product of the *MDR1/ABCB1* gene), MRP1 (*ABCC1*), and BCRP (*ABCG2*), are important as molecular causes of multidrug resistance (MDR) acquired by various types of human cancer cells. Since BCRP was isolated, we have characterized functional expression of BCRP in several types of human cancer cell lines (6–8).

In this chapter, we describe methods to detect peptide transport activity expressed in cancer cells in both in vitro and in vivo systems showing detailed protocols using a typical substrate for a peptide transporter, which can be utilized for activity of other SLC transporters expressed in cancer cells. In addition, we describe a protocol to measure BCRP-mediated efflux of substrate out of cells by retention assay. Furthermore, we present a useful methodology to evaluate transport activity utilizing membrane vesicles prepared from cancer cells, in which driving force for transporters are easily altered.

2. Materials

2.1. Cell Culture

1. HT1080, MKN-45, and MCF-7/AdrVp cells are obtained from American Type Culture Collection, Japanese Collection of Research Bioresources Cell Bank, and Dr Ross's laboratory at University of Maryland at Baltimore, respectively.
2. Cell culture medium: RPMI1640 and improved MEM medium. In general, cell culture medium is supplemented with 10% heat-inactivated fetal bovine serum.
3. Improved MEM.

4. Phosphate buffered saline (PBS): 150 mM NaCl, 5 mM KCl, 50 mM potassium phosphate (pH 7.5).
5. Antibiotics: Penicillin G and streptomycin are added to medium used at final concentrations of 100 U/mL and 100 μg/mL, respectively. Store at 4°C for several months before use.
6. Solution of trypsin (0.25%) with 0.38 g/L of ethylenediamine tetraacetic acid (EDTA)·4Na (Invitrogen, Carlsbad, CA). Store at 4°C for several months.
7. Phosphate buffered saline (PBS, pH 7.4).
8. 24-well cell culture plate (1.9 cm^2/well).

2.2. Intracellular Accumulation Study (In Vitro)

1. Transport buffer (Modified Hanks' balanced salt solution): 0.952 mM $CaCl_2$, 5.36 mM KCl, 0.441 mM KH_2PO_4, 0.812 mM $MgSO_4$, 136.7 mM NaCl, 0.385 mM Na_2HPO_4, 25 mM D-glucose, 10 mM HEPES/Tris, pH 7.4 (*see* **Note 1**). 10X buffer can be stored at –20°C for a long time. This buffer is also used as washing buffer.
2. Solubilization solution: 1% Triton X. This solution is prepared freshly.
3. Radiolabeled substrate (e.g., [^{14}C]glycylsarcosine (Gly-Sar) if you are studying PEPT1 – *see* **Table 5.1**).
4. Silicon layer: Liquid paraffin and silicon oil (SH550, available from Toray Dow Corning, Tokyo, Japan) are mixed at the ratio of 10 to 2.17, density of which is exactly 1.029. One day before the experiment, 100 μL of the mixture of liquid paraffin and silicon is overlaid onto 50 μL of 3 M KOH in a 0.4 mL microcentrifuge tube. This tube is spun at >13,000×*g* for 5 s using a benchtop microcentrifuge. Store the solution at 4C until use.
5. Liquid scintillation cocktail.
6. Multipurpose scintillation counter (e.g., LS 6500, Beckman Coulter, Inc. Fullerton, CA).

2.3. Intracellular Accumulation Study (In Vivo)

1. Female athymic mice (5 weeks old, CAnN.Cg-*Foxn1^{nu}*/Crl, available from Charles River).
2. Matrigel™ (BD Biosciences, Franklin Lakes, NJ).
3. Sevoflurane.
4. Radiolabeled Gly-Sar: [^{11}C]Gly-Sar (Procedures for chemical synthesis of [^{11}C]Gly-Sar is reported in (4, 28)).
5. γ-counter: Wallac Wizard3 (Long Island Scientific, East Setauket, NY).

Table 5.1
SLC transporters expressed in cancer cells and their established substrates and inhibitors

Transporter	Protein	Gene	Native substrate	Mechanism[a]	Cancer[b]	Established substrate/drug	Inhibitor
Oligopeptide	PEPT1	*SLC15A1*	Di- or tripeptide	C/H^+	Pancreas (AsPC-1) (3)	Gly-Sar (1) Ubenimex (Bestatin) (2)	Self[c]
Amino acid	LAT1	*SLC7A5*	H,M,L,I, V,F,Y,W,Q[d]	E/Amino acid	Leukemia (9) Colon (10) Glioma (11) Esophagus Bladder (T24) (12)	Amino acid (L) Melphalan (12)	BCH (13)
	ASCT2	*SLC1A5*	A,S,C,T,Q	C/Na^+	Liver (14)		
	$ATB^{0,+}$	*SLC6A14*	K,R,A,S,C, T,N,Q,H,C	C/Na^+& Cl^-	Colon (15) Cervix (16)	Carnitine (17) D-serine(18) NOX inhibior (19)	
	xCT	*SLC7A11*	Ci,E		Lung (A549) (20)	Amino acid (E) (20)	(S)-4-carboxy-phenylglycine (20)
Organic ion	OCT1	*SLC22A1*	Organic cation	F	Colon (21)	TAE Oxaliplatin (21) Imatinib (22)	
	CT2	*SLC22A16*	Carnitine/ Betain	F	Leukemia (23)	Doxorubicin (23)	

Table 5.1 (continued)

Nucleoside	CNT1	*SLC28A1*	Nucleoside	C/Na^+	Gynecologic tumor (24)	Cytarabine (ara-C)	
	ENT1	*SLC29A1*	Nucleoside, Inosine	F	Pancrea (25) Gynecologic tumor (24)	Cytarabine (ara-C)	(S)- (4-Nitrobenzyl) -6-thioinosine (NBMPR) Dipiridamole
Folate	RFC	*SLC19A1*	Folate	E/OH^-		Methotrexate	
Monocarboxylate	MCT1	*SLC16A1*	Lactate, pyruvate	C/H^+	Colon (26) Neuroblastoma (27)	Lactate	Lonidamine (27) α-Cyano-4-OH-cinnamate
Glucose	GLUT1	*SLC2A1*	Glucose	F		Fluoro-2-deoxy-*D*-glucose (FDG)	

[a]C, cotransporter, E, exchanger, F, facilitated transporter.

[b]Cancer, cancer cell lines or cancerous tissues are listed which are known or reported to overexpress transporters.

[c]Self, self-inhibition (using unlabeled substrate compound) has been reported to evaluate transport of the substrates listed because no compounds are known to block activity of the transporter specifically.

[d]Amino acids are shown in one letter code.

6. Planar positron imaging system (PPIS): IPS-1000-6XII (Hamamatsu Photonics K.K., Hamamatsu, Japan).

2.4. Preparation Plasma Membrane Vesicles

1. Phosphate buffered saline (PBS): The same described as in **Section 2.1**.
2. Cell suspension buffer: 10 mM NaCl, 1.5 mM $MgCl_2$, 10 mM Tris–HCl, pH 7.4. Store at 4°C. Add PMSF at final concentration of 0.02 mM before use.
3. 100 mM EDTA solution (100X, pH 7.4).
4. Sucrose cushion: 35% (w/v) sucrose, 10 mM Tris–HCl, pH 7.4.
5. Sucrose dilution buffer: 250 mM sucrose, 10 mM Tris–HCl, pH 7.5.
6. Vesicle storage buffer: 100 mM KCl, 100 mM mannitol, 20 mM HEPES-Tris, pH 6.5–8.5, or Mes-Tris, pH 5.0–6.5. Store at 4°C. Add PMSF at final concentration of 0.02 mM before use.
7. High-pressure chamber for nitrogen cavitation: Nitrogen Bomb Parr 4635 (Parr Instrument Company, Moline, IL).

2.5. Transport Study/Vesicle

1. Vesicle transport buffer: 100 mM KCl, 100 mM mannitol, 20 mM HEPES/Tris, pH 7.4. This buffer can be used to stop the reaction. However, in case a highly adsorptive radiolabeled substrate such as cationic compound (e.g., [^{3}H]TEA) is used, an excess amount of unlabeled substrate should be added to the buffer to reduce background radioactivity due to nonspecific binding to membrane filter.
2. Membrane filter: MF-Millipore membrane (HAWP, 0.45 μM, Millipore, Billerica, MA).

3. Methods

In general, influx transport activity of a membrane transporter is evaluated by measuring transport rate of a substrate compound in cells plated on tissue culture plates. Alternatively, the activity is measured using plasma membrane vesicles prepared from cultured cells. Transport rates are experimentally obtained by measuring an intracellular or intravesicular accumulation of a substrate compound for a certain period of time. In order to evaluate the activity of a transporter protein endogenously expressed in cancer cells, it is important (1) to select a specific substrate for the transporter and (2) to find out an experimental condition (as control) to estimate simple diffusion of the substrate in the presence of a selective inhibitor for the transporter or absence of its driving force out of transport buffer if it is known (e.g., replacing Na^+ with choline or *N*-methyl-d-glucamine (NMDG) or altering pH). To help readers to find out such conditions, in **Table 5.1**, established substrates and inhibitors for SLC transporter proteins are listed with transport mechanism if it is known. Furthermore, plasma membrane vesicles can be used to determine accurate rate and inhibition constants for both influx and efflux transporters because the method produces a mixture of both outside-out and inside-out vesicles. Therefore accumulation in inside-out vesicles can be used for efflux transporter studies, and accumulation in outside-out vesicles can be measured for influx transporters – and no need to separate the inside-out and outside-out vesicles.

To quantify the intracellular accumulation of a tested substrate, it is suggested to use radiolabeled or fluorescent compounds, if applicable, because the accumulation is easily measured by utilizing scintillation counter or flow cytometry. Otherwise, to quantify intracellular accumulation of the substrate, an appropriate analytical method should be applied, such as high-performance liquid chromatography or liquid chromatography in combination with mass spectrometry (e.g., LC/MS/MS).

3.1. Preparation of Cells and Seeding for Experiments

1. For *intracellular accumulation (in vitro)*: *For adherent cells*, 2 days before an experiment, plate cells of interest at a density such that the cells become semi-confluent when used. For example, HT1080 cells are usually plated at a 5.0×10^4 cells/cm^2 onto multiwell tissue culture plates in RPMI1640 medium supplemented with 10% FBS for uptake experimenta (*see* **Note 2**). For retention assays to evaluate efflux transporters such as BCRP, MCF-7/AdrVp cells overexpressing BCRP are plated at a density of $5.0–7.5 \times 10^4$ cells/cm^2onto multiwell tissue culture plate in IMEM supplemented with 10% FBS. *For floating cells*, cells of interest are cultured in a plastic flask (T-75 is preferably used) in an appropriate medium. Take care to ensure that the cell density does not exceed over 1×10^6 cells/mL.
2. *For intracellular accumulation (in vivo):* MKN-45 cells are grown to 95% confluence in RPMI1640 medium supplemented with 10% FBS. For xenotransplantation, trypsinized single-cell suspension is centrifuged at $90{\times}g$ at 4°C for 5 min to obtain cell pellet. The cell pellet is suspended directly in ice-cold PBS, which is followed by addition of the same volume of ice-cold liquid Matrigel™ on ice. Adult athymic mice at age 5 weeks are injected subcutaneously into right side of hind leg with 1×10^6 MKN45 cells and into left side of hind leg with the same volume of vehicle alone as control. MKN-45 cells are grown for 2 weeks to tumor

with a size of 300–500 mm^3. The mice at age 7 weeks are then used for evaluation in vivo [^{11}C]Gly-Sar accumulation to tumors.

3. *For membrane vesicle preparation:* Cells of interest are usually plated onto 150-cm^2 tissue culture flasks (10–15 flasks are usually required) to prepare plasma membrane vesicles. When cells become semi-confluent, cells are scraped off using a Teflon® cell scraper.

3.2. Intracellular Accumulation Study (In Vitro)

3.2.1. Adherent Cells

1. Culture medium is aspirated off, and then cells are washed gently but quickly with 1 mL of pre-warmed transport buffer (at 37°C for 20 min) twice.
2. (*Uptake experiment for peptide transporters*) A 250 μL of pre-warmed transport buffer containing 0.02–40 mM Gly-Sar plus 1 μCi/mL [^{14}C]Gly-Sar is placed onto cells. Then, the plate is transferred to a rack set in a benchtop water bath (37°C) and cells are incubated at 37°C for 0.5–120 min. The time course of intracellular accumulation of 20 μM [^{14}C]Gly-Sar in HT1080 and IMR90 (normal fibroblast cells, which do not have peptide transport activity) cells is shown in **Fig. 5.1**. Transport activity of [^{14}C]Gly-Sar was evaluated by the inhibitory effect of unlabeled Gly-Sar on tracer uptake in various types of human cancer cells (**Table 5.2**). Reduction in tracer uptake is considered to be carrier-mediated transport. When incubation ends, transport buffer containing substrate is aspirated off and then cells are gently washed off a 1 mL of ice-cold transport buffer twice.
3. (*Retention experiment for efflux transporter, e.g., BCRP*) In order to evaluate efflux of substrate, retention of fully loaded substrate into cells is measured. In case of cells expressing an efflux transporter such as BCRP, 250 μL of transport buffer containing radiolabeled substrate (e.g., 0.5–100 μM unlabeled MX plus 2.5 μCi/mL [^{3}H]mitoxantrone, *see* **Note 3**) with a specific inhibitor at appropriate concentration (e.g., 5 μM of FTC for BCRP, **Note 4**) is added to the cells in culture medium, and then the cells are incubated for 3 h at 37°C to load the substrate into the cells ($t = 0$). Cells are washed free of substrate by rinsing with ice-cold transport buffer twice, then the 250 μL of pre-warmed fresh buffer containing no substrate is added.
4. Immediately after washing, a 100 μL of solubilization solution is added directly to the cells, and then cells are gently shaken for 30 min on an orbital shaker on benchtop (*see* **Note 5** and **Section 3.2.3**).

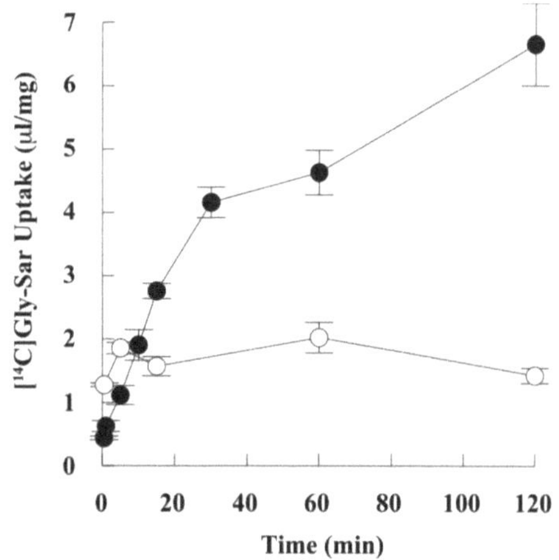

Fig. 5.1. Time course of intracellular accumulation of [^{14}C]Gly-Sar was measured in HT1080 (human fibrosarcoma, *closed circle*) and IMR-90 (normal fibrosarcoma, *open circle*) cells. Cells were incubated in the presence of 20 μM of [^{14}C]Gly-Sar at 37°C at pH 6.0. Each point represents the mean value (n = 3 or 4) with SEM. [**Fig.** 5.1 was redrawn using our original data published in (1)].

Table 5.2
Inhibition of [^{14}C]Gly-Sar uptake with unlabeled Gly-Sar

	Control	+ Unlabeled Gly-Sar (20 mM)	
Cell Line	**μL/mg protein**	**μL/mg protein**	**% of Control**
MKN-45 (Stomach)	8.3 ± 0.6	0.4 ± 0.5	5.3 ± 5.7
MG-63 (Bone)	12.9 ± 0.9	2.1 ± 0.3	16.2 ± 2.0
HT1080 (Connective tissues)	3.8 ± 0.4	1.0 ± 0.4	27.4 ± 10.3
T24 (Bladder)	5.7 ± 0.7	2.0 ± 0.2	34.6 ± 3.9
SK-MEL28 (Skin)	2.8 ± 0.4	1.9 ± 0.1	69.1 ± 4.0
T47D (Breast)	7.0 ± 0.4	5.1 ± 0.1	72.2 ± 1.0
Mia-Paca2 (Pancreas)	7.4 ± 0.2	5.8 ± 0.3	77.5 ± 3.4
COLO320DM (Colon)	8.1 ± 0.3	6.5 ± 0.2	80.2 ± 2.3
Nakajima (Squamous)	4.6 ± 0.2	3.8 ± 0.3	82.5 ± 0.3
BT20 (Breast)	3.4 ± 0.3	3.2 ± 0.2	94.3 ± 5.2

Table is redrawn from our original data published in (2).

Intracellular accumulation of [^{14}C]Gly-Sar was measured in various cancer cells by transport study. Cells were incubated in the presence of [^{14}C]Gly-Sar (16.7 μM) at 37°C at pH 6.0 for 15 min in the absence or presence of 20 mM unlabeled Gly-Sar. Then influx rate is calculated as uptake clearance of Gly-Sar (μL/mg protein) at 15 min. Each value represents the mean of three to four experiments with SEM.

*Significantly different ($P < 0.05$) from the control (uptake in the absence of unlabeled Gly-Sar).

3.2.2. Floating Cells

1. Floating cells (e.g., leukemia or lymphoma cell lines) are finally collected into a 50 mL polypropylene conical tube and spun at $200 \times g$ at room temperature for 5 min. The pellet is washed twice with PBS and then resuspended at a concentration of 2×10^6 cells/mL with transport buffer.
2. A 250 μL aliquot of cell suspension (containing 5×10^5 cells) is transferred into a 1.7-mL microcentrifuge tube and then pre-incubated for 20 min at 37°C in a water bath.
3. A 250 μL aliquot of transport buffer containing twofold concentration of substrate is added to the cell suspension in the same tube, and then incubated for an appropriate period of time at 37°C in a water bath.
4. When incubation ends, 400 μL of reaction mixture is transferred onto a silicon layer prepared in a 0.4 mL microtube. The tube is spun at $10{,}000 \times g$ for 1 min at room temperature to allow only live cells get to KOH phase through the layer because live cells have greater density than silicon layer. The tubes sit for 3 h on benchtop at room temperature until cells are completely solubilized in KOH. The tip part of the tube that contains solubilized cells is cut with a sharp blade and then transferred into a scintillation tube (*see* **Section 3.2.3**).

3.2.3. Quantification and Analysis

1. Four mL of liquid scintillation cocktail is added to the resultant cell lysates to determine radioactivity using a multipurpose scintillation counter.
2. Kinetic parameters are calculated for carrier-mediated transport by nonlinear least-squares regression analysis such as MULTI (29). Data obtained are fit to the Michaelis–Menten equation: $V = V_{max} \times S/(K_m + S) + k_d \times S$, where V is transport velocity, S is concentration of substrate, V_{max} is the maximum velocity of transport, K_m is Michaelis–Menten constant, and k_d is constant of diffusion. The constant of diffusion, k_d, should be estimated from uptake experiment, where the transporter function is abolished, such as at 4C or under the condition of driving force free.

3.3. Intracellular Accumulation (In Vivo)

1. The radiolabeled compound (e.g., [^{11}C]Gly-Sar; 0.7–29.8 MBq) dissolved in 150 μL of PBS is injected into each tumor-bearing mouse at age 7 weeks through the tail vein that had been injected MKN-45 cells at age 5 weeks (*see* **Section 3.1**).
2. A mice anesthetized with sevoflurane is positioned with the spine on an acrylic plate and placed on the midplane between the two opposing detectors arranged in a horizontal mode in PPIS system (IPS-1000-6XII).

3. Radiolabeled tracer (e.g., [^{11}C]Gly-Sar or [^{18}F]FDG) at a dose of 5 MBq (about 100 μL dissolved in PBS) is intravenously injected into each mouse from the tail vein. Whole-body imaging of biodistribution patterns of each tracer is determined is acquired with a 1-min timeframe interval for 60 min, and 6 summation images were created every 10 min (**Fig. 5.2**). To evaluate whether the intracellular accumulation is due to transport activity of interest, the image may be obtained from animals which are injected with an excess amount of unlabeled tracer (e.g., Gly-Sar) prior to the tracer injected (e.g., [^{11}C]Gly-Sar).

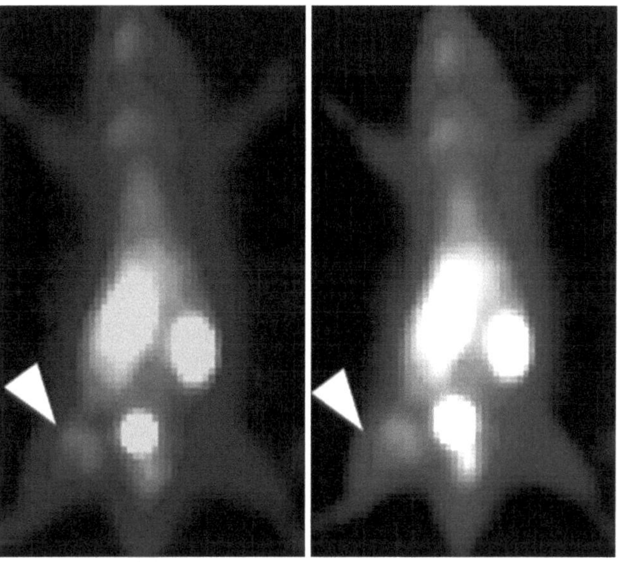

Fig. 5.2. Whole-body PPIS imaging of [^{11}C]Gly-Sar in tumor-bearing mice. Mice, inoculated subcutaneously with MKN-45 cells on right side of hind limb and anesthetized with sevoflurane, were positioned spine down on an acrylic plate and placed midplane between the two opposing detectors arranged in a horizontal mode. [^{11}C]Gly-Sar (5 MBq) was injected intravenously into each mouse via tail vein. Mice were imaged by PPIS with [^{11}C]Gly-Sar, and accumulated images were created from 0 to 10 min (left) and from 0 to 20 min (right) after injection. The *arrowhead* indicates tumor. [**Fig.** 5.2 was redrawn using our original data published in (4)].

4. To evaluate tumor tissue/blood distribution (K_p value) of the tested tracer, the animals are euthanized at designated time after the injection; tumor and blood are rapidly removed and weighed, and the radioactivity is measured with a γ-counter.
5. Tracer uptake by various organs is quantified as dimensionless standardized uptake values (SUVs) using the formula: SUV = (decay-corrected Bq per gram of tissue/Bq injected per gram of body weight) or %ID/g (decay-corrected Bq per tissue/Bq injected per gram of body weight).

3.4. Preparation of Plasma Membrane Vesicles

1. Cells grown to semi-confluence are washed twice with PBS at room temperature, harvested by using a cell scraper, and collected in 50-mL conical tubes.
2. The cell suspension is centrifuged at 4,000×*g* for 10 min at 4°C. If you have more than one tube of cell suspension, the pellets can be resuspended in small volume of PBS, combined in a single 50-mL tube, and then the cell suspension is pelleted under the same conditions as in step 1. The resultant pellet is suspended in 25 mL of cell suspension buffer and placed on ice for 30 min to allow sufficient time for cooling.
3. The cooled 25-mL cell suspension is transferred into a 100-mL glass beaker, and then the beaker is placed into a disruption bomb (Parr 4635) and equilibrated at 700 psi by nitrogen gas with gentle stirring. This procedure has to be done at 4°C.
4. Twenty minutes later, the homogenate is collected into a 50-mL conical tube through nozzle of the disruption bomb as the pressure is released. The content is homogenized with a teflon-glass homogenizer (20 strokes) at 4°C.
5. Immediately after the homogenate is collected into the 50-mL conical tube, a one-hundredth volume (0.5 mL) of 100 mM disodium EDTA solution is added to the homogenate to achieve a final concentration of 1 mM.
6. The homogenate is centrifuged at 1,000×*g* for 10 min at 4°C to remove nuclei and any intact cells.
7. The supernatant (approximately 5 mL/tube) is layered over a 3-mL sucrose cushion dispensed in a 12-mL round-bottom ultracentrifuge tube (NALGENE®, polycarbonate, up to 5,000×*g*), and then centrifuged in a Sorvall HB-4 rotor (Sorvall-Du Pont) for 30 min at 16,300×*g*.
8. The white layer of plasma membrane vesicles lying on the sucrose buffer cushion is collected using a fine gauge needle (e.g., 18G) without disturbing the layer (*see* **Note 6**) and put into a 12-mL ultracentrifuge tube.
9. The plasma membrane vesicles are diluted with several milliliters of sucrose dilution buffer and then are centrifuged at 45,000×*g* at 4°C for 45 min.
10. The pellet is suspended in vesicle storage buffer by passage through a fine gauge needle (e.g., 25G, *see* **Note** 7).
11. Vesicles are stored in 1-mL aliquots preferably in liquid nitrogen, but –80°C deep freezer can be alternatively used for a temporary storage of vesicles up to a couple of months (*see* **Note 8**).

3.5. Transport Study/Vesicle (Rapid Filtration Method)

1. After the vesicles are pre-incubated for 30 min at 25°C, 10 μL of the membrane vesicle suspended in storage buffer (*see* **Section 3.4**), containing 20 μg of protein, is mixed with a 90 μL of vesicle transport buffer containing 1.1 × fold concentration of the test compound: radiolabeled tracer usually is used for this assay.
2. At an appropriate period of time, the transport study is terminated by diluting the reaction mixture with 1 mL of ice-cold stop buffer (4°C).
3. The diluted mixture is immediately filtered through a membrane filter (Millipore, HAWP, 0.45 μm pore size, *see* **Note 9**).
4. Filter is quickly washed free of test compound four times with ice-cold stop buffer.
5. The filter is transferred into a 6-mL scintillation vial with 4 mL of liquid scintillation cocktail and left on bench at room temperature for an hour until it becomes dissolved.
6. The radioactivity retained on the membrane filter was quantitated with a liquid scintillation counter. The background radioactivity is evaluated by simply mixing an exact amount of test compound solution (90 μL) with diluted membrane vesicles (about 1 mL), followed immediately by filtration.

4. Notes

1. Lowering pH should be achieved by replacing HEPES with MES or other type of buffer. To evaluate Na^+ and Cl^- dependency of transport activity, NaCl is usually replaced with choline chloride or NMDG, and Na-gluconate, respectively.
2. Collagen (type I)-coated polystyrene tissue culture dishes may be helpful for cells that easily detach off non-coated dish.
3. Since mitoxantrone is fluorescent, its intracellular accumulation may be alternatively detected by flow cytometry, in lieu of using radioisotopes. Cells cultured in a 25-cm^2 flask are used per one point of flow cytometry assay. In general, cells are exposed to 10 μM or higher mitoxantrone to load them. Transporter (e.g., BCRP) expressing cells should also be incubated with a specific transporter inhibitor such as FTC to ensure maximal substrate loading of cells. To measure retention of intracellular accumulation of mitoxantrone, at the intended time intervals, cells are trypsinized off of the plates with 0.25% trypsin/EDTA solution as quickly as possible and then cells are collected into a 15-mL conical tube.

The collected cells are washed with ice-cold transport buffer solution twice, and finally cells are resuspended in a 300 μL of PBS. The cell suspension is kept at 4°C to avoid active efflux of intracellular drug into buffer. Intracellular mitoxantrone content is determined by using flow cytometry. For other fluorescent substrates for BCRP, *see* (30).

4. For other efflux transporters (e.g., P-glycoprotein or MRPs), see respective inhibitors and substrates in (31, 32).
5. Aliquot of solubilized protein can be utilized to estimate protein amount to normalize transport activity.
6. First by using a 1-mL microtip attached to vacuum pump, suction away lipid layer on top of tube. Then, carefully use a 200-μL eppendorf pipette to transfer the cloudy white interface layer consisting of plasma membrane vesicles.
7. The resultant pellet is usually very firm so that it should be passed through a fine gauge needle for at least 10 times. This should be operated on ice or in a cold room.
8. Before vesicles are stored, it is recommended to measure the protein content of the resultant suspension by protein assay, e.g., Bradford method.
9. If the test compound used is anticipated to have nonspecific binding to a membrane filter, the filter should be presoaked overnight with transport buffer containing an excess amount of unlabeled test compound and the washing buffer also must contain an excess amount of the compound. For example, such a procedure would be used for a cationic substrate such as [^{3}H]TEA, which is a typical radiolabeled substrate for organic cation transporters. In case of use of [^{3}H]TEA, the filter is presoaked with 5 mM of unlabeled TEA and the same amount of TEA should be added to the stop solution to avoid adsorption of [^{3}H]TEA to the surface of the filter.

Acknowledgements

We thank Dr. Akira Tsuji (at Kanazawa University, Japan) for his permission to use the data shown as **Fig. 5.2** and Dr. Ikumi Tamai (at Kanazawa University, Japan) for his thoughtful suggestions in describing this chapter.

References

1. Nakanishi, T., Tamai, I., Sai, Y., Sasaki, T., and Tsuji, A. (1997) Carrier-mediated transport of oligopeptides in the human fibrosarcoma cell line HT1080, *Cancer Res.* **57**, 4118–4122.
2. Nakanishi, T., Tamai, I., Takaki, A., and Tsuji, A. (2000) Cancer cell-targeted drug delivery utilizing oligopeptide transport activity. *Int. J. Cancer* **88**, 274–280.

3. Gonzalez, D.E., Covitz, K.-M.Y., Sadee, W., and Mrsny, R.J. (1998) An oligopeptide transporter is expressed at high levels in the pancreatic carcinoma cell lines AsPc-1 and Capan-2. *Cancer Res.* **58**, 519–525.
4. Mitsuoka, K., Miyoshi, S., Kato, Y., Murakami, Y., Utsumi, R., Kubo, Y., Noda, A., Nakamura, Y., Nishimura, S., and Tsuji, A. (2008) Cancer detection using a pet tracer, 11c-glycylsarcosine, targeted to h+/peptide transporter. *J. Nucl. Med.* **49**, 615–622.
5. Nakanishi, T. (2007) Drug transporters as targets for cancer chemotherapy. *Cancer Genomics Proteomics* **4**, 241–254.
6. Doyle, L.A., Yang, W., Abruzzo, L.V., Krogmann, T., Gao, Y., Rishi, A.K., and Ross, D.D. (1998) A multidrug resistance transporter from human MCF-7 breast cancer cells. *Proc. Natl. Acad. Sci. U S A* **95**, 15665–15670.
7. Ross, D.D. (2000) Novel mechanisms of drug resistance in leukemia. *Leukemia* **14**, 467–473.
8. Nakanishi, T., Doyle, L.A., Hassel, B., Wei, Y., Bauer, K.S., Wu, S., Pumplin, D.W., Fang, H.-B., and Ross, D.D. (2003) Functional characterization of human breast cancer resistance protein (BCRP, ABCG2) Expressed in the Oocytes of Xenopus laevis. *Mol. Pharmacol.* **64**, 1452–1462.
9. Yanagida, O., Kanai, Y., Chairoungdua, A., Kim, D.K., Segawa, H., Nii, T., Cha, S.H., Matsuo, H., Fukushima, J.-i., Fukasawa, Y., Tani, Y., Taketani, Y., Uchino, H., Kim, J.Y., Inatomi, J., Okayasu, I., Miyamoto, K.-i., Takeda, E., Goya, T., and Endou, H. (2001) Human L-type amino acid transporter 1 (LAT1): characterization of function and expression in tumor cell lines. *Biochimica et Biophysica Acta.* **1514**, 291–302.
10. Wolf, D.A., Wang, S., Panzica, M.A., Bassily, N.H., and Thompson, N.L. (1996) Expression of a highly conserved oncofetal gene, TA1/E16, in human colon carcinoma and other primary cancers: homology to schistosoma mansoni amino acid permease and caenorhabditis elegans gene products. *Cancer Res.* **56**, 5012–5022.
11. Nawashiro, H., Otani, N., Shinomiya, N., Fukui, S., Ooigawa, H., Shima, K., Matsuo, H., Kanai, Y., and Endou, H. (2006) L-type amino acid transporter 1 as a potential molecular target in human astrocytic tumors. *Int. J. Cancer* **119**, 484–492.
12. Kim, D.K., Kanai, Y., Choi, H.W., Tangtrongsup, S., Chairoungdua, A., Babu, E., Tachama, K., Anzai, N., Iribe, Y., and Endou, H. (2002) Characterization of the system L amino acid transporter in T24 human bladder carcinoma cells BBA *Biomembrane* **1565**, 112–122.
13. Kanai, Y., Segawa, H., Miyamoto, K.-i., Uchino, H., Takeda, E., and Endou, H. (1998) Expression cloning and characterization of a transporter for large neutral amino acids activated by the heavy chain of 4F2 antigen (CD98). *J. Biol. Chem.* **273**, 23629–23632.
14. Bode, B.P., Fuchs, B.C., Hurley, B.P., Conroy, J.L., Suetterlin, J.E., Tanabe, K.K., Rhoads, D.B., Abcouwer, S.F., and Souba, W.W. (2002) Molecular and functional analysis of glutamine uptake in human hepatoma and liver-derived cells. *Am. J. Physiol. Gastrointest. Liver Physiol.* **283**, G1062–1073.
15. Gupta, N., Miyauchi, S., Martindale, R.G., Herdman, A.V., Podolsky, R., Miyake, K., Mager, S., Prasad, P.D., Ganapathy, M.E., and Ganapathy, V. (2005) Upregulation of the amino acid transporter ATB0,+ (SLC6A14) in colorectal cancer and metastasis in humans. Biochimica et Biophysica Acta (BBA). *Mol. Basis Dis.* **1741**, 215–223.
16. Gupta, N., Prasad, P.D., Ghamande, S., Moore-Martin, P., Herdman, A.V., Martindale, R.G., Podolsky, R., Mager, S., Ganapathy, M.E., and Ganapathy, V. (2006) Upregulation of the amino acid transporter ATB0,+ (SLC6A14) in carcinoma of the cervix. *Gynecologic Oncology* **100**, 8–13.
17. Nakanishi, T., Hatanaka, T., Huang, W., Prasad, P.D., Leibach, F.H., Ganapathy, M.E., and Ganapathy, V. (2001) Na+- and Cl^--coupled active transport of carnitine by the amino acid transporter ATB0,+ from mouse colon expressed in HRPE cells and Xenopus oocytes. *J. Physiol. (Lond)* **532**, 297–304.
18. Hatanaka, T., Huang, W., Nakanishi, T., Bridges, C.C., Smith, S.B., Prasad, P.D., Ganapathy, M.E., and Ganapathy, V. (2002) Transport of -Serine via the Amino Acid Transporter ATB0,+ Expressed in the Colon. *Biochem. Biophys. Res. Comm.* **291**, 291–295.
19. Hatanaka, T., Nakanishi, T., Huang, W., Leibach, F.H., Prasad, P.D., Ganapathy, V., and Ganapathy, M.E. (2001) Na+- and Cl^--coupled active transport of nitric oxide synthase inhibitors via amino acid transport system B0,+. *J. Clin. Invest.* **107**, 1035–1043.
20. Huang, Y., Dai, Z., Barbacioru, C., and Sadee, W. (2005) Cystine-Glutamate Transporter SLC7A11 in Cancer Chemosensitivity and Chemoresistance. *Cancer Res.* **65**, 7446–7454.

21. Zhang, S., Lovejoy, K.S., Shima, J.E., Lagpacan, L.L., Shu, Y., Lapuk, A., Chen, Y., Komori, T., Gray, J.W., Chen, X., Lippard, S.J., and Giacomini, K.M. (2006) Organic cation transporters are determinants of oxaliplatin cytotoxicity. *Cancer Res.* **66**, 8847–8857.
22. White, D.L., Saunders, V.A., Dang, P., Engler, J., Zannettino, A.C. W., Cambareri, A.C., Quinn, S.R., Manley, P.W., and Hughes, T.P. (2006) OCT-1-mediated influx is a key determinant of the intracellular uptake of imatinib but not nilotinib (AMN107): reduced OCT-1 activity is the cause of low in vitro sensitivity to imatinib. *Blood* **108**, 697–704.
23. Okabe, M., Unno, M., Harigae, H., Kaku, M., Okitsu, Y., Sasaki, T., Mizoi, T., Shiiba, K., Takanaga, H., Terasaki, T., Matsuno, S., Sasaki, I., Ito, S., and Abe, T. (2005) Characterization of the organic cation transporter SLC22A16: A doxorubicin importer. *Biochem. Biophys. Res. Comm.* **333**, 754–762.
24. Farr, X., Guillen-Gomez, E., Sanchez, L., Hardisson, D., Plaza, Y., Lloberas, J., Casado, F.J., Palacios, J., and Pastor-Anglada, M. (2004) Expression of the nucleoside-derived drug transporters hCNT1, hENT1 and hENT2 in gynecologic tumors. *Int. J. Cancer* **112**, 959–966.
25. Spratlin, J., Sangha, R., Glubrecht, D., Dabbagh, L., Young, J.D., Dumontet, C., Cass, C., Lai, R., and Mackey, J.R. (2004) The absence of human equilibrative nucleoside transporter 1 is associated with reduced survival in patients with gemcitabine-treated pancreas adenocarcinoma. *Clin. Cancer Res.* **10**, 6956–6961.
26. Koukourakis, M.I., Giatromanolaki, A., Harris, A.L., and Sivridis, E. (2006) Comparison of metabolic pathways between cancer cells and stromal cells in colorectal carcinomas: a metabolic survival role for tumor-associated stroma. *Cancer Res.* **66**, 632–637.
27. Fang, J., Quinones, Q.J., Holman, T.L., Morowitz, M.J., Wang, Q., Zhao, H., Sivo, F., Maris, J.M., and Wahl, M.L. (2006) The H+-linked monocarboxylate transporter (MCT1/SLC16A1): a potential therapeutic target for high-risk neuroblastoma. *Mol. Pharmacol.* **70**, 2108–2115.
28. Nabulsi, N.B., Smith, D.E., and Kilbourn, M.R. (2005) [11C]Glycylsarcosine: synthesis and in vivo evaluation as a PET tracer of PepT2 transporter function in kidney of PepT2 null and wild-type mice. *Bioorg. Med. Chem.* **13**, 2993–3001.
29. Yamaoka, K., Tanigawara, Y., Nakagawa, T., and Uno, T. (1981) A pharmacokinetic analysis program (MULTI) for microcomputer. *J. Pharmacobiodyn.* **4**, 879–885.
30. Doyle, L.A., and Ross, D.D. (2003) Multidrug resistance mediated by the breast cancer resistance protein BCRP (ABCG2). *Oncogene* **22**, 7340–7358.
31. Ambudkar, S.V., Kimchi-Sarfaty, C., Sauna, Z.E., and Gottesman, M.M. (2003) P-glycoprotein: from genomics to mechanism. *Oncogene* **22**, 7468–7485.
32. Deeley, R., Westlake, C., and Cole, S. (2006) Transmembrane transport of endo- and xenobiotics by mammalian ATP-binding cassette multidrug resistance proteins. *Physiol. Rev.* **86**, 849–899.

Chapter 6

Analysis of Expression of Drug Resistance-Linked ABC Transporters in Cancer Cells by Quantitative RT-PCR

Anna Maria Calcagno and Suresh V. Ambudkar

Abstract

Quantitative real-time PCR (qRT-PCR) boasts many advantages over microarrays. For instance, very low amounts of total RNA are required to yield highly accurate and reproducible detection of transcript levels. As a consequence, qRT-PCR has become a popular technique for assessing gene expression levels and is now the gold standard. In this chapter, qRT-PCR using two distinct chemistries, SYBR Green and TaqMan, are described. We compare ABC transporter levels in various drug-resistant cancer cell lines by employing each method. SYBR Green yields reproducible results; nevertheless, TaqMan chemistry is superior to SYBR Green, as it displays higher specificity and sensitivity. Gene expression analysis by qRT-PCR is a powerful technique and shows potential as a diagnostic tool for predicting drug response in cancer patients.

Key words: Quantitative real-time PCR (qRT-PCR), SYBR green, TaqMan, ABC transporters, gene expression profiling.

1. Introduction

Gene expression analysis has become a valuable scientific tool. Microarrays and now quantitative real-time PCR (qRT-PCR) are commonly used to profile gene expression patterns in both cell lines and tumors (1, 2). qRT-PCR allows the user to monitor the reaction in real time as opposed to PCR, which relies on the end-point analysis (3). With the development of more sensitive and selective methodologies, qRT-PCR requires very little input material and provides accurate and reproducible detection of gene levels (4).

Q. Yan (ed.), *Membrane Transporters in Drug Discovery and Development*, Methods in Molecular Biology 637,
DOI 10.1007/978-1-60761-700-6_6, © Springer Science+Business Media, LLC 2010

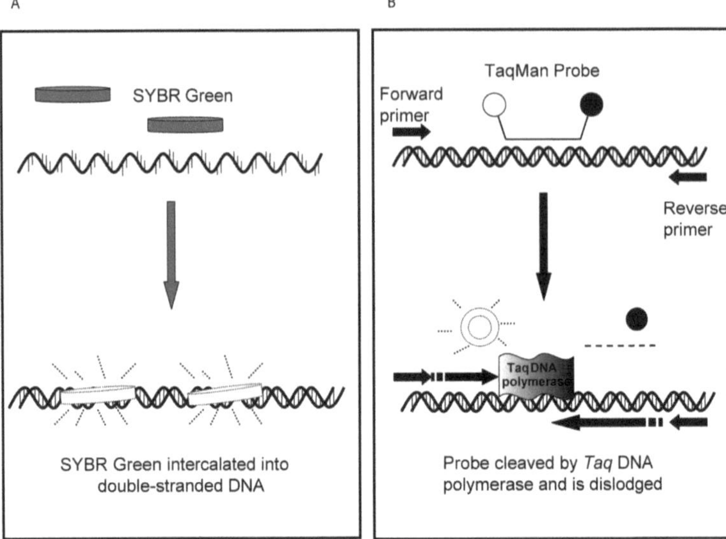

Fig. 6.1. Comparison of qRT-PCR chemistries. **(a)**Schematic showing SYBR Green qRT-PCR functions. SYBR Green will not bind to single-stranded DNA, yet it binds to double-stranded DNA. In the bound form, it exhibits fluorescence, allowing for the PCR reaction to be monitored in real time. **(b)**Characterization of TaqMan qRT-PCR. The dual-labeled probe binds between the two primers, but the closeness of the quencher dye prevents the reporter dye from fluorescing. As the Taq polymerase cleaves the probe with its 5′ nuclease activity, the reporter dye is no longer blocked by the quencher dye. The fluorescence intensity is proportional to the amount of PCR products generated.

SYBR Green is an intercalating dye which binds to double-stranded DNA product during the PCR reaction. It binds to the minor groove of DNA (5) (**Fig. 6.1a**); however, it can also bind to primer dimers, causing false increases in fluorescence. This will lead to overestimation of the amount of actual product. In contrast, TaqMan chemistry has greater selectivity due to the use of both primers and a specific probe that has a reporter dye and quencher dye attached at opposite ends. The TaqMan probe hybridizes between the two PCR primers; no fluorescence is seen when the quencher dye is located near the reporter dye (6). When the Taq DNA polymerase cleaves the reporter dye from the probe with its 5′–3′ exonuclease activity, fluorescence intensity increases proportional to the amount of PCR product formed (**Fig. 6.1b**).

Multidrug resistance (MDR) is a major cause of chemotherapy failure. Although MDR can be attributed to various mechanisms, the predominant mechanism is the overexpression of ATP-binding cassette (ABC) transporters (7). Of the 48 ABC transporter genes in the human genome, there are at least 12 transporters that function as drug efflux pumps (8). Members of this superfamily can efflux an assortment of substrates, including ions, sugars, amino acids, lipids, toxins, and anticancer drugs. The three dominant ABC transporters that contribute to MDR are ABCB1, ABCC1, and ABCG2 (9). Remarkably, these three transporters demonstrate great overlap in substrate specificity although their sequence homology varies. Microarray analysis for profiling ABC transporter genes has been utilized by several investigators (10–13). Microarrays, however, demonstrate poor probe specificity for genes in families possessing high homology such as ABC transporters (14), whereas qRT-PCR can more readily quantify and identify individual genes in such a homologous superfamily (2, 15). qRT-PCR has become known as the gold standard because of its sensitivity and reproducibility with very low amounts of sample; thus, it shows great utility in the clinic (16). Although both SYBR Green and TaqMan chemistries can quantitate ABC transporter expression, TaqMan provides greater sensitivity and selectivity. In the future, gene expression analysis using qRT-PCR of ABC transporters could aid in the diagnosis of MDR in clinical samples and help predict response to treatment.

We describe here the materials and methods that can be employed to analyze the gene expression of drug resistance-linked ABC transporters. To illustrate the various steps involved, we will refer to a sample unpublished study performed in our laboratory.

2. Materials

2.1. Cell Culture and RNA Isolation

1. In our sample study, the cervical epidermal carcinoma cell line KB-3-1 and its drug-resistant sublines, KB-V1, KB-C1, and KB-8.5–11, were generous gifts of Michael M. Gottesman (Laboratory of Cell Biology, National Cancer Institute, NIH, Bethesda, MD) (17).
2. These cell lines were cultured in Dulbecco's Modified Eagle's Medium (DMEM) with 10% (v/v) fetal calf serum supplemented with 2 mM glutamine and 100 units of penicillin/streptomycin/ml at 37°C in 5% CO_2 humidified air.
3. Vinblastine and colchicine were dissolved in DMSO and added to the medium of the resistant sublines at the appropriate concentrations.
4. Phosphate buffered saline.
5. RNeasy isolation kit and the RNase-free DNase kit (Qiagen, Valencia, CA). A table-top microcentrifuge with a rotor radius of 7 cm was used in the RNA isolation step.
6. Beta-mercaptoethanol and ethanol.

2.2. RNA Quantification and RNA Integrity Determination

1. The purified RNA was quantified using the Nanodrop ND-1000 Spectrophotometer (Wilmington, DE).
2. The integrity of the RNA was verified using the Agilent 2100 Bioanalyzer (Palo Alto, CA) with the Eukaryote total RNA Nano assay, which contains all necessary reagents for the assay.

2.3. SYBR Green qRT-PCR

1. The LightCycler RNA Master SYBR Green Kit and LightCycler 480 instrument (Roche Biochemicals, Indianapolis, IN) were utilized in our sample study.
2. Primers (**Table 6.1**) were designed using the LightCycler Probe Design Software 2.0 (Roche Biochemicals). Plasma membrane calcium ATPase 4 (PMCA4) was used as the reference gene (18) (*see* **Note 1**).

2.4. TaqMan qRT-PCR

1. A High-Capacity cDNA Reverse Transcription Kit with RNase Inhibitor, TaqMan Universal PCR Master Mix, optical adhesive covers, ABI 7900HT, and TaqMan Gene Expression Assays (**Table 6.2**).

3. Methods

3.1. RNA Isolation

1. RNA is isolated from the starting concentration of 200,000 cells/well grown in 6-well plates for 72 h to characterize the gene expression of select ABC transporters in parental KB-3-1 cells and three multi-step selected KB sublines. RNA samples should be isolated in at least duplicate.
2. The medium and any detached cells are first removed from the wells, and the cell monolayer is washed with PBS.
3. RNA isolation is performed on the cells that remain attached using the Qiagen RNeasy kit, as follows: Add 600 μl of Buffer RLT with beta-mercaptoethanol to each well. This buffer is prepared by adding 10 μl of beta-mercaptoethanol for each milliliter of Buffer RLT used. Use a pipette tip to first scrape the cells from the wells and then to transfer the contents of each well into an RNase-DNase-free, sterile microcentrifuge tube.
4. Add 600 μl of 70% ethanol to this lysate and pipette up and down several times to ensure complete mixing. Add 700 μl of this mixture to the mini-column and centrifuge 15 s at $\geq 7{,}000 \times g$ using a microcentrifuge. Pull the mini-column

Table 6.1
List of Primers used for Sybr Green RT-PCR (18)

Gene	Position of primer	Forward oligo sequence	Reverse oligo sequence
ABCA2	238–684	CATCCCCCTGGTGCTGTTCTT	GCTTGGGCCGTGCTATTGG
ABCA3	437–939	GCCCTCTTTACACTCAGTTTTCA	GACGAGCAGTTGTCGTACCTAAT
ABCA4	3,380–3,555	TCTGGGATCTGCTCCTGAAGTATCG	GGTTAAGTACAAGCCTGTGCCAAAG
ABCB1	834–1,086	GCCTGGCAGCTGGAAGACAAATAC	ATGGCCAAAATCACAAGGGTTAGC
ABCB11	2,204–2,371	CTTCCATCCGGCAACGCT	CACTGAATTTCAGAATCCTCCTAACTGGG
ABCC1	1,119–1,670	AGTGGAACCCCTCTCTGTTTAAG	CCTGATACGTCTTGGTCTTCATC
ABCC2	872–1,027	AATCAGAGTCAAAGCCAAGATGCC	TAGCTTCAGTAGGAATGATTTCAGGAGCAC
ABCC3	2,572–2,725	TCCTTTGCCAACTTTCTCTGCAACTAT	CTGGATCATTGTCTGTCAGATCCGT
ABCC4	3,880–4,124	TGATGAGCCGTATGTTTTGC	CTTCGGAACGGACTTGACAT
ABCC5	3,692–3,864	AGAGGTGACCTTTGAGAACGCA	CTCCAGATAACTCCACCAGACGG
ABCC11	3,025–3,560	CCACGGCCCTGCACAACAAG	GGAATTGCCAAAAGCCACGAACA
ABCG2	266–646	CCGCGACAGTTTCCAATGACCT	GCCGAAGAGCTGCTGAGAACTGTA
PMCA4	2,309–2,465	ATCTGCATAGCTTACCGGGACT	TGCCAGCTTGTTTGCATTTGGCAATA

Table 6.2
TaqMan Assays used for qRT-PCR

Gene symbol	Assay ID	Gene name
ABCB1	Hs00184491_ml	ATP-binding cassette, sub-family B (MDR/TAP), member 1
ATP2B4	Hs00608066_ml	ATPase, Ca++ transporting, plasma membrane 4

The TaqMan assays were obtained from Applied Biosystems (Foster City, CA). For each reaction, 40 ng cDNA for each cell line was used (See **Fig. 6.3**).

from the collection tube and discard the flow-through. Return the mini-column to the collection tube. Add the remaining lysate (500 μl) to the mini-column and spin 15 s at ≥ 7,000×*g* and discard flow-through.

5. Gently add 350 μl of RW1 buffer to the filter of the mini-column and centrifuge 15 s at ≥ 7,000×*g* and discard flow-through. Add 80 μl of the DNase/RDD mixture onto the column and incubate for 15 min at room temp. The DNase mixture is prepared by adding 10 μl of DNase stock to 70 μl Buffer RDD per tube of RNA. Mix by gently inverting(*see* **Note 2**). Add 350 μl of RW1 buffer to the filter of the mini-column at the completion of the DNase step. Centrifuge 15 s at ≥7,000×*g*, then discard flow-through as before. Transfer the mini-column into a new collection tube and pipette 500 μl of RPE buffer. Centrifuge 15 s at ≥ 7,000×*g*, then discard flow-through.
6. Add another 500 μl of RPE buffer and centrifuge 2 min at 13,000×*g*. Discard both the flow-through and the collection tube. Put the column in a 1.5 ml tube w/lid and add 50 μl of RNase-free water and centrifuge 1 min at 7,000×*g* to elute the RNA (*see* **Note 3**).
7. Store RNA samples at –80°C until ready to use.

3.2. RNA Quantification and RNA Integrity Determination

1. After isolating the RNA, it is then quantified using the NanoDrop 1000. The NanoDrop is turned on and the nucleic acid tab is selected. To initialize the instrument, place 2 μl of distilled water on the lower pedestal and bring down the top lever before clicking to initialize. The setting is changed to quantitate RNA. Clean the pedestal and place 2 μl of RNase-free water on the pedestal using a Kimwipe. Click on blank. The pedestal is cleaned following each new sample. The sample is measured by placing 2 μl on the pedestal, closing the lever, and clicking on measure.
2. The integrity of the RNA is verified using the Agilent 2100 Bioanalyzer with the Eukaryote total RNA assay. Allow the reagents to come to room temperature for 30 min prior to beginning. The RNA 6,000 ladder remains at –20°C. The

gel (550 μl) is filtered for 10 min using the microcentrifuge at ~1,500×*g*. Aliquot 65 μl of filtered gel into microcentrifuge tubes. Store additional aliquots for later use at 4°C (*see* **Note 4**). Add 1 μl of dye to 65 μl of filtered gel and vortex for 10 s. Centrifuge the gel-dye mixture for 10 min at 13,000×*g*.

3. While the gel-dye mixture is spinning, prepare the RNA samples for testing as follows: Place 2 μl of a sample into an individual small PCR tube and heat at 70°C for 2 min (*see* **Note 5**). Another small PCR tube should contain 2 μl of RNA 6000 ladder and should be heated with the other RNA samples.
4. Clean the electrodes by placing 350 μl of RNase Zap in the electrode cleaner chip 1 and insert into the Bioanalyzer for 1 min. Remove this chip. Load the second electrode cleaner chip with 350 μl of RNase-free water and insert into the Bioanalyzer for 10 s. Remove this chip and keep the instrument open for 10 s to allow the electrodes to dry. Close the instrument until the filled RNA Nano chip is ready to insert.
5. Assemble the priming station by locking the syringe into the top of the priming station. Place the RNA Nano chip into the priming station set on the "C" setting. Add 9 μl of gel-dye mixture to the well marked with "G" and a dark circle around it. Close the priming station for 30 s. Slowly release the pressure and open the priming station. Place 9 μl of gel-dye mixture into the two remaining wells marked with "G." Place 5 μl of marker into each of the 12 wells and the ladder well. Place 1 μl of RNA 6000 ladder into the well marked as "ladder." Add 1 μl of the first sample to a well marked by the number 1. Add 1 μl of the next sample to the well marked 2. Continue until all wells are filled. Twelve samples can be analyzed per run. If fewer than 12 samples are analyzed, add an additional 1 μl of marker to the wells which do not contain sample.
6. Vortex the filled Lab-Chip for 1 min at 2,400 rpm in the special IKA vortexer provided by Agilent. Place the Lab-Chip in the Bioanalyzer and run the instrument. Total RNA with a RNA integrity (RIN) number above 7 can be utilized in qRT-PCR. The Agilent software assigns a RIN number, which presents a standardized RNA integrity evaluation, to each sample run on the Bioanalyzer (19).

3.3. SYBR Green qRT-PCR Using the One-Step Method

1. Following quantitation and analysis of the total RNA, the RNA is diluted to the desired concentration. For SYBR Green qRT-PCR, an optimal range is 75–150 ng/μl. Each reaction uses 2 μl of total RNA. The Roche RNA Master SYBR Green kit and total RNA are used for the one step method.

2. The total volume for each reaction is 20 μl.

RNase-free water	4.2 μl
Mn acetate	1.3 μl
SYBR Green Master Mix	7.5 μl
Forward primer 2 μM stock	2.5 μl
Reverse primer 2 μM stock	2.5 μl

Then add 2 μl of the appropriate total RNA at the desired concentration.

3. The RT-PCR reaction is performed on 300 ng total RNA with 250 nM specific primers under the following conditions on the LightCycler 480: reverse transcription (20 min at 61°C), one cycle of denaturation at 95°C for 30 s, and PCR reaction of 45 cycles with denaturation (15 s at 95°C), annealing (30 s at 58°C), and elongation (30 s at 72°C with a single fluorescence measurement). The PCR reaction was followed by a melting curve program (65–95°C with a heating rate of 0.1°C per second and a continuous fluorescence measurement) and then a cooling program at 40°C.
4. Negative controls consisting of no-template (water) reaction mixtures should be run with all reactions. PCR products can also be run on agarose gels to confirm the formation of a single product of the desired size.
5. Crossing points for each transcript are determined using the second derivative maximum analysis with an arithmetic baseline adjustment. Data are normalized to expression of only a single convenient reference gene. We used PMCA4. Data are presented as a comparison of gene expression for the resistant sublines relative to that for the parental cells using the delta–delta Ct method. **Figure 6.2** displays the unpublished gene expression profile for each of the KB-resistant sublines relative to the parental KB-3-1 obtained in our sample study using SYBR Green q-RT-PCR.

3.4. TaqMan qRT-PCR

1. The TaqMan assay requires a reverse transcription step to be performed using the High-Capacity cDNA Reverse Transcription Kit with RNase Inhibitor. The quantitated RNA is utilized to generate cDNA. In a 20 μl reaction, up to 2 μg of RNA can be reverse transcribed. In our own study, 400 ng of RNA in a total volume of 10 μl was used in a PCR tube (dilute up to 10 μl with RNase-free water), and mixed with a 10 μl 2X reverse transcription mix given below:

10X RT Buffer	2.0 μl
25X dNTP Mix (100 mM)	0.8 μl
10X RT Random Primers	2.0 μl

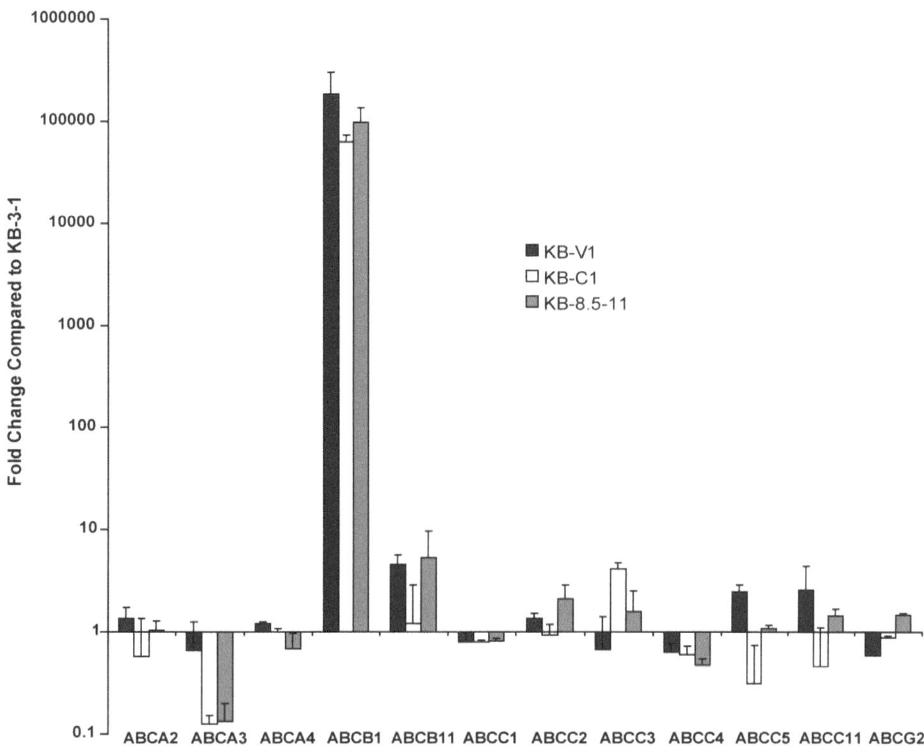

Fig. 6.2. Analysis of ABC transporter expression in multi-step selected KB sublines using SYBR Green qRT-PCR. The delta–delta CT method was used to determine the fold change of ABC transporter gene expression in the multi-step selected KB sublines (17), compared to their parental line, KB-3-1. 300 ng of Total RNA used per reaction. Data were normalized to expression of the reference gene, plasma membrane Ca^{2+} ATPase 4 (PMCA4) (18). The values represent the mean and standard deviation ($n = 2$). The 12 ABC transporters known to transport drugs are indicated on the *x*-axis. (Calcagno AM and Ambudkar SV, unpublished data).

MultiScribe reverse transcriptase	1.0 μl
RNase inhibitor	1.0 μl
RNase-free water	3.2 μl
Total	10 μl

Mix by pipetting and briefly centrifuge to ensure homogeneous mixing. The PCR tube with the 20 μl is placed in a thermal cycler for reverse transcription with the following temperatures and times: Step 1. 10 min at 25°C; Step 2. 120 min at 37°C; Step 3. 5 s at 85°C; Step 4. Hold at 4°C (*see* **Note 6**).

2. The step above will yield cDNA to be employed in the qRT-PCR.

For qRT-PCR using TaqMan assays on a 96 well plate, the following are mixed together.

cDNA from Step 3.3.1 (40 ng cDNA)	2 μl
TaqMan Universal PCR Master Mix	25 μl

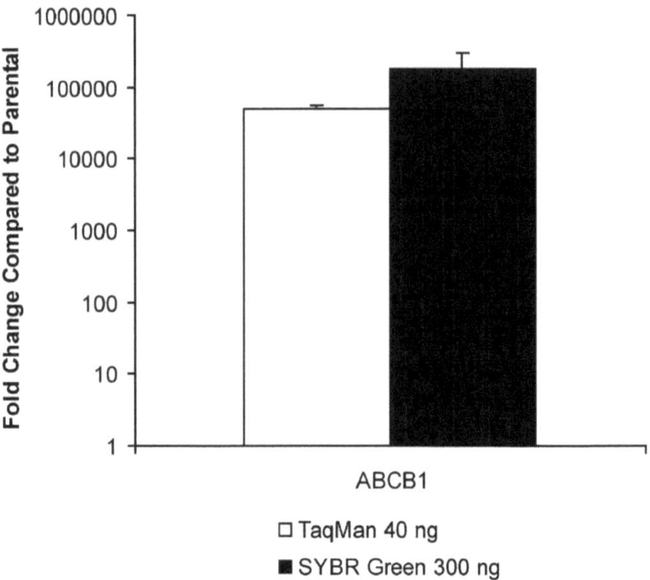

Fig. 6.3. Evaluation of ABCB1 gene expression in KB-V1 cells using SYBR Green and TaqMan qRT-PCR. The delta–delta CT method was used to determine the fold change for ABCB1 gene expression in KB-V1 cells compared to the parental KB-3-1. The SYBR Green method utilized 300 ng of total RNA, whereas the TaqMan used 40 ng of cDNA. Both were normalized to the equivalent amounts of PMCA4 as the reference gene. The values represent the mean and standard deviation ($n = 2$). (Calcagno AM and Ambudkar SV, unpublished data).

20X TaqMan Gene Expression Assay Mix	2.5 μl
RNase-free water	20.5 μl
TOTAL	50 μl

The desired genes can be evaluated on a 96-well plate. After each reaction is loaded on a 96-well plate, the plate is sealed with an optical adhesive cover. The plate is centrifuged at ~500×*g* for 1 min in a centrifuge with adapters for a 96-well plate to ensure proper mixing of the reagents.

3. The plate is run on the ABI 7900HT with the following settings:
   Step 1. 2 min at 50°C; Step 2. 10 min at 95°C, and then 40 cycles of Step 3. 15 s at 95°C and then 1 min at 60°C.

4. Crossing points for each transcript are determined using the SDS software. Data are normalized to the expression of a reference gene, PMCA4. In our sample study, data are presented as a comparison of ABCB1 gene expression for the resistant KB-V1 subline relative to that for the parental KB-3-1 cells using the delta–delta Ct method (**Fig. 6.3** and **Table 6.3**).

Table 6.3
Comparison of CT values for ABCB1 and PMCA4 using SYBR Green and TaqMan for qRT-PCR

	ABCB1 CT value KB-3-1	ABCB1 CT value KB-V1	PMCA4 CT value KB-3-1	PMCA4 CT value KB-V1
SYBR Green (300 ng)	37.0 ± .49	19.8 ± .30	22.4 ± .64	22.4 ± .22
TaqMan (40 ng)	32.7 ± .05	16.2 ± .03	27.0 ± .05	26.1 ± .03

The values represent the mean CT value and the standard deviation ($n = 2$).

4. Notes

1. All primer sets should be tested prior to use to ensure that only a single product of the correct size is amplified for all ABC transporter primer sets. Each product is run on a DNA gel to determine the product sizes.
2. The RNase-free DNase Set was from Qiagen Catalog number 79254 and contains DNase and Buffer RDD. This is an optional step; however, it seems to produce the best quality RNA.
3. The kit calls for two elution spins; however, one elution with a larger volume such as 50 μl at 7,000×*g* for 1 min yields good results.
4. These aliquots expire 30 days following the filtering step. Store at 4°C. Usually 7 additional aliquots are made with one filtering step. Once the dye is added to the 65 μl tube of filtered gel, this can be used for only 24 h. This mixture must be centrifuged prior to using in the RNA Nano Chip.
5. Only 1 μl sample will be loaded on the RNA Nano Chip; however, 2 μl are placed in the tube for heat denaturing to ensure that enough remains for loading.
6. For this step, 100% conversion of RNA to cDNA is assumed. 400 ng cDNA would be present in the final volume of 20 μl; therefore, each μl would contain 20 ng cDNA.

Acknowledgments

We thank Mr. George Leiman for editorial assistance. This research was supported by the Intramural Research Program of the National Institutes of Health, National Cancer Institute, Center for Cancer Research.

References

1. Staunton, J. E., Slonim, D.K., Coller, H.A., et al. (2001) Chemosensitivity prediction by transcriptional profiling. *Proc. Natl. Acad. Sci. U S A* **98,** 10787–10792.
2. Szakacs, G., Annereau, J., Lababidi, S., et al. (2004) Predicting drug sensitivity and resistance: profiling ABC transporter genes in cancer cells. *Cancer Cell* **6,** 129–137.
3. Wong, M.L. and Medrano, J.F. (2005) Real-time PCR for mRNA quantitation. *Biotechniques.* **39,** 75–85.
4. Gyorffy, B., Surowiak, P., Kiesslich, O., et al. (2006) Gene expression profiling of 30 cancer cell lines predicts resistance towards 11 anticancer drugs at clinically achieved concentrations. *Int. J. Cancer* **118,** 1699–1712.
5. Wittwer, C.T., Herrmann, M.G., Moss, A.A., and Rasmussen, R.P. (1997) Continuous fluorescence monitoring of rapid cycle DNA amplification. *Biotechniques* **22,** 130–131, 134–138.
6. Overbergh, L., Giulietti, A., Valckx, D., Decallonne, R., Bouillon, R., Mathieu, C. (2003) The use of real-time reverse transcriptase PCR for the quantification of cytokine gene expression. *J. Biomol. Tech.* **14,** 33–43.
7. Gottesman, M., Fojo, T., and Bates, S. (2002) Multidrug resistance in cancer: role of ATP-dependent transporters. *Nat. Rev. Cancer* **2,** 48–58.
8. Gottesman, M. and Ambudkar, S.V. (2001) Overview: ABC transporters and human disease. *J. Bioenerg. Biomembr.* **33,** 453–458.
9. Calcagno, A.M., Kim, I.W., Wu, C.P., Shukla, S., Ambudkar, S.V. (2007) ABC drug transporters as molecular targets for the prevention of multidrug resistance and drug-drug interactions. *Curr. Drug Deliv.* **4,** 324–333.
10. Annereau, J.P., Szakacs, G., Tucker, C.J., et al. (2004) Analysis of ATP-binding cassette transporter expression in drug-selected cell lines by a microarray dedicated to multidrug resistance. *Mol. Pharmacol.* **66,** 1397–1405.
11. Gillet, J.P., Efferth, T., Steinbach, D., et al. (2004) Microarray-based detection of multidrug resistance in human tumor cells by expression profiling of ATP-binding cassette transporter genes. *Cancer Res.* **64,** 8987–8993.
12. Huang, Y., Anderle, P., Bussey, K.J., et al. (2004) Membrane transporters and channels: role of the transportome in cancer chemosensitivity and chemoresistance. *Cancer Res.* **64,** 4294–4301.
13. Liu, Y., Peng, H., and Zhang, J.T. (2005) Expression profiling of ABC transporters in a drug-resistant breast cancer cell line using AmpArray. *Mol. Pharmacol.* **68,** 430–438.
14. Lee, J.K., Bussey, K.J., Gwadry, F.G., et al. (2003) Comparing cDNA and oligonucleotide array data: concordance of gene expression across platforms for the NCI-60 cancer cells. *Genome Biol.* **4,** R82.
15. Langmann, T., Mauerer, R., Zahn, A., et al. (2003) Real-time reverse transcription-PCR expression profiling of the complete human ATP-binding cassette transporter superfamily in various tissues. *Clin. Chem.* **49,** 230–238.
16. Bustin, S.A. and Mueller, R. (2005) Real-time reverse transcription PCR (qRT-PCR) and its potential use in clinical diagnosis. *Clin. Sci. (Lond)* **109,** 365–379.
17. Shen, D., Cardarelli, C., Hwang, J., et al. (1986) Multiple drug-resistant human KB carcinoma cells independently selected for high-level resistance to colchicine, adriamycin, or vinblastine show changes in expression of specific proteins. *J. Biol. Chem.* **261,** 7762–7770.
18. Calcagno, A.M., Chewning, K.J., Wu, C.P., Ambudkar, S.V. (2006) Plasma membrane calcium ATPase (PMCA4): a housekeeper for RT-PCR relative quantification of polytopic membrane proteins. *BMC Mol. Biol.* **7,** 29.
19. Schroeder, A., Mueller, O., Stocker, S., et al. (2006) The RIN: an RNA integrity number for assigning integrity values to RNA measurements. *BMC Mol. Biol.* **7,** 3.

Chapter 7

Fluorescence Studies of Drug Binding and Translocation by Membrane Transporters

Frances J. Sharom, Ronghua Liu, and Balpreet Vinepal

Abstract

Resistance to multiple drugs is a serious limitation to chemotherapy treatment of human cancers. In addition, many clinically useful drugs show limited uptake in the intestine and cannot gain access to the brain. Three multidrug efflux pumps of the ABC superfamily (P-glycoprotein/ABCB1, MRP1/ABCC1, and BCRP/ABGG2) are responsible for most drug transport out of mammalian cells. P-glycoprotein is the best characterized of the ABC drug transporters. However, the lipophilic nature of its substrates has made it difficult to directly quantitate drug binding to the protein by classical biochemical methods, and the measurement of drug transport rates has also proved challenging. In recent years, fluorescence spectroscopic approaches have proved very useful in overcoming these problems. This chapter focuses on the use of fluorescence tools to quantitate the affinity of binding of various drugs to purified P-glycoprotein and to measure its drug transport activity in reconstituted proteoliposomes in real time. The ability of various drugs to inhibit P-glycoprotein mediated transport can also be assessed using this approach.

Key words: ABC superfamily, multidrug efflux pump, P-glycoprotein, fluorescence quenching, binding affinity, transport rate, real-time assay.

1. Introduction

1.1. Drug Efflux Transporters of the ABC Superfamily

The ATP-binding cassette (ABC) superfamily is one of the largest groups of protein found in both eukaryotes and prokaryotes (1–3). Most ABC proteins are membrane transporters and several are involved in the ATP-dependent efflux of drugs, thus causing the cells that express them to display multidrug resistance (MDR). The most important mammalian ABC proteins known to export drugs are P-glycoprotein (Pgp; ABCB1), multidrug resistance-associated protein (MRP1; ABCC1), and breast cancer resistance

Q. Yan (ed.), *Membrane Transporters in Drug Discovery and Development*, Methods in Molecular Biology 637,
DOI 10.1007/978-1-60761-700-6_7,

protein (BCRP; ABCG2) (4–7). All carry out the ATP-driven export from cells of chemically unrelated compounds, including many drugs used clinically in the treatment of cancers and other medical conditions.

This chapter will describe the application of fluorescence approaches to study the drug binding and drug transport functions of the Pgp multidrug efflux pump, which is a 170-kDa protein expressed in the plasma membrane of many different MDR tumor cells (8). Pgp is the best characterized and most intensively studied of the ABC drug efflux pumps and is known to export hundreds of amphipathic drugs, natural products, peptides, and lipids (9). The protein is believed to contribute to resistance to chemotherapeutic drugs in 50% of human cancers (10). Pgp plays an important role in normal physiology as a mechanism for dealing with both endogenous and exogenous toxic agents. It is expressed at the apical surface of epithelial cells in the gut, where it prevents absorption of many clinically administered drugs as well as xenobiotics. Pgp is also found at the apical surface of the capillary endothelial cells that make up the blood–brain barrier, where it greatly limits entry of drugs into the brain.

Pgp consists of two similar halves, each comprising six membrane-spanning helices and one NBD (nucleotide-binding domain) on the cytosolic side (*see* **Fig. 7.1**, top panel). The NBDs couple the energy of ATP hydrolysis to the transport of drug substrates and contain three highly conserved motifs: the Walker A and Walker B sequences found in other ATP/GTP-hydrolyzing proteins and the Signature C motif, which is unique to ABC proteins (11). Drugs that are Pgp substrates are generally lipophilic and appear to gain access to the transporter from the cytoplasmic membrane leaflet after partitioning into the lipid bilayer (12, 13). Drug transport by Pgp is active, driven by ATP hydrolysis, and generates a drug concentration gradient across the membrane, as shown by studies using functionally reconstituted protein (14–16). Compounds known as Pgp modulators can inhibit drug transport by directly binding to the transporter. Some modulators appear to be transport substrates (thus competing with drugs for translocation), while others may bind tightly to the protein and prevent drug substrates from gaining access to the substrate binding pocket. Verapamil, cyclosporin A, and PSC833 are commonly used Pgp modulators for in vitro studies.

1.2. Fluorescence Spectroscopy

The use of fluorescence spectroscopic approaches to study membrane transporters has been steadily increasing over the past decade. Fluorescence techniques are of high sensitivity and typically require only small amounts of protein, especially when low-volume microcuvettes are used. Fluorescence instrumentation is commonly available and relatively inexpensive, and several different aspects of membrane protein function can be studied using

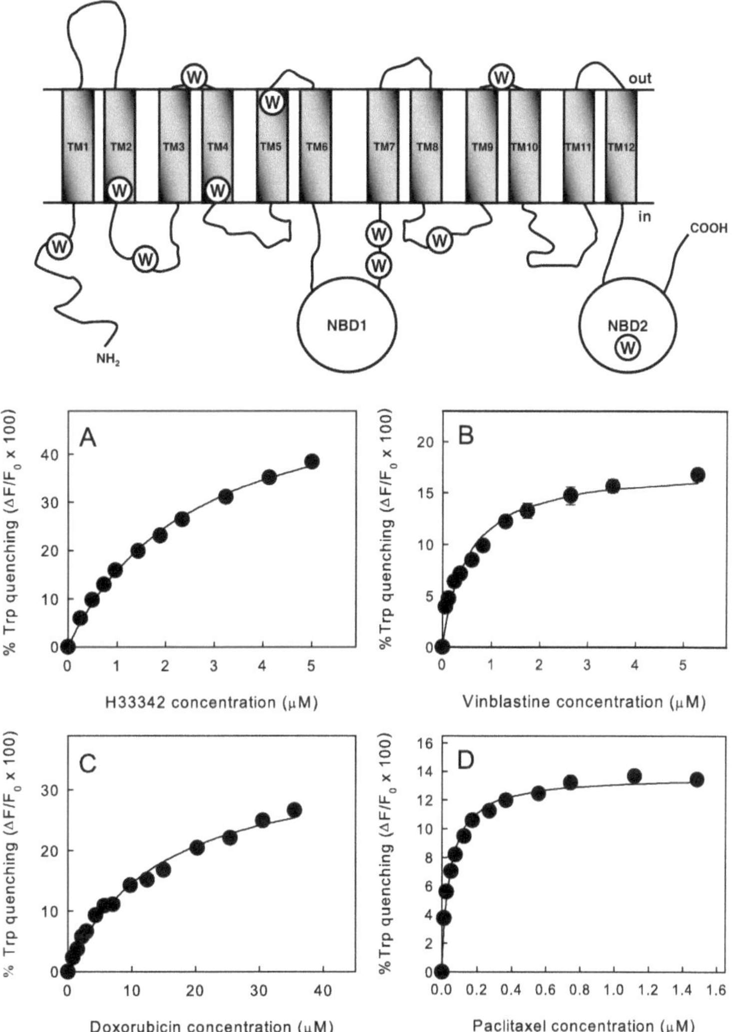

Fig. 7.1. The *top panel* shows a topological cartoon indicating the locations of the 11 Trp residues in Pgp which are responsible for its intrinsic fluorescence. The *bottom panels* show saturable concentration-dependent quenching of Pgp Trp fluorescence by various drugs. Purified Pgp in buffer A with 2 mM CHAPS was titrated at 22°C with increasing concentrations of various drugs: (**a**) H33342 [reproduced from (12) with permission from The American Chemical Society], (**b**) vinblastine [reproduced from (22) with permission from The American Chemical Society], (**c**) doxorubicin [reproduced from (31) with permission from The American Chemical Society], and (**d**) paclitaxel. Fluorescence emission at 330 nm was recorded at 22°C following excitation of Trp residues at 290 nm. Fitting of the quenching data to a binding equation allowed estimation of K_d, the dissociation constant for binding of drug.

complementary approaches. The reader is referred to the most recent edition of the authoritative book by Lakowicz (17) for more details of the principles and practice of modern fluorescence spectroscopy as applied to protein structure and function.

Reviews providing an overview of the application of fluorescence approaches to Pgp are also available (18–20).

1.3. Drug Binding Affinity Measured by Quenching of Intrinsic Protein Tryptophan Fluorescence

It has proved challenging to measure the binding of drugs to Pgp by classical biochemical methods because of their lipophilic nature. Fluorescence spectroscopic approaches have proved invaluable in demonstrating that drugs directly interact with the protein and in quantitating the affinity of binding (21, 22). Most membrane transporters contain Trp residues, which are useful spectroscopic probes, since they are highly sensitive to the local polarity of their environment and the presence of neighboring fluorophores. The locations of the 11 Trp residues in Pgp are shown in **Fig. 7.1** (top panel); eight of them are located in the N-terminal half of the protein (one in the N-terminal tail, three in the TM regions, one in a cytoplasmic loop, one in an extracellular loop, and two in the linker region immediately downstream of NBD1), and the remaining three residues are in the C-terminal half (one in a cytoplasmic loop, one in an extracellular loop, and one in NBD2). When drug substrates bind to Pgp, the intrinsic Trp fluorescence is quenched in a concentration-dependent, saturable manner (**Fig. 7.1a–d**). A quenching titration in which increasing amounts of drug are added to purified Pgp can be utilized to quantitate substrate binding, by fitting of the experimental data to a binding equation using standard graphics/fitting software. This allows estimation of the dissociation constant, K_d, for binding of the drug to the transporter. **Table 7.1** shows some K_d values obtained for a variety of drugs and peptides; the binding affinities cover a very broad range, from 46 nM to 75 μM. The Trp quenching approach requires only small amounts of purified protein (typically 3 μg for a complete titration) and is a very useful way to estimate drug binding affinity. This method may also be applied to the other ABC drug efflux pumps, MRP1 and ABCG2.

1.4. Real-Time Drug Transport Assays Using Fluorescent Substrates

Early studies of Pgp-mediated drug transport, using either plasma membrane vesicles from MDR cells or reconstituted proteoliposomes, revealed that the transport process was relatively fast, reaching equilibrium in only a few minutes (14, 16, 23). Fixed time-point rapid filtration assays can be used to measure the equilibrium uptake of drug, but do not have the time resolution to determine the initial rate of drug transport. In addition, these methods are time-consuming and use relatively large amounts of transport protein. The development of real-time assays for drug transport, using fluorescent substrates, allowed continuous monitoring of the transport process (15, 24, 25). Only a single small sample of membrane vesicles or proteoliposomes is needed to record a progress curve and estimate the initial rate of transport.

Several fluorescent dyes are good transport substrates for Pgp, including various rhodamine dyes, H33342 (Hoechst 33342), LDS-751, and acetoxy esters of Ca^{2+}- and pH-sensitive dyes (calcein-AM, Fura-2-AM). The two dyes that will be employed in the methods described here are H33342 and TMR (tetramethylrosamine, a rhodamine dye); both are relatively high-affinity substrates for Pgp, with K_d values of 2.6 and 0.7 μM, respectively (*see* **Table 7.1**). In these experiments, the fluorescent dye and ATP (together with an ATP-regenerating system) are added to the exterior of well-sealed membrane vesicles or reconstituted proteoliposomes containing the transporter. The NBDs located on the exterior membrane surface (i.e., from inward-facing transporter proteins) hydrolyze ATP and transport the dye into the vesicle interior.

In the case of H33342, this compound displays very low fluorescence in aqueous solution, but its emission intensity is greatly enhanced in a hydrophobic environment (**Fig. 7.2**). When the dye is added to membrane vesicles or proteoliposomes containing Pgp, it partitions into the bilayer interior, thus displaying high levels of fluorescence. On addition of ATP (together with an ATP-regenerating system), Pgp actively expels the dye from

Table 7.1
K_d values for drug binding to purified Pgp estimated by quenching of the intrinsic Trp fluorescence

Drug	K_d (μM)
Colchicine	74.9
Doxorubicin	14.4
Quinine	12.5
Pepstatin A	9.5
Quinidine	7.8
H33342	2.6
LY294002	1.6
Daunorubicin	1.4
Valinomycin	0.72
TMR	0.70
Vinblastine	0.50
Cyclosporin A	0.30
PSC833	0.081
Paclitaxel	0.046

Values for the dissociation constant, K_d, were estimated by fitting of the corrected Trp fluorescence quenching data (*see* **Fig. 7.1**) to an equation for a single binding site (*see* **Section 2.5**).

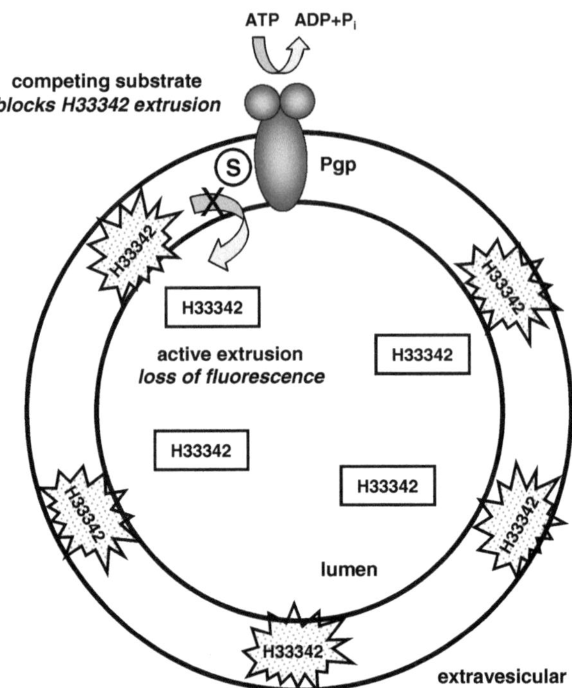

Fig. 7.2. Cartoon illustrating the principles of the real-time assay for Pgp-mediated transport of the fluorescent substrate H33342 in reconstituted proteoliposomes. The drug shows greatly enhanced fluorescence when located within the hydrophobic interior of the lipid bilayer. ATP-powered expulsion of the drug from the membrane to the aqueous solution by Pgp results in a time-dependent loss of fluorescence intensity. This can be followed in real time to give an initial rate of drug transport. The inclusion of other competing substrates in the assay results in a decrease in the rate of H33342 transport.

the bilayer into the aqueous phase in the vesicle lumen, resulting in a time-dependent loss of fluorescence intensity (**Figs. 7.2** and **7.3a**). Rhodamine dyes such as TMR typically lose their fluorescence following active transport by Pgp into the lumen of a vesicle or proteoliposome, likely as a result of self-quenching at higher concentrations. Thus the rate of loss of fluorescence intensity of TMR is a direct measure of the rate of transport of the dye into the vesicle lumen (**Fig. 7.3b**). In both cases, the rate of passive diffusion of dye out of the vesicle lumen, down the concentration gradient generated by Pgp, soon matches the rate at which the dye is pumped inward. At this time, a steady state is reached, and the fluorescence intensity of the dye levels off and eventually reaches a plateau. The initial rate of transport of the dye is measured from the first 20 s of data collected after initiation of transport by addition of ATP.

Real-time transport assays using H33342 and TMR can also be used to quantitate the effect of other compounds (substrates and modulators) on Pgp-mediated drug transport. Increasing

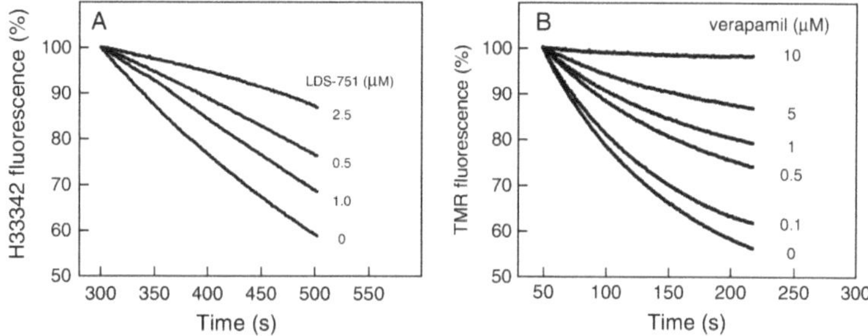

Fig. 7.3. (**a**) Real-time fluorescence trace for transport of H33342 by Pgp in reconstituted proteoliposomes and its inhibition by increasing concentrations of LDS-751. (**b**) Real-time fluorescence trace for transport of TMR by Pgp in reconstituted proteoliposomes and its inhibition by increasing concentrations of verapamil.

concentrations of the test compound are added into the assay, the fluorescence traces are recorded after ATP addition (**Fig. 7.3,a,b**), and the initial rate of transport is measured for each. A plot of the initial rate vs. concentration of competitor is plotted (**Fig. 7.4**) and used to determine the IC_{50} value for the inhibition of transport (**Table 7.2**). Both substrates and modulators have an effect on transport of H33342 and TMR (*see* **Figs. 7.3** and **7.4**). It should be noted that not all Pgp substrates and modula-

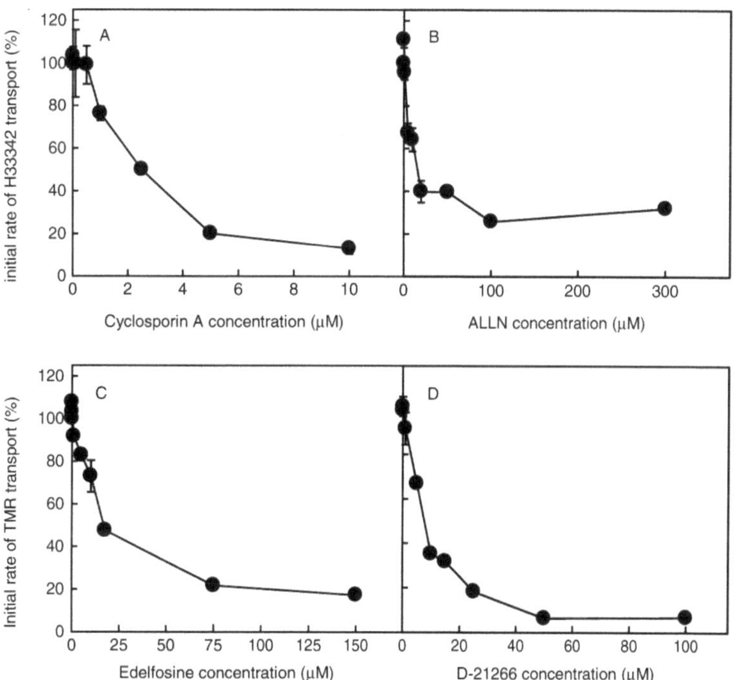

Fig. 7.4. Concentration-dependent inhibition of H33342 transport by (**a**) cyclosporin A and (**b**) *N*-acetyl-leucyl-leucyl-norleucinal (ALLN). Concentration-dependent inhibition of TMR transport by (**c**) edelfosine and (**d**) D21266.

tors inhibit drug transport. Pgp contains at least two linked transport sites which display complex allosteric interactions (26). Some compounds are able to stimulate transport of dye, and others display biphasic behavior, with stimulation observed at low concentrations and inhibition at higher concentrations (15). These transport sites reflect the existence within Pgp of a large flexible substrate binding pocket with sub-sites for different drugs, which may interact with each other both sterically and allosterically to a larger or smaller extent. Because of this complex relationship, a given drug when used as a competitor of TMR and H33342 transport may show different IC_{50} values for the two dyes. For example, LDS-751 inhibits transport of both dyes with a similar IC_{50} value, whereas edelfosine displays a low IC_{50} value for TMR but is over 17-fold less effective at inhibiting H33342 transport (*see* **Table 7.2**). In contrast, the peptide ALLN is 14-fold more effective at inhibiting H33342 transport compared to TMR transport.

Table 7.2
IC_{50} values for inhibition of TMR and H33342 transport by various substrates and modulators

Compound	TMR transport IC_{50} (μM)	H33342 transport IC_{50} (μM)
LDS-751	0.7	0.9
Verapamil	1.0	12
Cyclosporin A	1.6	2.3
Valinomycin	12	35
Miltefosine	12	105
Edelfosine	12	>200
D21266	15	47
Pepstatin A	15	75
ALLN	153	11

Values for IC_{50} were estimated from inhibition plots similar to those shown in **Fig. 7.4**, which were derived from real-time fluorescence data of the type displayed in **Fig. 7.3**.

2. Materials

2.1. Preparation of Purified Pgp

1. Plasma membrane vesicles are prepared from an MDR cell line that overexpresses Pgp. A useful source of these vesicles is the Chinese hamster ovary cell line CH^RB30 (27), which

displays high levels of Pgp in the plasma membrane. These vesicles may be stored frozen at –80°C for up to 3 months.

2. Buffer A: 20 mM HEPES, 0.1 M NaCl, 5 mM $MgCl_2$, 1 mM dithioerythritol (DTE), pH 7.5 (*see* **Note 1**).
3. Buffer A containing 25 mM 3-[(3-cholamidopropyl)-dimethylammonio]-1-propanesulfonate (CHAPS), 15 mM CHAPS, and 2 mM CHAPS.
4. Column of concanavalin A-Sepharose 4B (Sigma-Aldrich) with a bed volume of ~7 mL.

2.2. Reconstitution of Pgp into Proteoliposomes for Transport Studies

1. Purified Pgp in buffer A containing 2 mM CHAPS.
2. Stock solutions of the desired phospholipids in $CHCl_3$–MeOH (4:1 v/v). Suitable lipids are egg phosphatidylcholine (PC) or dimyristoylphosphatidylcholine (DMPC). These lipids should be of high purity and free of fluorescent contaminants.
3. Buffer A with no CHAPS, and buffer A containing 200 mM CHAPS.
4. Sephadex G-50 column (1 × 15 cm, ~10 mL bed volume) equilibrated in buffer A.

2.3. Transport of TMR or H33342 by Pgp in Reconstituted Proteoliposomes

1. Transport buffer: 10 mM HEPES, 250 mM sucrose, 5 mM $MgCl_2$, pH 7.5.
2. ATP solution (50 mM) in transport buffer.
3. An ATP-regenerating system: 30 μg/mL creatine kinase and 3.5 mM creatine phosphate in transport buffer.
4. Working solutions of TMR or H33342 in transport buffer (*see* **Section 2.4**).
5. Working solutions of drugs to be tested as transport inhibitors, in transport buffer (*see* **Section 2.4**).

2.4. Preparation of Drugs and Fluorescent Transport Substrates

1. Because Pgp drug substrates are typically amphipathic and of limited aqueous solubility, stock solutions of drugs are prepared in DMSO at a concentration 100- to 500-fold higher than the highest concentration required in the experiments (typically 1–100 mM). The final concentration of DMSO in the Pgp sample solution after dilution should be <1% (v/v). Drug stock solutions may be stored at –20°C for ~12 months, depending on the stability of the individual drug (follow the supplier's advice). Just prior to use, a working solution of the drug is prepared in either buffer A (for binding experiments) or transport buffer (for transport inhibition experiments) (*see* **Note 2**).
2. Stock solutions of 1 mM TMR (DMSO) or 5 mM H33342 (water) are made up and stored at –20°C (TMR) or 4°C

(H33342) in the dark. Just prior to use, working solutions of both fluorescent substrates are prepared by dilution of the stock solutions into transport buffer; 25 μM TMR or 87.5 μM H33342.

2.5. Fluorescence Data Collection and Correction

The experiments in this chapter are best carried out using an analytical fluorimeter with a quartz microcuvette (0.3 cm path length) for drug binding experiments. Transport experiments may be carried out using either a 0.3 cm path length microcuvette or a standard cuvette of 1.0 cm path length. The emission intensity of many fluorophores is temperature-dependent, so accurate temperature control of the sample cuvette compartment is essential.

Fluorescence data from Trp quenching titrations must be corrected for the inner filter effect which arises when molecules are present in solution that absorb the exciting radiation and the emitted fluorescence radiation. If there are significant numbers of scattering particles in solution (such as detergent micelles), scattering must also be corrected for by subtracting a background fluorescence reading taken on a sample that is identical, but with no fluorescent probe present. During a quenching titration, the sample becomes progressively diluted, and this must also be taken into account. Fluorescence emission intensities may be simultaneously corrected for all of these effects using the following equation:

$$F_{\mathrm{icor}} = (F_{\mathrm{i}} - B)\left(\frac{V_i}{V_0}\right) 10^{0.5b(A_{\lambda ex}+A_{\lambda em})}$$

where F_{icor} is the corrected value of the fluorescence intensity at a given point in the titration, F_{i} is the experimentally measured fluorescence intensity, B is the background fluorescence intensity (caused mainly by scattering), V_0 is the initial volume of the sample, V_i is the volume of the sample at a given point in the titration (V_i/V_0 is the dilution factor), b is the path length of the optical cell in cm, and $A_{\lambda ex}$ and $A_{\lambda em}$ are the absorbances of the sample at the excitation and emission wavelengths, respectively. For more details of all these corrections, the reader is referred to Lakowicz (1).

3. Methods

3.1. Pgp Purification

The MDR Chinese hamster ovary cell line, CH^{R}B30, is a useful source of purified Pgp, which makes up approximately 15% of the protein in plasma membrane vesicles isolated from these cells. Plasma membrane is treated with 25 mM CHAPS to extract many membrane proteins, followed by pelleting of the insoluble material, which contains most of the Pgp. Re-extraction of the pellet with 15 mM CHAPS at a higher detergent:protein weight ratio results in solubilization of the bulk of the Pgp in the soluble supernatant. Pgp is then further purified to remove contaminant glycoproteins by lectin affinity chromatography. The resulting protein preparation can be used directly for fluorescence binding experiments or reconstituted into proteoliposomes for transport experiments.

1. 30 mg of plasma membrane vesicle protein is thawed and sedimented at 164,000×*g* for 30 min at 4°C (70.1 Ti rotor, ultracentrifuge). The membrane pellet is resuspended in 3 mL of buffer A containing 25 mM CHAPS at a final protein concentration of 10 mg/mL and incubated on ice for 15 min. The insoluble pellet is collected by centrifugation at 164,000×*g* for 15 min at 4 C, resuspended in 5–7 mL of buffer A containing 15 mM CHAPS (final protein concentration ~1 mg/mL) by passing repeatedly through a 26 gauge needle, and incubated on ice for at least 1 h with periodic passes through the needle. The sample is pelleted at 15,000×*g* for 15 min (microcentrifuge), and the resulting supernatant containing partially purified Pgp is retained.
2. Partially purified Pgp is further purified by affinity chromatography on a 7 mL concanavalin A Sepharose 4B column equilibrated in buffer A/2 mM CHAPS. Typically, 2.5 mL of partially purified Pgp (2.5–3.0 mg of protein) is loaded onto the column at 4°C at a flow rate of 0.2 mL/min. On washing the column with the same buffer, the Pgp peak elutes in the run-through fractions and can be detected by absorbance at 280 nm, followed by measurement of its ATPase activity using a colorimetric assay for released P_i (28) and by Western blotting, if required (29). Typically, 12 fractions of volume 0.5 mL are collected, and fractions 5–8 are pooled. The final product consists of 90–95% pure Pgp, at a concentration of ~0.2 mg/mL, in buffer A containing 2 mM CHAPS. The Pgp preparation should be kept on ice and used within 24 h or frozen at –80°C (*see* **Note 3**). The column is regenerated by washing with ethanol (*see* **Note 4**).

3.2. Determination of Drug Binding Affinity by Quenching of Pgp Intrinsic Trp Fluorescence

1. Quenching experiments are performed by successively adding 1 μL aliquots of drug solution to 60 μL of purified Pgp solution (50 μg/mL) in buffer A containing 2 mM CHAPS, using a quartz microcuvette (0.3 cm path length) equilibrated at the required temperature (typically 22°C). For the suggested mixing procedure, *see* **Note 5**.
2. After each addition, the Pgp sample is excited at 290 nm and the steady-state fluorescence emission is measured at 330 nm

(the emission maximum for Pgp Trp residues). A 305 nm cutoff filter may be placed at the emission monochromator to reduce the stray light.

3. The fluorescence intensities are corrected for dilution, scattering, and the inner filter effect. For each drug, a control titration should be carried out with 30 μM NATA (*N*-acetyltryptophanamide) to assess the nonspecific quenching of Trp fluorescence by the compound. A control titration should also be carried out with DMSO alone, to confirm that there are no solvent-related effects. The experimental data for each drug are computer-fitted to the following equation (*see* **Fig. 7.1a–d**):

$$\left(\frac{\Delta F}{F_0} \times 100\right) = \frac{\left(\frac{\Delta F_{\text{max}}}{F_0} \times 100\right) \times [S]}{K_\text{d} + [S]}$$

where $(\Delta F/F_0 \times 100)$ is the percent quenching (percent change in fluorescence relative to the initial value) following addition of drug at a concentration $[S]$ and K_d is the dissociation constant. Fitting is carried out using nonlinear regression with the Marquardt–Levenberg algorithm (SigmaPlot, Systat Software, Chicago IL), and values of K_d are extracted (*see* **Table 7.1**).

3.3. Reconstitution of Pgp into Proteoliposomes for Transport Studies

1. Purified Pgp is reconstituted into the desired phospholipids using rapid detergent removal by gel filtration chromatography. For transport kinetic studies, typically 125 μg of Pgp is reconstituted into 5 mg of lipid; DMPC or egg PC may be used.
2. The lipid stock solution in $CHCl_3$–MeOH is dispensed into a small glass tube, and the solvent is evaporated to dryness under a stream of N_2 gas. The dried lipid is then pumped under vacuum for 1 h to remove all traces of solvent.
3. The dried lipid is dissolved at a concentration of 10 mg/mL in buffer A containing 200 mM CHAPS, with warming to 37°C and occasional vortexing, to give an optically clear solution. Purified Pgp in buffer A containing 2 mM CHAPS is added, typically 125 μg in a volume of ~1 mL, and the mixture is incubated on ice for 30 min with periodic mixing.
4. The lipid–protein mixture is passed at 4°C through a Sephadex G-50 column (1 × 15 cm, equilibrated with buffer A) and eluted with buffer A, collecting ~20 fractions of ~1 mL volume. Turbid fractions containing proteoliposomes can be located visually (or the absorbance of the fractions

may be read at 280 nm); typically two to three fractions close to the void volume of the column are pooled.

5. The Pgp proteoliposomes should be kept on ice until ready for use. Their protein content may be measured using a modified Lowry method (30) and the ATPase activity using a colorimetric method (28). The final lipid:protein ratio of the proteoliposomes is approximately 40:1 (w/w).

3.4. Real-Time Fluorescence Assay for Transport of TMR and H33342 and Inhibition by Drugs and Modulators

1. Samples for transport measurements in a 1.0 cm path length cuvette consist of 175 μL of transport buffer, 50 μL of proteoliposomes containing approximately 10 μg of Pgp, and either 10 μL of 25 μM TMR in transport buffer (1 μM final concentration) or 10 μL of 87.5 μM H33342 in transport buffer (3.5 μM final concentration). When using a small-volume microcuvette, the reagent volumes should be adjusted appropriately. Samples are preincubated for 5 min at 27°C before the start of the experiment.
2. The sample is transferred to a thermostatted cuvette, mixed 15–20 times using a plastic pipette with a long tip, and then allowed to equilibrate for 300 s to allow stabilization of the fluorescence signal. Fluorescence emission intensity is monitored at excitation/emission wavelengths of 550/575 nm for TMR and 355/450 nm for H33342, respectively.
3. After 300 s equilibration, TMR or H33342 transport is initiated by the addition of 25 μL of 50 mM ATP and an ATP-regenerating system (30 μg/mL creatine kinase, 3.5 mM creatine phosphate) to the sample cuvette using a long-tipped pipette, to give a final ATP concentration of 4.8 mM. The sample is mixed at least 10 times before placing the cuvette back into the fluorimeter.
4. Transport of TMR or H33342 is monitored for an additional 100–200 s and the data points collected over the first 20 s are used to estimate the initial rate of transport (% change in the fluorescence intensity per second).
5. To quantitate the effect on TMR and H33342 transport of other drugs that are Pgp substrates, 25 μL of a particular drug at the desired concentration is substituted for 25 μL of transport buffer in the sample. A transport curve is collected at increasing drug concentrations, as shown in **Fig. 7.3a,b**.
6. A plot of the initial rate of TMR or H33342 transport with increasing drug concentration (*see* **Fig. 7.4**) can be used to estimate the value of IC_{50} for the competing drug (*see* **Table 7.2**).

4. Notes

1. Many membrane transport proteins may be stabilized by the use of sulfhydryl reagents, such as DTE or dithiothreitol.
2. Adsorption of highly hydrophobic/water-insoluble drugs onto the surface of glass or plastic sample tubes may take place during prolonged storage (several hours) of dilute solutions. Such drugs are best diluted to the required concentration just prior to use and used promptly.
3. Purified Pgp may be stored frozen at –80°C for a period of up to 1 year with only a 10% loss of ATPase and transport activity after thawing.
4. The concanavalin A-Sepharose column is regenerated by washing with 50 mL of 20% (v/v) ethanol in buffer A, followed by storage in 0.1 sodium acetate, 1 M NaCl, 1 mM $CaCl_2$, 1 mM $MgCl_2$, 1 mM $MnCl_2$, pH 6.0.
5. It is very important to keep the microcuvette in a fixed position inside the fluorimeter sample compartment during the entire titration process. Any shift in the position of the microcuvette may cause a significant change in the fluorescence emission intensity. Therefore, after each addition of aliquot of drug, the Pgp sample is mixed without touching the sample cuvette or removing it from the instrument. This is done by gently drawing up and ejecting the sample repeatedly using a small micropipette tip, taking care to avoid the introduction of any air bubbles.

Acknowledgments

Research in the author's laboratory on the P-glycoprotein multidrug efflux pump is supported by an operating grant from the Canadian Cancer Society.

References

1. Dassa, E. and Schneider, E. (2001) The rise of a protein family: ATP-binding cassette systems. *Res. Microbiol.* **152**, 203.
2. Davidson, A.L., Dassa, E., Orelle, C., and Chen, J. (2008) Structure, function, and evolution of bacterial ATP-binding cassette systems. *Microbiol. Mol Biol. Rev.* **72**, 317–364.
3. Dean, M., Rzhetsky, A., and Allikmets, R. (2001) The human ATP-binding cassette (ABC) transporter superfamily. *Genome Res.* **11**, 1156–1166.
4. Sharom, F.J. (2008) ABC multidrug transporters: structure, function and role in chemoresistance. *Pharmacogenomics* **9**, 105–127.

5. Deeley, R.G., Westlake, C., and Cole, S.P. (2006) Transmembrane transport of endo- and xenobiotics by mammalian ATP-binding cassette multidrug resistance proteins. *Physiol. Rev.* **86**, 849–899.
6. Calcagno, A.M., Kim, I.W., Wu, C.P., Shukla, S., and Ambudkar, S.V. (2007) ABC drug transporters as molecular targets for the prevention of multidrug resistance and drug-drug interactions. *Curr. Drug Deliv.* **4**, 324–333.
7. Litman, T., Druley, T.E., Stein, W.D., and Bates, S.E. (2001) From MDR to MXR: new understanding of multidrug resistance systems, their properties and clinical significance. *Cell. Mol. Life Sci.* **58**, 931–959.
8. Alvarez, M., Paull, K., Monks, A., Hose, C., Lee, J.S., Weinstein, J., Grever, M., Bates, S., and Fojo, T. (1995) Generation of a drug resistance profile by quantitation of mdr-1/P-glycoprotein in the cell lines of the National Cancer Institute Anticancer Drug Screen. *J. Clin. Invest.* **95**, 2205–2214.
9. Sharom, F.J. 2007. Multidrug resistance protein: P-glycoprotein. In *Drug Transporters: Molecular Characterization and Role in Drug Disposition* (G. You and M.E. Morris, eds.). John Wiley & Sons, Hoboken, NJ, pp. 223–262.
10. Gottesman, M.M. (1993) How cancer cells evade chemotherapy: sixteenth Richard and Hinda Rosenthal Foundation Award Lecture. *Cancer Res.* **53**, 747–754.
11. Holland, I.B., and Blight, M.A. (1999) ABC-ATPases, adaptable energy generators fuelling transmembrane movement of a variety of molecules in organisms from bacteria to humans. *J. Mol. Biol.* **293**, 381–399.
12. Qu, Q., and Sharom, F.J. (2002) Proximity of bound Hoechst 33342 to the ATPase catalytic sites places the drug binding site of P-glycoprotein within the cytoplasmic membrane leaflet. *Biochemistry* **41**, 4744–4752.
13. Lugo, M.R. and Sharom, F.J. (2005) Interaction of LDS-751 with P-glycoprotein and mapping of the location of the R drug binding site. *Biochemistry* **44**, 643–655.
14. Sharom, F.J., Yu, X., and Doige, C.A. (1993) Functional reconstitution of drug transport and ATPase activity in proteoliposomes containing partially purified P-glycoprotein. *J. Biol. Chem.* **268**, 24197–24202.
15. Lu, P., Liu, R., and Sharom, F.J. (2001) Drug transport by reconstituted P-glycoprotein in proteoliposomes. Effect of substrates and modulators, and dependence on bilayer phase state. *Eur. J. Biochem.* **268**, 1687–1697.
16. Sharom, F.J., Yu, X., DiDiodato, G., and Chu, J.W.K. (1996) Synthetic hydrophobic peptides are substrates for P-glycoprotein and stimulate drug transport. *Biochem. J.* **320**, 421–428.
17. Lakowicz, J.R. (2006) *Principles of Fluorescence Spectroscopy*. Springer Science+Business Media, New York.
18. Sharom, F.J., Liu, R., Qu, Q., and Romsicki, Y. (2001) Exploring the structure and function of the P-glycoprotein multidrug transporter using fluorescence spectroscopic tools. *Seminars Cell Dev. Biol.* **12**, 257–266.
19. Sharom, F.J., Liu, R., Romsicki, Y., and Lu, P. (1999) Insights into the structure and substrate interactions of the P-glycoprotein multidrug transporter from spectroscopic studies. *Biochim. Biophys. Acta* **1461**, 327–345.
20. Sharom, F.J. (2006) Shedding light on drug transport: structure and function of the P-glycoprotein multidrug transporter (ABCB1). *Biochem. Cell Biol.* **84**, 979–992.
21. Liu, R., and Sharom, F.J. (1996) Site-directed fluorescence labeling of P-glycoprotein on cysteine residues in the nucleotide binding domains. *Biochemistry* **35**, 11865–11873.
22. Liu, R., Siemiarczuk, A., and Sharom, F.J. (2000) Intrinsic fluorescence of the P-glycoprotein multidrug transporter: sensitivity of tryptophan residues to binding of drugs and nucleotides. *Biochemistry* **39**, 14927–14938.
23. Doige, C.A., and Sharom, F.J. (1992) Transport properties of P-glycoprotein in plasma membrane vesicles from multidrug-resistant Chinese hamster ovary cells. *Biochim. Biophys. Acta* **1109**, 161–171.
24. Shapiro, A.B., Corder, A.B., and Ling, V. (1997) P-glycoprotein-mediated Hoechst 33342 transport out of the lipid bilayer. *Eur. J. Biochem.* **250**, 115–121.
25. Kolaczkowski, M., van der Rest, M., Cybularz-Kolaczkowska, A., Soumillion, J.P., Konings, W.N., and Goffeau, A. (1996) Anticancer drugs, ionophoric peptides, and steroids as substrates of the yeast multidrug transporter Pdr5p. *J. Biol. Chem.* **271**, 31543–31548.
26. Shapiro, A.B., and Ling, V. (1997) Positively cooperative sites for drug transport by P-glycoprotein with distinct drug specificities. *Eur. J. Biochem.* **250**, 130–137.
27. Doige, C.A., and Sharom, F.J. (1991) Strategies for the purification of P-glycoprotein from multidrug-resistant Chinese hamster ovary cells. *Protein Expr. Purif.* **2**, 256–265.

28. Doige, C.A., Yu, X., and Sharom, F.J. (1992) ATPase activity of partially purified P-glycoprotein from multidrug-resistant Chinese hamster ovary cells. *Biochim. Biophys. Acta* **1109**, 149–160.
29. Sharom, F.J., Yu, X., Chu, J.W.K., and Doige, C.A. (1995) Characterization of the ATPase activity of P-glycoprotein from multidrug-resistant Chinese hamster ovary cells. *Biochem. J.* **308**, 381–390.
30. Peterson, G.L. (1977) A simplification of the protein assay method of Lowry et al. which is more generally applicable. *Anal. Biochem.* **83**, 346–356.
31. Qu, Q., Chu, J.W., and Sharom, F.J. (2003) Transition state P-glycoprotein binds drugs and modulators with unchanged affinity, suggesting a concerted transport mechanism. *Biochemistry* **42**, 1345–1353.

Chapter 8

A Model for Transport Studies of the Blood–Brain Barrier

Dennis J. Bobilya

Abstract

The blood–brain barrier (BBB) is the cellular structure between the blood flowing through the brain and the parenchymal tissues of the brain. This physiological barrier is formed by the endothelial cells of the capillary walls. It exquisitely regulates the passage of substances into and out of the brain. Astrocytes (astroglial cells) signal the endothelial cells to adopt BBB characteristics. An in vitro BBB model can be very useful for the study of the nutrition, physiology, and pharmacology of the brain. We took advantage of numerous advances made by previous researchers in this field to develop a co-culture BBB model. Capillary endothelial cells and astrocytes are isolated from the brains of miniature swine and grown on permeable membranes suspended between two chambers of media: analogous to the capillary lumen and the interstitium of the brain, respectively. The endothelial cell isolation procedure includes mechanical and enzymatic digestion of the brain tissue followed by separation of the capillary fragments, based on size and density, from other brain cells. Astrocytes are purified from these "other" cells. The endothelial cells of the capillary fragments proliferate in culture flasks and are then seeded onto the upper surface of a polycarbonate semi-permeable membrane suspended between two chambers of fluid. Astrocytes are seeded on the underside of the membranes. Their close proximity enables the astrocytes to communicate with the endothelial cells and encourage their expression of BBB characteristics without disrupting the endothelial cell monolayer. Transport studies across the monolayer can be conducted by introducing test compounds into the media on one side and observing its appearance on the other side. Mechanisms of transport can also be studied.

Key words: Blood–brain barrier, capillary endothelial cells, microvascular, astrocytes, neurophysiology, pharmacology.

1. Introduction

The blood–brain barrier (BBB) is the anatomical and physiological interface between the brain and the blood components of the vascular system. It is composed of the specialized endothelial cells that constitute the walls of the capillary vessels in parts of the brain

Q. Yan (ed.), *Membrane Transporters in Drug Discovery and Development*, Methods in Molecular Biology 637, DOI 10.1007/978-1-60761-700-6_8, © Springer Science+Business Media, LLC 2010

(1). The BBB functions to regulate the passage of substances into and out of the bloodstream perfusing the brain. It tightly controls homeostasis in the brain by permitting necessary compounds to enter, while preventing the passage of toxins. Due to the complex physiological and biochemical environment that exists in vivo and the difficulties surrounding manipulations in situ in the brain, an in vitro model can be a valuable tool for researchers in the biological, medical, and pharmacological sciences. While a cell culture model can never be expected to respond exactly as similar tissue would in a living brain, it does enable experimental inquiries into transport mechanisms and the relative influences of specific alterations in environmental conditions and physiological stimuli on the transport rate of nutrients and substances across the BBB. It also permits screening of potential pharmacological compounds or procedures with the potential of permeating the blood–brain barrier. Therefore, we have developed procedures for constructing a reliable and reproducible in vitro model of the BBB.

Brain capillary endothelial cells have been difficult to isolate and cultivate in the laboratory. We built upon the significant advances made by our predecessors in the establishment of BBB models (2–5). Brain tissue is digested mechanically and enzymatically. Capillary fragments are separated out based on their size and density and seeded into culture flasks. The capillary endothelial cells proliferate in culture under optimized conditions that include the use of platelet-poor horse serum, heparin, and endothelial cell growth supplements (6–8). Astrocytes (astroglial cells) adjacent to the capillaries in the brain promote the development of BBB characteristics, e.g., tight junctions, in the endothelial cells by releasing paracrine substances that induce the endothelial cells to adopt the unique characteristics of the BBB (9, 10). Our BBB model advances upon the pioneering work of Gaillard et al. (11), by co-culturing astrocytes and brain capillary endothelial cells from the same species and potentially from the same animal. Astrocytes are grown on the bottom of a permeable membrane and brain capillary endothelial cells on the top of the membrane. This way the astrocytes can communicate with the endothelial cells through the permeable membrane as they would in vivo without disrupting the monolayer. Transport across the BBB model includes passage through the endothelial cell monolayer, the permeable membrane, and the astrocyte cell layer.

2. Materials

2.1. Animals

1. Young (2–4 months old) pig or other suitable source of brain tissue (*see* **Notes 1** and **2**).

2. Anesthesia and surgical tools for euthanizing and decapitating the donor animal. We anesthetized the animal with an intramuscular injection of 40 mg ketamine HCl and 2.2 mg xylazine per kg bodyweight and then euthanized by exsanguination. This avoided the injection of a more stringent euthanasia compound that might adversely affect cells that form the walls of blood vessels.
3. Hand-held rotary bone saw or other instrument to open and remove the cranial plate.

2.2. Isolation of Astrocytes and Capillary Endothelial Cells for Co-culture

1. Sodium bicarbonate.
2. MEM: minimum essential medium Eagle with Earle's salts and L-glutamine, plus 2.2 g/L bicarbonate. Prepare from powder in deionized water. Store at 4°C for up to 1 month after sterile filtering.
3. MEM w/HEPES: MEM supplemented with 25 mM HEPES. Prepare from powder in deionized water. Store at 4°C for up to 1 month after sterile filtering.
4. FBS: Fetal bovine serum. Store at 4°C for up to 1 month; otherwise, at –70°C.
5. Gentamicin sulfate. The 50 mg/mL source is stored at 4°C and 500 μL is added to each 500 mL bottle of MEM and MEM w/HEPES.
6. Amphotericin B. Prepared in advance to a concentration of 10 mg/mL, frozen at 4°C in 250 μL aliquots for additions to 500 mL bottles of MEM and MEM w/HEPES.
7. CM: Collection medium; 2% FBS in MEM w/HEPES, 50 μg/mL gentamicin, 5 μg/mL amphotericin B, buffer to pH 7.4, and sterile filter. Store at 4°C for up to 1 month after sterile filtering.
8. Collagenase: (EC 3.4.24.3), Type IA, 270 U/mg. Must be made fresh within a few hours of usage and held on ice. Collagenase will auto-digest.
9. BSA: Bovine serum albumin.
10. 25% BSA in CM: Add 25 g of BSA to CM to a final volume of 100ml and allow it to dissolve with gentle mixing overnight at 4°C. This concentration of BSA can be slow to go into solution. After the BSA is thoroughly dissolved, sterile filter it. If sterile, can be stored for a month at 4°C (*see* **Note 3**).
11. Nylon screens: 20, 30, 60, and 149 μm pore sizes.
12. Glass beads, 0.5 mm diameter.
13. Separation columns: We attach the nylon screens using rubber bands to the bottom of glass columns that are 12 mm diameter (10 mL glass test tubes with their bottoms

removed). The separation column with the 20 μm screen includes glass beads (40 mm height) suspended by the screen. The columns need to be prepared in advance and sterilized by autoclaving.

14. PPHS: Platelet-poor horse serum. Store at 4°C for up to 1 month; otherwise, at –70°C.
15. Heparin: Heparin sulfate: 170 USP/mg. Prepared in advance to be 10 mg/mL in MEM and stored at –20°C in small aliquots of 100–500 μL for inclusion in PGM and SGM at 1%.
16. ECGS: Endothelial cell growth supplement. ECGS can also be prepared according to the method of Maciag et al. (12). Prepare in advance to be 5 mg/mL in MEM and store at –80°C for long-term storage. Small aliquots of 100–500 μL can be frozen for a few months at –20°C for inclusion in PGM and SGM at 1%. Loses activity if stored unfrozen for more than 2 days. It is best to thaw fresh aliquots each time new medium is made.
17. PGM: Primary growth medium: 15% PPHS in MEM w/HEPES and 2.2 mg/mL bicarbonate, buffered to pH 7.4, and sterile filtered. Plus, 100 μg/mL heparin, 50 μg/mL ECGS, 50 μg/mL gentamicin, and 5 μg/mL amphotericin B. Make fresh as needed.
18. Fibronectin. Freeze dried source is rehydrated to a concentration of 20 μg/mL in sterile deionized water. Can be stored at 4°C for months.
19. AGM.: Astrocyte growth medium: 10% FBS in MEM, plus 50 μg/mL gentamicin, 5 μg/mL amphotericin B, buffer to pH 7.4, and sterile filter.
20. SGM: Secondary growth medium: 13% PPHS and 2% FBS in MEM, plus 100 μg/mL heparin, 50 μg/mL ECGS, 50 μg/mL gentamicin, and 5 μg/mL amphotericin B (PGM with 2% of the PPHS replaced with FBS).

2.3. Construction of the Blood–Brain Barrier Model

1. PBS: phosphate buffered saline: 8.00 g NaCl, 1.15 g Na_2HPO_4 (dibasic), 0.20 g KCl, 0.20 g KH_2PO_4 (monobasic)/L, buffer to pH 7.4, and sterile filter.
2. EDTA: Ethylenediaminetetraacetic acid.
3. PBS w/EDTA: PBS buffer with 0.2% EDTA (5.37 mM). EDTA will go into solution quickly if the pH of the PBS is made acidic. Return to pH 7.4 and sterile filter.
4. Trypsin. Prepared in advance to be 2.5% in PBS and stored at –20°C in 200 μL aliquots for inclusion in trypsin/EDTA at 2% of final volume.

5. Trypsin/EDTA: PBS buffer with 0.1% EDTA and 0.05% trypsin. Must be made fresh within an hour of usage and held on ice. Trypsin will auto-digest.
6. Transwell™ polycarbonate filters (Corning, Inc., Corning, NY). We prefer the 12-well Transwell™ plates from Costar with 0.4 μm pores and 12 mm diameter membranes with a surface area of 1 cm^2.

2.4. Measuring Transport Rate Across the Blood–Brain Barrier

1. Voltohmmeter: Endohm chamber and EVOM resistance meter.
2. Representative substance that can be reliably and accurately measured in the amounts expected to pass through the BBB. Usually, labeling the molecule with a fluorescent or radioactive tag is most useful.
3. Orbital shaker.

3. Methods

3.1. Animals

1. Anesthetize and euthanize (according to your ACUC institutional guidelines) the donor animal (*see* **Note 2**).
2. Decapitate the animal and transfer the head to a sterile environment such as a laminar flow hood. Clean the surface of the cranium with Betadine or other suitable sanitizing agent. Using a scalpel and forceps, peel back the skin from the cranium.
3. Using a rotary bone saw, remove the cranial plate to reveal the brain. Peel back the meninges membrane to expose the cerebral cortex.
4. Extract approximately 20 g of grey matter from the brain's cerebral cortex and transfer it to a petri dish containing 20 mL of cold CM (*see* **Note 4**).

3.2. Isolation of Astrocytes and Capillary Endothelial Cells

1. Mechanically disburse the cerebral tissue into 1–2 mm^3 by mincing with scalpels in a petri dish with cold (4°C) collection medium (CM). Then, transfer the minced tissue to a 50 mL test tube using a large bore (25 mL) pipet. Repeatedly (3–4X) aspirate and eject the tissue for further disbursal.
2. Wash the tissue by centrifuging at 400×*g* for 5 min at 4°C. Aspirate and discard the supernatant, leaving about 15–20 mL of packed tissue. Add fresh collection medium to nearly fill the tube (~45 mL mark). Resuspend the tissue by

aspirating repeatedly (3–4X) with a 25 mL pipet. Centrifuge as before and then repeat this wash again.

3. Digest the tissue enzymatically by resuspending it in an equal volume (1:1) of CM containing 1.0 mg/mL collagenase Type IA (270 IU/mL). The effective final concentration is ~135 IU/mL. Incubate in a water bath for 60 min at 37°C, hand mixing by inversion every 10 min. At 20 min intervals during the incubation, remove the tube and aspirate the cell clumps three to four times against the test tube wall with a large bore 25 mL pipet to break them up. Stop the incubation by adding cold CM to inhibit the collagenase activity and place the tube on ice (*see* **Note 5**).
4. Wash the tissue three times in cold CM by centrifugation at 400×*g* for 5 min at 4°C. When resuspending the tissue for washing, repeatedly aspirate it with a 25 mL pipet to facilitate further dispersion.
5. Discard the supernatant and resuspend the tissue with 25 mL of 25% BSA in CM by repeatedly aspirating with a 25 mL pipet. Centrifuge at 1,000×*g* for 15 min at 4°C. The BSA gradient will permit small clumps of cells to penetrate, while resisting the passage of the larger fattier tissue particles. The pellet contains single cells, microvessels, and other small multicellular clumps. The floating cake contains mostly larger tissue particles, but also some smaller particles (including capillaries) that still need to be removed. So, repeat this process five to six times, until most of the capillary fragments have been separated out (*see* **Note 6**).
6. Transfer the supernatant, including the floating cake, to a new 50 mL tube using a 25 mL pipet without disturbing the pellet. Vigorously mix the floating cake throughout the BSA gradient by repeatedly aspirating with a 25 mL pipet. Centrifuge at 1,000×*g* for 15 min at 4°C.
7. Resuspend the pellet with about 4 mL of fresh 25% BSA in CM and transfer it to a fresh 50 mL collecting tube and place this in a holding container with ice water.
8. Repeat the previous three steps until the resulting pellet is negligible (typically five to six times), consolidating the pellets together into the collecting tube. Then, as a final wash repeat these steps with the BSA solution of consolidated pellets. The resulting pellet should be void of any large clumps of tissue matter.
9. Resuspend the final pellet in 30 mL of cold CM. May require gentle vortexing.
10. Pass the suspension through successive screens with decreasing pore sizes (149, 60, and 30 μm) to sequentially

decrease the maximum size of the particles in the suspension (*see* **Note 7**). Collect the filtrate into 50 mL test tubes. After passage of the suspension through the screen, wash the screen and tube with CM into the filtrate. The volume will increase with each successive filtering. The final volume will be 50–100 mL and contain only fragments that are smaller than 30 μm in diameter. The retentates are discarded.

11. Pass the filtrate through a premoistened separation column formed of glass beads suspended by a 20 μm screen. Capillary fragments between 20 and 30 μm in diameter will be retained on the glass beads and screen (*see* **Note 8**). Thoroughly flush out the single cells and nuclei (non-capillary fragments) with an additional 150 mL of CM. Collect the filtrate in 50 mL tubes and set aside on ice; this will be used in step #20 to isolate astrocytes.
12. Release the screen from the column and allow the glass beads and screen to fall into an 80 mL sterile glass beaker with 10–20 mL of CM. Release the capillary fragments from the screen by agitating it or spraying it with CM from a pipet, bringing the final volume up to about 30–40 mL (*see* **Note 9**). Remove the screen from the beaker and discard. Release the capillary fragments from the glass beads by swirling the container vigorously for 5–10 s. After waiting 5 s for the beads to settle (but not the capillaries) pour off most of the CM into a 50 mL tube for collection. Repeat this step six to seven times. The total volume of capillaries in CM should be approximately 200 mL in five to six tubes.
13. Centrifuge the tubes at 400×*g* for 5 min at 4°C. Discard the supernatant.
14. Resuspend and combine the pellets into a final volume of 20 mL of PGM (*see* **Note 10**).
15. Transfer 10 mL of the suspension of capillary fragments into each of the two fibronectin-coated (2 μg/cm^2) plastic T75 (75 cm^2 surface area) tissue culture flask and incubate at 37°C, 5% CO_2, and 95% humidity. Do not disturb the flask for 18 h to enable cell attachment. Change the medium daily (*see* **Note 11**) and observe the cells for capillary endothelial cell characteristics (*see* **Note 12**). Use PGM for the first three changes. Then, use SGM thereafter. These brain capillary endothelial cells can be used to construct the blood–brain barrier model after approximately 5–7 days in culture, when they are approximately 75–80% confluent.
16. Centrifuge the filtrate (from step #15) that was retained from passage through the glass beads and 20 μm screen at

400×g for 5 min at 4°C. Discard the supernatant. Resuspend the pellets into 20 mL of AGM. Transfer 10 mL into each of two T75 tissue culture flasks. Incubate at 37°C, 5% CO_2, and 95% humidity. Change the medium every 2–3 days. Observe daily for the development of astroglial culture (*see* **Note 12**).

17. Harvest the cells (neurons, glial cells) by trypsinization when the culture is ~80% confluent (8–9 days after isolation) and transfer them to two T75 tissue culture flasks (1:1). Incubate for 30 min, then change the medium. Astrocytes attach more quickly than neurons, so this selective attachment enhances the astrocyte culture. These astrocytes can be subcultured into a blood–brain barrier model anytime after 2 days in culture. Alternatively, they can be further propagated by subculture into new flasks by trypsinization prior to confluency.

3.3. Construction of the Blood–Brain Barrier Model

1. The surface of semi-permeable membranes (*see* **Note 13**) is covered with an attachment factor (*see* **Note 14**).
2. Subculture the astrocytes (at passage one to three) by trypsinization onto the bottom side of polycarbonate filter inserts (*see* **Note 15**). This is accomplished by inverting the filters and positioning them upside-down. Seed the cells at a concentration of 50,000 cells per filter (1 cm^2 surface area) in 0.5 mL volume of AGM. Allow the astrocytes to adhere for 15 min (*see* **Note 16**) and then invert the insert and return it to its culture plate. Feed the cells with AGM (0.5 mL in the lumenal [upper] compartment, 1.5 mL in the ablumenal [lower] compartment of a 12-well plate). The lumenal chamber is analogous to the lumen of the capillary. The ablumenal chamber is analogous to the interstitial fluid of the brain.
3. Approximately 24 h later, subculture the brain capillary endothelial cells from passage one by trypsinization and seed them onto polycarbonate membranes at 50,000 cells/cm^2. Feed all the cells with SGM (0.5 mL in the lumenal compartment, 1.5 mL in the ablumenal compartment). When changing the media, every 2–3 days, only replace half of the ablumenal chamber's medium (0.75 mL) so that some of the astrocyte-conditioned medium remains behind. The co-culture model will be in an optimal state for the measurement of transport across the blood–brain barrier 5–7 days after the capillary cells were seeded. The monolayer of capillary endothelial cells growing on the polycarbonate membrane is analogous to the capillary wall in the brain's blood vessels.

3.4. Measuring the Rate of Transport Across the Blood–Brain Barrier

1. Assess BBB integrity on a daily basis by measuring transendothelial electrical resistance (TEER) with a voltohmmeter (*see* **Notes 17** and **18**). Record TEER (Ω cm^2) of the BBB model (polycarbonate membrane with co-culture of astrocytes and endothelial cells) with an Endohm chamber connected to an EVOM resistance meter. Compare TEER of the BBB models with that of a polycarbonate membrane without cells.
2. Measure TEER immediately before and immediately after the transport study (*see* **Note 19**). Introduce the substance under investigation into the media in the lumenal chamber to measure transport from the capillary into the brain. This can be accomplished by replacing the medium in the lumenal chamber or by adding a very small volume with the test compound. A radioactive- or otherwise-labeled molecule may be used if your assay for the compound is not sufficiently sensitive. Alternatively transport in the opposite direction could be performed by introducing the substance of interest into the ablumenal chamber and measuring its appearance in the lumenal chamber.
3. Incubate the BBB models for 60 min on an orbital shaker at 30 rpm (at 37°C, 5% CO_2, and 95% humidity). The agitation mimics the movement of blood through the vessel and avoids the development of an un-stirred layer near the endothelial cell wall. (The incubation time will vary depending on the anticipated rate of transport across the BBB.)
4. Stop the reaction by removing the insert with the cells and lumenal medium from the well. Collect the medium from the ablumenal chamber, or a representative fraction, for analysis.
5. Calculate the rate of transport as moles/(min × cm^2) or percent of the administrated compound per unit of time.
6. Additionally, the endothelial cells on the permeable membrane can be dissolved in a suitable medium for collection and analysis.

4. Notes

1. This BBB model is a co-culture model with astrocytes and capillary endothelial cells from pig brain. The astrocytes and endothelial cells could potentially be isolated from the same brain and then reconstituted into a co-culture model. However, the astrocytes usually require longer cultivation than the endothelial cells before they are suitable for use

in the co-culture model. Astrocytes retain their suitability through multiple passages in culture, unlike the capillary endothelial cells. Therefore, we usually plan to combine endothelial cells from a primary with astrocytes isolated previously, typically the previous week.

2. We chose Yucatán minipigs (*Sus scrofa*) as the donor species because we had a resident herd and one pig would yield enough tissue for approximately 48 experimental units (BBB models) with common genetic characteristics after only one passage. We avoided rodents because the number of animals that would need to be pooled together to obtain enough brain tissue would increase the genetic variability of the cells. Or, the cells would need to be propagated through multiple passages. In our experience, the brain capillary endothelial cells lose some of their BBB characteristics, e.g., tight junctions, with each passage. Cells from older pigs tended to become senescent more quickly in culture then cells from younger pigs, so use of adult animals should be avoided if possible. Nevertheless, these procedures should be adaptable to any species, including humans.

3. It can be a challenge to incorporate 25% BSA into CM. We would sprinkle the BSA on top of the CM the previous day and refrigerate overnight. By morning, after some gentle mixing, it would be soluble enough for sterile filtering.

4. When removing the cerebral cortex tissue, use caution to avoid contaminating your sample of grey matter with the epithelial cells of the intracerebral ventricles (white matter). Best results have been achieved by removing the anterior region of the brain to a depth of about 0.5 cm. If any portion of the ventricles is excised, this should be cut out of the collected tissue and discarded.

5. The collagenase separation may require some calibration. The goal is to free the capillary fragments from the parenchymal tissue, while keeping intact capillary fragments composed of 5–15 endothelial cells each. Of course there will be a wide range of sizes. The larger and smaller fragments will be removed later on. The yield will depend upon how many of the appropriately sized fragments remain afterward. Temperature, duration, agitation, and collagenase activity of the incubation affect the outcome. We incubated at 37°C and stopped the incubation by adding ice-cold CM. We incubated for 60 min, but agitated the tube by hand every 10 min and vigorously aspirated the tissue every 20 min with a 25 mL pipet. Different sources and lots of collagenase will typically have different enzymatic activities; we standardized our activity

to 135 IU/mL, taking into account the dilution effect of the tissue to which the collagenase is being added. The collagenase should be made up fresh daily and kept ice cold until it is warmed to 37°C just prior to being added to the tissue.

6. After the tissue has been mechanically and enzymatic dispersed, most of the parenchymal tissue is removed by centrifugation in BSA gradient. Separation of the supernatant from the pellet can be difficult because of the floating cake of fatty tissue. The pellet contains the capillary fragments, as well as cellular debris, erythrocytes, and other individual cells. The supernatant contains a floating cake of fatty tissue above the BSA gradient. The floating cake will need to be "washed" of their capillary fragments by repeating the centrifugation step and combining the pellets. Use a 25 mL pipet to vacuum the floating cake from the top, working your way down through the gradient to the pellet at the bottom. We found that six or seven washes were appropriate, as indicated by the shrinking size of the pellet with each successive “wash.”
7. The three filtration steps are intended to gradually reduce the maximum size of the particles in the filtrate, while maximizing their quantity. A goal is to avoid clogging the filters. First, pipet the suspension through a premoistened separation column with a 149 μm nylon screen to remove large arterioles and venules from the capillaries. Collect the filtrate into sterile 50 mL test tubes. Begin by mixing the suspension well, let larger fragments settle for 5–7 s, then transfer the suspension. Pipet from the top first and work your way down. This will result in delivery of smaller fragments through the screen first and delay its clogging until most of these have passed through. Follow with about 30 mL of collection medium to wash the retentate on the screen. Two screens may be required to filter the entire suspension, since sometimes one screen will clog up with myelin. Discard filter and retained matter. Repeat this with the 60 μm and 30 μm screens. It is recommended that microscope slides be prepared from small aliquots of each of the subsequent preparations to observe the progress of this procedure.
8. Cells and other matter smaller than capillary fragments are removed by passing the cell suspension through a column (40 mm column of 0.5 mm diameter beads in a 12 mm diameter glass tube supported by a 20 mm screen). Pipet the cell suspension through the beads from their 50 mL tubes in the order they were collected off the previous screen. As always, mix the suspension, let it settle for a few

seconds, then transfer the upper portion of the suspension to the column before the lower portion (with larger particles). This will allow most of the smallest fragments to pass through the column before the larger fragments begin to impede passage.

9. Separating the capillary fragments from the screen and beads can be a challenge. Some fragments can adhere tenaciously. We found that holding the screen with a forceps in the beaker (above the CM and beads) and vigorously washing it with CM from a 10 mL pipet releases most of the fragments (this can be confirmed by microscopic examination). We found that cutting the screen in half enabled it to be washed most effectively. The capillary fragments are loosened from the beads by swirling them vigorously in the beaker. Then, allow the beads to settle for a few seconds and decant off the CM before the capillary fragments settle out. Fresh CM (20–30 mL) should be added, swirled, and decanted six to seven times. Eventually, nearly all the capillary fragments should be freed from the beads and in the collection tubes for centrifugation and pooling.
10. Our procedure for combining the pellets of capillary fragments together into 20 mL of PGM was to add 5 mL of PGM to the first tube, resuspend the fragments, then transfer this suspension into the second tube and resuspend those fragments, repeating this for all the tubes. The final volume is brought up to 10 mL with additional PGM. Occasionally, gentle vortexing is necessary to resuspend the fragments. Vortex vigorously if necessary, since visible clumps of fragments will not survive in culture. At this point, it is very important to make a microscope slide of a drop of this suspension to visualize the yield. The vast majority of the components should be capillary fragments of 5–15 cells per fragment, with very few single cells or larger fragments. Bring the final volume up to 20 mL.
11. When the medium is changed for the first time from both the capillary and the astrocyte cultures, wash the surface gently with MEM to remove cellular debris and nonviable cells before applying new growth medium.
12. There are a variety of methods to confirm the nature of the isolated cells. These methods will not be described in detail here. Visually, capillary endothelial cells grow into a confluent monolayer of elongated, cigar-shaped cells. The borders between the cells are poorly defined, reflecting the tightness of the interendothelial junctions. Pinocytotic vesicles will be undetectable by phase contrast microscopy. Endothelial cells express the Factor VIII related antigen, which can be measured immunofluorescently by the

procedure of Hoyer et al. (13). Endothelial cells will also take up acetylated low-density lipoprotein (ac-LDL) by endocytosis and can be tested with the DiI-Ac-LDL method of Voyta et al. (14). Astrocytes express a number of specific enzymes and other proteins, such as the glial fibrillary acidic protein (GFAP), which can be measured by immunofluorescence to confirm the characteristics of these cells.

13. We prefer the 12-well Transwell™ plates from Costar with 12 mm diameter membranes with 0.4 μm pores and a surface area of 1 cm^2. We found that larger pore sizes yielded poor cell attachment, while smaller pore sizes limited the passage of macromolecules. In transport studies, we want the BBB cells to be the barrier and not a permeable membrane.
14. An attachment factor needs to be applied to the top surface of the filters prior to their use. Typically a few hours before the astrocytes are added. We found that collagen and fibronectin seemed to work equally well, but usually used fibronectin (2 $\mu g/cm^2$) because it was more physiologically and anatomically relevant. We applied 100 μL of a 20 μg/mL solution.
15. Our procedure for trypsinization is as follows: wash the cells twice with PBS and once with PBS w/EDTA. Apply 2–3 mL of trypsin/EDTA and remove, leaving behind a residual amount. Observe the cells using an inverted cell culture microscope as they detach from the flask. Collect the cells in growth medium.
16. Astrocyte attachment can be improved, if needed, by extending the incubation time after initial seeding up to 60 min before returning the membrane to its well in the culture plate. During this time, keep the cells wet by adding a drop of AGM every 5–10 min.
17. TEER, when measured properly, is the best indicator of blood–brain barrier integrity. Electrical resistance (Ω cm^2) reflects the cell layer's ability to resist the passage of a low electrical current, essentially reflecting the passage of small electrolytes (atoms). We measure TEER everyday to monitor the development of the barrier in our co-culture model. Typically, we see a steady rise in TEER over the first 5 days, followed by a plateau at days 5–8. Subsequently, TEER declines as the cells begin to overgrow the model. Therefore, we usually measure transport across the BBB models on days 6 or 7. While we prefer to work with models that have a TEER measurement greater than 100 Ω cm^2, we observe very little difference in the rates of transport of

most substances across the replicate models when TEER is greater than 30 Ω cm^2.

18. Inulin and albumin can also be suitable to measure blood–brain barrier integrity. However, these molecules represent the passage of very large macromolecules. The rate of ^{3}H-inulin transport should be under 0.5 % per hour for the in vitro BBB to be considered intact (15).

19. We measure TEER immediately before and immediately after the experimental procedure to measure transport of a substance across the BBB model. The "before" measurement indicates whether the model is suitable (sufficient barrier) to continue with the study. The "after" measurement is compared with the "before" value of the same unit to indicate the degree to which the treatment affected barrier function. For example, if the transport substance being investigated is toxic to the cells, this would have a deleterious effect on the barrier. Without knowing this, the data could be misinterpreted.

References

1. Reese, T. and Karnovsky, M. (1967) Fine structural localization of a blood-brain barrier to exogenous peroxidase. *J. Cell Biol.* **34**, 207–217.
2. Goetz, I.E., Warren, J., Estrada, C., Roberts, E., and Krause, D.N. (1985) Long term serial cultivation of arterial and capillary endothelium from adult bovine brain. In Vitro Cell. *Dev. Biol.* **21**, 172–180.
3. Bowman, P.D., du Bois, M., Dorovini-Zis, K., and Shivers, R.R. (1990) Microvascular endothelial cells from brain. In *Cell Culture Techniques in Heart and Vessel Research* (Piper, H.M., ed.), pp. 140–157, Springer Verlag, Berlin, Germany.
4. Abbott, N.J., Hughes, C.C.W., Revest, P.A., and Greenwood, J. (1992) Development and characterization of a rat brain capillary endothelial culture: towards an in vitro blood-brain barrier. *J. Cell Sci.* **103**, 23–37.
5. Mischeck, U., Meyer, J., and Galla, H.J. (1989) Characterization of gamma glutamyl transpeptidase activity of cultured endothelial cells from porcine brain capillaries. *Cell Tissue Res.* **256**, 221–226.
6. Wall, R.T., Harker, L.A., Quadracci, L.J., and Striker, G.E. (1978) Factors influencing endothelial cell proliferation in vitro. *J. Cell. Physiol.* **96**, 203–214.
7. Castellot, J.J., Jr., Addonizio, M.L., Rosenberg, R., and Karnovsky, M.J. (1981) Cultured endothelial cells produce a heparin like inhibitor of smooth muscle cell growth. *J. Cell Biol.* **90**, 372–379.
8. Thornton, S.C., Mueller, S.N., and Levine, E.M. (1983) Human endothelial cells: use of heparin in cloning and long-term serial cultivation. *Science* **222**, 623–625.
9. Raub, T.J., Kuentzel, S.L., and Sawada, G.A. (1992) Permeability of bovine brain microvessel endothelial cells in vitro: barrier tightening by a factor released from astroglioma cells. *Exp. Cell Res.* **199**, 330–340.
10. Gaillard, P., van der Sandt, I., Voorwinden, L., D.Vu, D., Nielsen, J., de Boer, A., and Breimer, D. (2000) Astrocytes increase the functional expression of P-glycoprotein in an in vitro model of the blood-brain barrier. *Pharm. Res.* **17**, 1198–1205.
11. Gaillard, P., Voorwinden, L., Nielsen, J., Ivanov, A., Atsumi, R., Engman, H., Ringbom, C., de Boer, A., and Breimer, D. (2001) Establishment and functional characterization of an in vitro model of the blood-brain barrier, comprising a co-culture of brain capillary endothelial cells and astrocytes. *Eur. J. Pharm. Sci.* **12**, 215–222.
12. Maciag, T., Cerundolo, J., Ilsley, S., Kelley, P.R., and Forand, R. (1979) An endothelial cell growth factor from bovine hypothalamus: identification and partial characterization. *Proc. Natl. Acad. Sci. U S A* **76**, 5674–5678.

13. Hoyer, L.W., de Los Santos, R.P., and Hoyer, J.R. (1973) Antihemophilic factor antigen: localization in endothelial cells by immunofluorescent microscopy. *J. Clin. Invest.* **52**, 2737–2744.
14. Voyta, J.C., Via, D.P., Butterfield, C.E., and Zetter, B.R. (1984) Identification and isolation of endothelial cells based on their increased uptake of acetylated-low density lipoprotein. *J. Cell Biol.* **99**, 2034–2040.
15. Dehouck, M.P., Meresse, S., Delorme, P., Fruchart, J.C., and Cecchelli, R. (1990) An easier, reproducible, and mass-production method to study the blood-brain barrier in vitro. *J. Neurochem.* **54**, 1798–1801.

Chapter 9

Genetic Variants in the Vesicular Monoamine Transporter 1 (*VMAT1*/SLC18A1) and Neuropsychiatric Disorders

Falk W. Lohoff

Abstract

Vesicular monoamine transporters (VMATs) are involved in the presynaptic packaging of monoaminergic neurotransmitters into storage granules. Upon an action potential, vesicles release their contents into the synaptic cleft via exocytosis. Since insufficient or excess release of neurotransmitter might alter neurochemical function and neurotransmission, VMATs are an important target for biological research in neuropsychiatric disorders. Two structurally related but pharmacologically distinct VMATs have been identified, encoded by separate genes, *VMAT1* (*SLC18A1*) and *VMAT2* (*SLC18A2*). Although it was reported initially that only VMAT2 is expressed in brain, recent studies indicate that VMAT1 is also expressed in brain, thus making both transporters plausible candidate genes for neuropsychiatric disorders. The gene encoding *VMAT1* is located on chromosome 8p21, a region implicated in linkage studies of schizophrenia, bipolar disorder, and anxiety-related phenotypes. Furthermore, several recent genetic case–control studies have documented an association between common missense variations in the *VMAT1* gene and susceptibility to bipolar disorder and schizophrenia. Variations in the *VMAT1* gene might affect transporter function and might be involved in the etiology of neuropsychiatric disorders. This chapter describes methods for genotyping three missense polymorphisms implicated in neuropsychiatric disorders (Thr4Pro, Thr98Ser, Thr136Ile) using TaqMan-based PCR and standard PCR approaches.

Key words: Vesicular monoamine transporter, psychiatric disorders, schizophrenia, bipolar disorder, depression, genetics, brain expression, reserpine, tetrabenazine, SLC18A1.

1. Introduction

Monoaminergic synaptic transmission requires two types of neurotransmitter transporters: plasma membrane transporters that remove neurotransmitters from the synaptic cleft to terminate signaling and vesicular monoamine transporters (VMATs) that

Q. Yan (ed.), *Membrane Transporters in Drug Discovery and Development*, Methods in Molecular Biology 637,
DOI 10.1007/978-1-60761-700-6_9, © Springer Science+Business Media, LLC 2010

package neurotransmitter molecules into presynaptic storage vesicles. Upon arrival of an action potential at the nerve terminal, vesicles release their contents into the synaptic cleft via exocytosis. Regulation and storage of newly synthesized and accumulating neurotransmitter molecules from the synapse is thus a critical part in monoamine signaling between neurons. Since insufficient or excess release of neurotransmitter might alter neurochemical function and neurotransmission, VMATs are an important target for biological research in neuropsychiatric disorders.

1.1. Physiology and Mechanism

Two structurally related but pharmacologically distinct VMATs have been identified, encoded by separate genes, *VMAT1* (*SLC18A1*) located on chromosome 8p21 and *VMAT2* (*SLC18A2*) located on chromosome 10q25 (1–3). Expression studies, mostly conducted in animal tissue, report a distinct distribution pattern of the two isoforms and differences of expression between species (2–4). It was reported initially that VMAT1, previously known as chromaffin granule amine transporter, is expressed exclusively in neurons of the peripheral nervous system, endocrine tissue, and chromaffin cells, while only the VMAT2 isoform, previously known as synaptic monoamine transporter, was thought to be expressed in brain (2–4); however, recent studies show that VMAT1 is expressed in rat brain (5) and human brain (6). Review of the public databases further supports expression of VMAT1 in human brain (Affymetrix GeneChip Human Genome U95 Set HG-U95A Accession # GDS181, Merck Rosetta Chip Accession # GDS833; expression profile information suggested by analysis of EST counts, UniGene Hs.158322; www.ncbi.nlm.nih.gov).

VMATs display a comparable size and molecular topography similar to other plasma membrane transporters, such as the dopamine transporter (DAT), serotonin transporter (SERT), and norepinephrine transporter (NET), with 12 transmembrane domains and both tails located in the interior (**Fig. 9.1**) (1). Nevertheless, VMAT physiology is distinct from plasma membrane transporters in that they use a proton gradient to transport substrates and they lack an extracellular compartment. Both proteins are able to transport monoamines (serotonin, dopamine, epinephrine, norepinephrine); however, they differ in their substrate preferences and affinities. VMAT1 shows higher affinity for serotonin (7), whereas VMAT2 is also able to transport histamine (3).

1.2. Pharmacological Characteristics

The activity of both transporters is inhibited irreversibly by reserpine, an indole alkaloid derived from the plant species *Rauwolfia serpentina* (8). The drug was introduced in 1954 as antipsychotic and antihypertensive medication. Reserpine was instrumental in the development of the “monoamine hypothesis” of affective

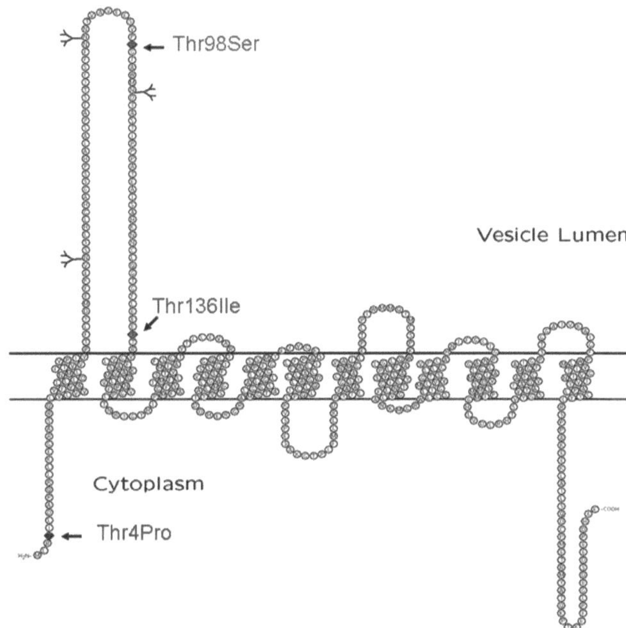

Fig. 9.1. Predicted structure of human *VMAT1*. *Highlighted* are the amino acid variations implicated in psychiatric disorders.

disorders (9) and continues to be used as research tool in rodent models of depression and anxiety (10). Clinical use of reserpine has been limited due to many drug interactions and side effects which include lethargy and clinical depression (11–13). On the other hand, more recent reports show that the causal relationship between the use of reserpine and depression should be reevaluated (14) and recent clinical trials of reserpine document that the drug was well tolerated and safe (15, 16). Reserpine has gained recent attention as a potential treatment for cocaine addiction. In the National Institute on Drug Abuse (NIDA)-sponsored Cocaine Rapid Efficacy and Safety Trial (CREST) 60 cocaine-dependent outpatient participated in a 12-week, double-blind, four-arm study that compared the safety and efficacy of reserpine (at a 0.50 mg dose), gabapentin, lamotrigine, and placebo (15, 17). Results indicated that the reserpine group significantly decreased their cocaine use as assessed by urine benzoylecgonine levels at endpoint when compared to baseline. A follow-up clinical trial failed to replicate the initial results, possibly due to differences in clinical variables and demographics (16).

Tetrabenazine is another VMAT blocker, with predominantly VMAT2 inhibitory properties (3). It has recently been approved in the United States for the treatment of choreiform movements in Huntington disease (18, 19). Tetrabenazine is structurally related to antipsychotics and has many similar side effects, including akathisia, Parkinsonism, dizziness, and depression.

Indirect evidence that VMAT1 might be involved in neuropsychiatric disorders stems from studies in vitro that show that lithium and valproate, effective pharmacotherapies for BPD, increase the expression of VMAT1 (20–22), suggesting that the *VMAT1* gene might be a target for therapeutic drug action.

1.3. Relevance to Neuropsychiatric Disorders

Dysregulation of dopamine and serotonin neurotransmission has been long postulated to play a role in the etiology of psychiatric disorders such as major depressive disorder (MDD), bipolar disorder (BPD), and schizophrenia (SZ) (23, 24), thus making VMATs plausible biological targets for neuropsychiatric disorders. In fact, several studies have implicated the VMAT2 transporter in Parkinson disease, addiction, and other psychiatric disorders (25–30).

Additional evidence that VMATs are involved in psychiatric disorders stems from positron emission tomography (PET) imaging studies. Binding of radiolabeled dihydrotetrabenazine, a catecholamine depleter with higher affinity for VMAT2 than VMAT1 (31, 32), was increased in thalamus and brainstem of BPD patients when compared to controls (26). Ventral brainstem binding was higher in BPD and SZ patients compared to controls (27). These experiments suggest that higher levels of VMAT expression may represent a trait-related abnormality in patients with BPD and SZ.

The VMAT1 transporter has gained recent attention in the field of neuropsychiatric research. This is based on the discovery that VMAT1 is in fact brain expressed and furthermore that genetic variants of VMAT1 are associated with several psychiatric phenotypes (6, 33, 34). *VMAT1* maps to chromosome 8p21 (2, 35). This genomic region has been implicated to harbor genes for BPD and SZ (36, 37). Remarkably, the linkage peaks for SZ and BPD seem to overlap in this genomic region, suggesting that chromosome 8p21 might contain shared susceptibility factors for both disorders. This is especially interesting since recent BPD linkage analyses found a peak on 8p in BPD patients with psychotic symptoms (38, 39), suggesting that the phenotype of psychosis might influence susceptibility to BPD and SZ. Furthermore, the two disorders have many epidemiological characteristics in common. Both disorders are chronic debilitating psychiatric disorders with a strong heritable component, affecting 1–2% of the general population (40, 41). Clinically, they share several characteristics including psychosis, mood symptoms, and cognitive deficits. While the therapeutic modalities for both disorders differ depending on the clinical course, bipolar mania and psychotic exacerbation of schizophrenia are treated effectively with antipsychotic drugs (42, 43), suggesting a shared common pathway in the etiology of some symptom dimensions.

Recently, *VMAT1* genetic polymorphisms have been associated with BPD, SZ, and anxiety-related personality traits

Table 9.1
Genetic association studies of *VMAT1* in bipolar disorder and schizophrenia

	Authors	Sample size cases	Variant	Ethnicity	*p* Value
Bipolar disorder	(6)	$n = 580$	Thr136Ile	European American	$p = 0.0006$
Schizophrenia	(44)	$n = 28$	Thr4Pro	European American	$p = 0.01$
	(45)	$n = 354$	Thr98Ser	Japanese	$p = 0.01$
	(46)	$n = 319$	Thr4Pro	Chinese	$p = 0.006$
	(34)	$n = 63$	Thr4P	European American	$p = 0.003$

(**Table 9.1**) (6, 33, 34, 44–46). Lohoff et al. (6) showed in a case–control association study of bipolar patients and normal controls that the SNP predicting the amino acid variant Thr136Ile in the *VMAT1* gene was associated with disease. Bly (44) showed that the Thr4Pro polymorphism was associated with SZ in a small sample of cases and controls of European descent, with the Pro allele conferring risk for SZ. This observation was recently confirmed by Chen et al. (46) in a Han Chinese population. Richards et al. (45) could not detect an association of the Thr4Pro polymorphism in their sample but provided evidence for an association of the Thr98Ser polymorphism in a Japanese population. Lohoff et al. (34) provided further evidence for an association between the Thr4Pro polymorphisms in the *SLC18A1* gene and SZ.

Although there are no data available that address the potential biological effects of the *VMAT1* Thr4Pro, Thr98Ser, and Thr136Ile polymorphisms, indirect evidence suggests potential relevance to serotonergic neurotransmission. The common Thr136Ile polymorphism is located in the first luminal domain of the transporter (**Fig. 1**). This region of the protein interacts with inhibitors and substrates (47) and mediates G-protein-dependent regulation of transmitter uptake (7). Deletion of the first intravesicular loop results in marked decrease in serotonin transport capacity (7). Subtle variation in the protein could thus affect serotonin and norepinephrine levels, consistent with the "serotonin hypothesis" of neuropsychiatric disorders (48–50).

Taken together, several lines of new evidence suggest that VMAT1 is an important target for neuropsychiatric disorders. First, VMAT1 is brain expressed; second, the gene encoding the protein lies in a genomic region where linkage studies have indicated a susceptibility locus for BPD and SZ; third, genetic association studies in BPD and SZ show that potentially functional missense variants are associated with disease; and fourth, VMAT1 is a drugable transporter target thus increasing the likelihood that new therapeutic drug development is feasible and could be successful.

2. Materials

The *VMAT1*/SLC18A1 gene is located on chromosome 8, 20,046,660–20,084,906 bp (www.genome.ucsc.edu). It encodes 525 amino acids and consists of 16 exons spanning 38,346 bp (NM_003053; www.ncbi.nlm.nih.gov). The three non-synonymous SNPs [Thr4Pro (exon 2), Thr98Ser (exon 3), Thr136Ile (exon 3)] that have been associated with psychiatric disorders can be genotyped via multiple methods. Described here are two genotyping methods for these SNPs, each of which uses different technologies and thus might be more applicable depending on the equipment in individual laboratories.

2.1. DNA Extraction from Leukocytes

1. Red blood cell (RBC) lysis solution (Qiagen, Hilden, Germany).
2. Cell lysis solution (Qiagen, Hilden, Germany).
3. Protein precipitation solution (Qiagen, Hilden, Germany).
4. 100% isopropanol.
5. 100% ethanol is diluted with dH_2O to a 70% ethanol solution and stored at 4°C.
6. 1X Tris-EDTA solution, pH 7.6.

2.2. TaqMan Genotyping

1. Applied Biosystems (ABI) Assay-on-Demand genotyping assay (ABI, Foster City, CA, USA; www.appliedbiosystems.com/), which consists of two unlabeled PCR primers and a mixture of TaqMan MGB probes labeled at the 5′ end with the fluorescent reporter dye VIC or FAM. Assay is prepared at 40X concentration (900 nM primers and 200 nM of the fluorescent probe). Details are described in **Table 9.2**.
2. 2X TaqMan Universal PCR Master Mix, No AmpErase UNG (universal concentration conditions: 4 mM MgCl2, 200 mM each dATP, dCTP, and dGTP; 400 mM dTTP, 1.25 U AmpliTaq Gold).

2.3. Polymerase Chain Reaction (PCR)

1. GeneAmp® 10X PCR buffer containing 15 mM $MgCl_2$ from ABI.
2. Integrated DNA Technologies (IDT) Custom-DNA-Oligos (IDT, San Jose, CA) was used to design "*VMAT1* set 1"

Table 9.2
TaqMan applied biosystems genotyping assays for *SLC18A1*

Amino acid variation	dBSNP ID	ABI ID	Allele	Codon	Codon	Major allele
Thr4Pro	rs2270641	C__22271506_10	[G/T]	T=Thr	G=Pro	Major Allele: T
Thr98Ser	rs2270637	C__2716008_1_	[C/G]	C=Thr	G=Ser	Major Allele: C
Thr136Ile	rs1390938	C__8804621_1_	[A/G]	G=Thr	A=Ile	Major Allele: G

primers. Prepare 0.1 mM of primer solution by diluting 25 nmol of primer pellet (*see* **Note 1**) in 250 μL dH_2O. Dilute 10-fold to generate working stock solution. Details are described in **Table 9.3**.

3. IDT Custom-DNA-Oligos was used to design "*VMAT1* set 2" primer set. Dilute primers in same method as described above. Details are described in **Table 9.3**.
4. GeneAmp® dNTP mix (ABI, Foster City, CA, USA): deoxynucleotide triphosphates dissolved in glass-distilled water and titrated to pH 7.0 with NaOH.

Table 9.3
RFLP assay for *SLC18A1*

Amino acid	Primer set	Codon	Observed band sizes
Thr4Pro	*VMAT1* set 1 Forward primer: 5′-GAA CAG TGA TAC AGA CAT GG-3′ Reverse primer: 5′-TCA CTG TTG AAC TCA AGC AG-3′	G=Pro T=Thr[a]	420 bp 211 bp+209 bp
Thr98Ser	*VMAT1* set 2 Forward primer: 5′-CAG TGG TAA GAT GAC ATC G-3′ Reverse primer: 5′-CCT CCT ACT CCT GTG TAC-3′	C=Thr G=Ser[a]	468 bp 226 bp+242 bp
Thr136Ile	*VMAT1* set 3 Forward primer: 5′-CAG TGG TAA GAT GAC ATC G-3′ Reverse primer: 5′-CCT CCT ACT CCT GTG TAC-3′	A=Ile G=Thr[a]	468 bp 129 bp+339 bp

Primer set 2 and 3 are identical.
[a]The digest enzyme recognizes and cuts this allele.

5. AmpliTaq® DNA polymerase, 5 U/μL: AmpliTaq® DNA polymerase (ABI, Foster City, CA, USA) is a 94 kDa ultra-pure, gelatin-free, thermostable, recombinant DNA polymerase obtained by expression of a modified form of the *Thermus aquaticus* Taq DNA polymerase gene in *Escherichia coli*.

2.4. Restriction Fragment Length Polymorphism (RFLP) Genotyping

1. BspE1 enzyme, 10,000 U/mL (New England Biolabs, Ipswich, MA).
2. Bfa1 enzyme, 5,000 U/mL (New England Biolabs, Ipswich, MA).
3. BsmA1 enzyme, 5,000 U/mL (New England Biolabs, Ipswich, MA).
4. 10X NEBuffer #3 from NEB: 50 mM Tris–HCl, 100 mM NaCl, 10 mM MgCl2, 1 mM dithiothreitol, pH 7.9 at 25°C.
5. 10X NEBuffer #4 (New England Biolabs, Ipswich, MA): 20 mM Tris-acetate 50 mM potassium acetate, 10 mM magnesium acetate, 1 mM dithiothreitol, pH 7.9 at 25°C.

2.5. Polyacrylamide Gel Electrophoresis (PAGE)

1. 5X buffers: 10X Tris-borate, pH 8.3, 2 mM Na_2EDTA (National Diagnostics, Atlanta, GA) is diluted 1:1 with dH_2O.
2. 40% acrylamide (*see* **Note 2**) and 99% *N*,*N*,*N*′,*N*′-tetramethylethylenediamine (TEMED, Sigma-Aldrich, St. Louis, MO).
3. 10% ammonium persulfate (APS): prepare 1 g of APS in 10 mL dH_2O. Store at 4°C.
4. 1X running buffer: 10X Tris-borate, pH 8.3, 2 mM Na_2EDTA is diluted 1:9 with dH_2O.
5. Molecular Weight Marker: Φ174 DNA/HaeIII marker (Promega, Madison, WI).
6. 6X loading dye: gel loading solution (Sigma-Aldrich).
7. 0.5 μg/mL ethidium bromide (EtBr) is prepared by diluting 250 μL of 2 mg/mL EtBr in 1 L of dH_2O.

3. Methods

3.1. DNA Isolation from Blood Sample – Qiagen Recommended Protocol

1. Add 24 mL RBC lysis solution to 8 mL of blood sample in 50 mL tube.
2. Allow solution to sit at room temperature for 10 min.
3. Centrifuge for 10 min at 2,000×*g* and discard supernate.

4. Add 8 mL of cell lysis solution.
5. Homogenize the solution in the 50 mL tube:
   a. Vortex the tube for 10 s.
   b. Pipette-mix the mixture with transfer pipette until homogenous.
   c. Incubate the tubes in 37°C water bath for 10 min (*see* **Note 3**).
6. Add 2.67 mL of protein precipitation solution.
7. Vortex tube for 20 s.
8. Centrifuge for 10 min at 2,000×*g*.
9. Transfer supernate to a new 50 mL tube, containing 8 mL of isopropanol, and invert 50 times or until DNA precipitates.
10. Centrifuge tubes for 3 min at 2,000×*g* – DNA should form pellet.
11. Carefully discard supernate, without discarding the DNA pellet.
12. Add 4 mL of 70% ethanol.
13. Centrifuge for 3 min at 2,000×*g*.
14. Carefully discard supernate without discarding the DNA pellet.
15. Air-dry DNA pellet.
16. After EtOH is completely evaporated, suspend DNA pellet in 1,000 μL of Tris-EDTA Solution.
17. Let the DNA dissolve in solution overnight at room temperature.
18. Transfer 1,000 μL DNA solution to appropriately labeled 2 mL DNA tubes.

3.2. TaqMan Genotyping

1. Set up TaqMan PCR with the following cocktail:

Reaction Components	*1×*
2X Master mix	2.500 μL
40X Assay	0.125 μL
dH_2O	0.375 μL

2. Add 2 μL DNA (1 ng/μL) to reaction well.
3. Thermal cycling conditions: 95°C hold for 10 min followed by 40 cycles of 92°C for 15 s and 60°C for 60 s.
4. Data analysis: Using SDS software, analyze data with allele discrimination program. Assign fluorescent readers for appropriate reporters to wells. Designate non-

template controls as NTCs. Run an allelic discrimination post-read.

3.3. RFLP Genotyping

3.3.1. Threonine 4 Proline Genotyping Protocol

1. Set up standard PCR with the following cocktail and primers:

Reaction Components	*1×*
dH_2O	40.5 μL
10X buffer	5.00 μL
VMAT1 set 1 forward primer	1.00 μL
VMAT1 set 1 reverse primer	1.00 μL
dNTPs (10 nM)	1.00 μL
Taq polymerase	0.50 μL
DNA template (20–50 ng/μL)	1.00 μL

See **Table 9.3** for primer sequences.

2. Run the PCR beginning with a 95°C hold for 5 min followed by 40 cycles of 95°C for 45 s, 54°C for 45 s, and 72°C for 45 s. End with a 72°C hold for 7 min.
3. Set up BspE1 digest reaction with the following cocktail:

Reaction Components	*1×*
dH_2O	3.50 μL
10X buffer (*New England Buffer #3*)	1.00 μL
BspE1 enzyme	0.50 μL
PCR product	5.00 μL

4. Run digest reaction in standard PCR machine set at 37°C for a 120 min hold.
5. Run digested product on a 5% PAGE. PAGE protocol elaborated below.
6. Analyze results:
   The original PCR product is 420 bp. BspE1 cuts only when recognizing the allele encoding threonine. If the DNA sample is homozygous for the allele encoding the proline amino acid, then there will be no cut, and a single 420 bp band will be observed. If the DNA sample is homozygous for the allele encoding the threonine amino acid, the BspE1 enzyme will cut the sequence, and two bands, one 211 bp and one 209 bp, will be observed. If the DNA sample is heterozygous, then the uncut 420 bp band will be observed along with the cut 211 and 209 bp bands.

3.3.2. Threonine 98 Serine Genotyping Protocol

1. Set up standard PCR with the following cocktail and primers:

Reaction Components	*1×*
dH_2O	40.5 μL
10X buffer	5.00 μL
VMAT1 set 2 forward primer	1.00 μL
VMAT1 set 2 reverse primer	1.00 μL
dNTPs (10 nM)	1.00 μL
Taq polymerase	0.50 μL
DNA template (20–50 ng/μL)	1.00 μL

See **Table 9.3** for primer sequences.

2. Run the PCR beginning with a 95°C hold for 5 min followed by 40 cycles of 95°C for 45 s, 58°C for 45 s, and 72°C for 45 s. End with a 72°C hold for 7 min.
3. Set up Bfa1 digest reaction with the following cocktail:

Reaction Components	*1×*
dH_2O	3.50 μL
10X buffer (*New England Buffer #4*)	1.00 μL
Bfa1 enzyme	0.50 μL
PCR product	5.00 μL

4. Run digest reaction in standard PCR machine set at 37°C for a 120 min hold.
5. Run digest product on a 5% PAGE. SDS-PAGE protocol elaborated below.
6. Analyze results:
   The original PCR product is 468 bp. Bfa1 cuts only when recognizing the allele encoding serine. If the DNA sample is homozygous for the allele encoding the threonine amino acid, then there will be no cut, and a single 468 bp band will be observed. If the DNA sample is homozygous for the allele encoding the serine amino acid, the Bfa1 enzyme will cut the sequence, and two bands, one 226 bp and one 242 bp, will be observed. If the DNA sample is heterozygous, then the uncut 468 bp band will be observed along with the cut 226 and 242 bp bands.

3.3.3. Threonine 98 Isoleucine Genotyping Protocol

1. Set up standard PCR with the following cocktail and primers:

Reaction Components	*1×*
dH_2O	40.5 μL
10X buffer	5.00 μL
VMAT1 set 3 forward primer	1.00 μL
VMAT1 set 3 reverse primer	1.00 μL
dNTPs (10 nM)	1.00 μL
Taq polymerase	0.50 μL
DNA template (20–50 ng/μL)	1.00 μL

See **Table 9.3** for primer sequences.

2. Run the PCR beginning with a 95°C hold for 5 min followed by 40 cycles of 95°C for 45 s, 58°C for 45 s, and 72°C for 45 s. End with a 72°C hold for 7 min.
3. Set up BsmA1 digest reaction with the following cocktail:

Reaction Components	*1×*
dH_2O	3.50 μL
10X buffer (*New England Buffer #3*)	1.00 μL
BsmA1 Enzyme	0.50 μL
PCR product	5.00 μL

4. Run digest reaction in standard PCR machine set at 37°C for a 120 min hold.
5. Run digest product on a 5% PAGE. PAGE protocol elaborated below.
6. Analyze results:
   The original PCR product is 468 bp. BsmA1 cuts only when recognizing the allele encoding threonine. If the DNA sample is homozygous for the allele encoding the isoleucine amino acid, then there will be no cut, and a single 468 bp band will be observed. If the DNA sample is homozygous for the allele encoding the threonine amino acid, the BsmA1 enzyme will cut the sequence, and two bands, one 129 bp and one 339 bp, will be observed. If the DNA sample is heterozygous, then the uncut 468 bp band will be observed along with the cut 129 and 339 bp bands.

3.4. PAGE

1. Prepare 5% gel by mixing the following reagents: 4.95 mL of dH_2O, 1.5 mL of 5X TBE buffer, 0.95 mL of 40% acrylamide, 87.5 μL of APS, and 4 μL of TEMED.
2. Pour liquid gel mixture into cast with appropriate gel comb. The gel should polymerize within 20 min.

3. Set up gel in Mini Protean® 3 Cell (Bio-Rad, Hercules, CA). Pour 1X running TBE buffer into middle chamber, allowing buffer to overflow into outer chamber.
4. Aliquot 10 μL of PCR product into new tubes. Dispense 2 μL of loading dye into each 10 μL aliquot of PCR. Pipette mix.
5. Load 10 μL of PCR product – loading dye mixture into wells. Designate one well, for which 2 μL of Φ174 DNA/HaeIII marker (100 ng/μL) will be loaded.
6. Complete assembly of Mini Protean® 3 Cell unit and connect to power supply. Set the power supply to 400 mA and run gel at 100 V. Stop the power supply when blue loading dye reaches the bottom of the gel.
7. Remove the gel from the Mini Protean® 3 Cell, and post-stain the gel with EtBr for 10 min.
8. Expose gel to Trans UV light waves and photograph.

4. Notes

1. Primers will be delivered as a pellet. While 25 nM of primers will often be provided, variation of this amount is common. Adjust dilution accordingly.
2. When unpolymerized, acrylamide is a neurotoxin, so much precaution must be taken to avoid exposure. Also, store acrylamide in 4°C.
3. If the mixture is still not homogenous after 10 min in the water bath, repeat vortexing, pipette-mixing, and water bath until mixture homogenized.

Acknowledgments

This work was supported by the Center for Neurobiology and Behavior, Department of Psychiatry, University of Pennsylvania. Financial support is gratefully acknowledged from National Institutes of Health grant K08MH080372. The author would like to thank Thomas Ferraro, Glenn Doyle, and Paul Bloch for their very helpful comments, suggestions, improvements, and corrections.

References

1. Liu, Y., Peter, D., Roghani, A., Schuldiner, S., Prive, G.G., Eisenberg, D., Brecha, N., and Edwards, R.H. (1992) A cDNA that suppresses MPP+ toxicity encodes a vesicular amine transporter. *Cell* **70**, 539–551.
2. Peter, D., Finn, J.P., Klisak, I., Liu, Y., Kojis, T., Heinzmann, C., Roghani, A., Sparkes, R.S., and Edwards, R.H. (1993) Chromosomal localization of the human vesicular amine transporter genes. *Genomics* **18**, 720–723.
3. Erickson, J.D., Schafer, M.K., Bonner, T.I., Eiden, L.E., and Weihe, E. (1996) Distinct pharmacological properties and distribution in neurons and endocrine cells of two isoforms of the human vesicular monoamine transporter. *Proc. Natl. Acad. Sci. U S A* **93**, 5166–5171.
4. Eiden, L.E., Schafer, M.K., Weihe, E., and Schutz, B. (2004) The vesicular amine transporter family (SLC18): amine/proton antiporters required for vesicular accumulation and regulated exocytotic secretion of monoamines and acetylcholine. *Pflugers Arch.* **447**, 636–640.
5. Hansson, S.R., Hoffman, B.J., and Mezey, E. (1998) Ontogeny of vesicular monoamine transporter mRNAs *VMAT1* and *VMAT2*. I. The developing rat central nervous system. *Brain Res. Dev. Brain Res.* **110**, 135–158.
6. Lohoff, F.W., Dahl, J.P., Ferraro, T.N., Arnold, S.E., Gallinat, J., Sander, T., and Berrettini, W.H. (2006) Variations in the vesicular monoamine transporter 1 gene (*VMAT1*/SLC18A1) are associated with bipolar I disorder. *Neuropsychopharmacology* **31**, 2739–2747.
7. Brunk, I., Blex, C., Rachakonda, S., Holtje, M., Winter, S., Pahner, I., Walther, D.J., and Ahnert-Hilger, G. (2006) The first luminal domain of vesicular monoamine transporters mediates G-protein-dependent regulation of transmitter uptake. *J. Biol. Chem.* **281**, 33373–33385.
8. Chen, F.E. and Huang, J. (2005) Reserpine: a challenge for total synthesis of natural products. *Chem. Rev.* **105**, 4671–4706.
9. Schildkraut, J.J. and Kety, S.S. (1967) Biogenic amines and emotion. *Science* **156**, 21–37.
10. Heslop, K.E. and Curzon, G. (1999) Effect of reserpine on behavioural responses to agonists at 5-HT1A, 5-HT1B, 5-HT2A, and 5-HT2C receptor subtypes. *Neuropharmacology* **38**, 883–891.
11. Goodwin, F.K. and Bunney, W.E., Jr. (1971) Depressions following reserpine: a reevaluation. *Semin. Psychiatry 3*, 435–448.
12. Bant, W.P. (1978) Antihypertensive drugs and depression: a reappraisal. *Psychol. Med.* **8**, 275–283.
13. Widmer, R.B. (1985) Reserpine: the maligned antihypertensive drug. *J. Fam. Pract.* **20**, 81–83.
14. Baumeister, A.A., Hawkins, M.F., and Uzelac, S.M. (2003) The myth of reserpine-induced depression: role in the historical development of the monoamine hypothesis. *J. Hist. Neurosci.* **12**, 207–220.
15. Berger, S.P., Winhusen, T.M., Somoza, E.C., Harrer, J.M., Mezinskis, J.P., Leiderman, D.B., Montgomery, M.A., Goldsmith, R.J., Bloch, D.A., Singal, B.M., and Elkashef, A. (2005) A medication screening trial evaluation of reserpine, gabapentin and lamotrigine pharmacotherapy of cocaine dependence. *Addiction* **100** (Suppl 1), 58–67.
16. Winhusen, T., Somoza, E., Sarid-Segal, O., Goldsmith, R.J., Harrer, J.M., Coleman, F.S., Kahn, R., Osman, S., Mezinskis, J., Li, S.H., Lewis, D., Afshar, M., Ciraulo, D.A., Horn, P., Montgomery, M.A., and Elkashef, A. (2007) A double-blind, placebo-controlled trial of reserpine for the treatment of cocaine dependence. *Drug Alcohol Depend.* **91**, 205–212.
17. Leiderman, D.B., Shoptaw, S., Montgomery, A., Bloch, D.A., Elkashef, A., LoCastro, J., and Vocci, F. (2005) Cocaine Rapid Efficacy Screening Trial (CREST): a paradigm for the controlled evaluation of candidate medications for cocaine dependence. *Addiction* **100** (Suppl 1), 1–11.
18. Hayden, M.R., Leavitt, B.R., Yasothan, U., and Kirkpatrick, P. (2009) Tetrabenazine. *Nature Rev.* **8**, 17–18.
19. Kenney, C., and Jankovic, J. (2006) Tetrabenazine in the treatment of hyperkinetic movement disorders. *Expert. Rev. Neurother.* **6**, 7–17.
20. Cordeiro, M.L., Gundersen, C.B., and Umbach, J.A. (2002) Lithium ions modulate the expression of *VMAT2* in rat brain. *Brain Res.* **953**, 189–194.
21. Cordeiro, M.L., Gundersen, C.B., and Umbach, J.A. (2004) Convergent effects of lithium and valproate on the expression of proteins associated with large dense core vesicles in NGF-differentiated PC12 cells. *Neuropsychopharmacology* **29**, 39–44.
22. Cordeiro, M.L., Umbach, J.A., and Gundersen, C.B. (2000) Lithium ions Up-regulate mRNAs encoding dense-core vesicle proteins in nerve growth factor-differentiated PC12 cells. *J Neurochem* **75**, 2622–2625.

23. Manji, H.K. and Lenox, R.H. (2000) Signaling: cellular insights into the pathophysiology of bipolar disorder. *Biol. Psychiatry* **48**, 518–530.
24. Tamminga, C.A. and Holcomb, H.H. (2005) Phenotype of schizophrenia: a review and formulation. *Mol. Psychiatry* **10**, 27–39.
25. Zheng, G., Dwoskin, L.P., and Crooks, P.A. (2006) Vesicular monoamine transporter 2: role as a novel target for drug development. *Aaps J. 8*, E682–692.
26. Zubieta, J.K., Huguelet, P., Ohl, L.E., Koeppe, R.A., Kilbourn, M.R., Carr, J.M., Giordani, B.J., and Frey, K.A. (2000) High vesicular monoamine transporter binding in asymptomatic bipolar I disorder: sex differences and cognitive correlates. *Am. J. Psychiatry* **157**, 1619–1628.
27. Zubieta, J.K., Taylor, S.F., Huguelet, P., Koeppe, R.A., Kilbourn, M.R., and Frey, K.A. (2001) Vesicular monoamine transporter concentrations in bipolar disorder type I, schizophrenia, and healthy subjects. *Biol. Psychiatry* **49**, 110–116.
28. Lin, Z., Walther, D., Yu, X.-Y., Li, S., Drgon, T., and Uhl, G.R. (2005) SLC18A2 promoter haplotypes and identification of a novel protective factor against alcoholism. *Hum. Mol. Genet.* **14**, 1393–1404.
29. Little, K.Y., Krolewski, D.M., Zhang, L., and Cassin, B.J. (2003) Loss of striatal vesicular monoamine transporter protein (*VMAT2*) in human cocaine users. *Am. J. Psychiatry* **160**, 47–55.
30. Wilson, J.M., Levey, A.I., Bergeron, C., Kalasinsky, K., Ang, L., Peretti, F., Adams, V.I., Smialek, J., Anderson, W.R., Shannak, K., Deck, J., Niznik, H.B., and Kish, S.J. (1996) Striatal dopamine, dopamine transporter, and vesicular monoamine transporter in chronic cocaine users. *Ann. Neurol.* **40**, 428–439.
31. DaSilva, J.N., Kilbourn, M.R., and Mangner, T.J. (1993) Synthesis of a [11C]methoxy derivative of alpha-dihydrotetrabenazine: a radioligand for studying the vesicular monoamine transporter. *Appl. Radiat. Isot.* **44**, 1487–1489.
32. DaSilva, J.N., Carey, J.E., Sherman, P.S., Pisani, T.J., and Kilbourn, M.R. (1994) Characterization of [11C]tetrabenazine as an in vivo radioligand for the vesicular monoamine transporter. *Nucl. Med. Biol.* **21**, 151–156.
33. Lohoff, F.W., Lautenschlager, M., Mohr, J., Ferraro, T.N., Sander, T., and Gallinat, J. (2008) Association between variation in the vesicular monoamine transporter 1 gene on chromosome 8p and anxiety-related personality traits. *Neurosci. Lett.* **434**, 41–45.
34. Lohoff, F.W., Weller, A.E., Bloch, P.J., Buono, R.J., Doyle, G.A., Ferraro, T.N., and Berrettini, W.H. (2008) Association between polymorphisms in the vesicular monoamine transporter 1 gene (*VMAT1*/SLC18A1) on chromosome 8p and schizophrenia. *Neuropsychobiology* **57**, 55–60.
35. Roghani, A., Welch, C., Xia, Y., Liu, Y., Peter, D., Finn, J.P., Edwards, R.H., and Lusis, A.J. (1996) Assignment of the mouse vesicular monoamine transporter genes, Slc18a1 and Slc18a2, to chromosomes 8 and 19 by linkage analysis. *Mamm. Genome 7*, 393–394.
36. Berrettini, W. (2003) Evidence for shared susceptibility in bipolar disorder and schizophrenia. *Am. J. Med. Genet.* **123C**, 59–64.
37. Berrettini, W. (2004) Bipolar disorder and schizophrenia: convergent molecular data. *Neuromolecular Med.* **5**, 109–117.
38. Park, N., Juo, S.H., Cheng, R., Liu, J., Loth, J.E., Lilliston, B., Nee, J., Grunn, A., Kanyas, K., Lerer, B., Endicott, J., Gilliam, T.C., and Baron, M. (2004) Linkage analysis of psychosis in bipolar pedigrees suggests novel putative loci for bipolar disorder and shared susceptibility with schizophrenia. *Mol. Psychiatry* **9**, 1091–1099.
39. Cheng, R., Juo, S.H., Loth, J.E., Nee, J., Iossifov, I., Blumenthal, R., Sharpe, L., Kanyas, K., Lerer, B., Lilliston, B., Smith, M., Trautman, K., Gilliam, T.C., Endicott, J., and Baron, M. (2006) Genome-wide linkage scan in a large bipolar disorder sample from the National Institute of Mental Health genetics initiative suggests putative loci for bipolar disorder, psychosis, suicide, and panic disorder. *Mol. Psychiatry* **11**, 252–260.
40. Craddock, N. and Jones, I. (1999) Genetics of bipolar disorder. *J. Med. Genet.* **36**, 585–594.
41. Smoller, J.W. and Finn, C.T. (2003) Family, twin, and adoption studies of bipolar disorder. *Am. J. Med. Genet. C. Semin. Med. Genet.* **123**, 48–58.
42. Moller, H.J. (2003) Bipolar disorder and schizophrenia: distinct illnesses or a continuum? *J. Clin. Psychiatry* **64** (Suppl 6), 23–27; discussion 28.
43. Sikich, L. (2008) Efficacy of atypical antipsychotics in early-onset schizophrenia and other psychotic disorders. *J. Clin. Psychiatry* **69** (Suppl 4), 21–25.
44. Bly, M. (2005) Mutation in the vesicular monoamine gene, SLC18A1, associated with schizophrenia. *Schizophr. Res.* **78**, 337–338.

45. Richards, M., Iijima, Y., Kondo, H., Shizuno, T., Hori, H., Arima, K., Saitoh, O., and Kunugi, H. (2006) Association study of the vesicular monoamine transporter 1 (*VMAT1*) gene with schizophrenia in a Japanese population. *Behav. Brain Funct.* **2**, 39.
46. Chen, S.F., Chen, C.H., Chen, J.Y., Wang, Y.C., Lai, I.C., Liou, Y.J., and Liao, D.L. (2007) Support for association of the A277C single nucleotide polymorphism in human vesicular monoamine transporter 1 gene with schizophrenia. *Schizophr. Res.* **90**, 363–365.
47. Sievert, M.K., and Ruoho, A.E. (1997) Peptide mapping of the [125I]Iodoazidoketanserin and [125I]2-N-[(3′-iodo-4′-azidophenyl)propionyl]tetrabenazine binding sites for the synaptic vesicle monoamine transporter. *J. Biol. Chem.* **272**, 26049–26055.
48. Senkowski, D., Linden, M., Zubragel, D., Bar, T., and Gallinat, J. (2003) Evidence for disturbed cortical signal processing and altered serotonergic neurotransmission in generalized anxiety disorder. *Biol. Psychiatry* **53**, 304–314.
49. Charney, D.S. and Deutch, A. (1996) A functional neuroanatomy of anxiety and fear: implications for the pathophysiology and treatment of anxiety disorders. *Crit. Rev. Neurobiol.* **10**, 419–446.
50. Handley, S.L. (1995) 5-Hydroxytryptamine pathways in anxiety and its treatment. *Pharmacol. & Ther.* **66**, 103–148.

Chapter 10

Equilibrium Binding and Transport by Vesicular Acetylcholine Transporter

Parul Khare, Aubrey R. White, Anuprao Mulakaluri, and Stanley M. Parsons

Abstract

The method describes production and the selection of neurosecretory PC12A123.7 cells stably transfected with human vesicular acetylcholine transporter (hVAChT). Transfected cells provide postnuclear supernatant used to characterize equilibrium binding of the neurotransmitter acetylcholine (ACh), the pH dependence for transport of ACh, and the rate behavior for dissociation of the allosteric, high-affinity inhibitor vesamicol. Retention of radiolabeled ACh or vesamicol, mediated by hVAChT in synaptic-like microvesicles of postnuclear supernatant, is measured using filter assays. The procedure for regression analysis of data also is described.

Key words: Vesicular acetylcholine transporter, transport, acetylcholine, vesamicol, VAChT, transfectant, PC12A123.7 cells, blasticidin S, major facilitator superfamily.

1. Introduction

This chapter describes expression and characterization of human vesicular acetylcholine transporter (hVAChT) by biochemical means. VAChT in cholinergic nerve terminals stores acetylcholine (ACh) in synaptic vesicles by exchanging vesicular protons for cytoplasmic ACh (1). Vacuolar ATPase supplies the required protons to vesicular lumen. Rand and his colleagues cloned the gene encoding VAChT in 1993 (2). VAChT is a member of the major facilitator superfamily of secondary active transporters and almost certainly possesses 12 transmembrane helices (3–6). Because most mammalian proteins integral to membranes cannot be isolated

Q. Yan (ed.), *Membrane Transporters in Drug Discovery and Development*, Methods in Molecular Biology 637,
DOI 10.1007/978-1-60761-700-6_10, © Springer Science+Business Media, LLC 2010

from a natural source, they often are studied after expression in a heterologous host.

Cloned VAChT expresses well in PC12$^{A123.7}$ cells, which are derived from the PC12 cell line that commonly is used to study neurosecretory phenomena (7). The cells arose from rat. They grow rapidly and, unlike the parental PC12 cells, do not express endogenous VAChT. The current chapter updates a previous publication in this series that utilized transient transfection to express rat VAChT in PC12$^{A123.7}$ cells in order to obtain the saturation curves for ACh transport and binding of the allosteric, inhibitory compound vesamicol (8). Here we describe procedures used to produce and select clonal lines of PC12$^{A123.7}$ cells that stably express high levels of human VAChT (hVAChT, **Sections 3.4** and **3.6**). We detail procedures used to grow biochemical quantities of the clones (**Section 3.2**) and to prepare postnuclear supernatant containing synaptic-like microvesicles (**Section 3.5**). We also describe experiments to determine the rate of dissociation of vesamicol from its complex with VAChT (**Section 3.8**), the pH dependence of ACh transport (**Section 3.9**), and the binding of ACh to VAChT at equilibrium (**Section 3.10**). Quantitative analysis of the resulting data by regression is described (**Section 3.11**), and typical results are interpreted and discussed (**Section 3.12**).

2. Materials

2.1. Twenty-Channel Filter Manifold

Construction of a vacuum filtration manifold supporting 20 filters of 1.3-cm diameter is detailed in the earlier version of this chapter (chapter 8). Twenty stopcocks apply vacuum to each filtration channel individually. Hoefer, Inc., manufactures and sells a similar manifold supporting 10 filters of 2.5-cm diameter (http://www.hoeferinc.com/product.asp?ProdID=224). However, 2.5-cm filters do not lie flat on the bottom of standard liquid scintillation spectrometry vials, making our custom-built manifold, which supports a larger number of smaller diameter filters, superior for the current application.

2.2. Growth of PC12$^{A123.7}$ Cells

You should buy tissue culture grade reagents. Prepare and use growth media, including highly purified water, under sterile conditions in a laminar flow cabinet.

1. Laminar flow cabinet.
2. CO_2 incubator.

3. Waste trap: vacuum tubing, one-hole rubber stopper, and 4-L vacuum flask. Add bleach to collected waste before disposal.
4. 2-, 5-, 10-, 25-, and 50-mL sterile, cotton-plugged serological pipettes.
5. Disposable bottle-top filters with 0.10-μm polyethersulfone membrane.
6. T-25 and T-175 flasks: treated polystyrene cell culture flasks with vent cap (0.2-μm filter) and canted neck. T-25 has 25 $cm^2$and T-175 has 175 cm^2 of growth surface, respectively.
7. Fetal bovine serum.
8. Horse serum.
9. Pen/Strep (10,000 units penicillin/mL and 10,000 μg streptomycin/mL).
10. Incomplete medium (1 L): Dissolve 15.6 g D-MEM/F-12 nutrient mixture and 1.2 g $NaHCO_3$ in 900 mL of distilled water. Adjust the pH to 7.4 with 1 M NaOH, bring to 1 L with water, filter sterilize, and store at 4°C until used.
11. Complete medium (1 L): Mix 840 mL of incomplete medium with 50 mL fetal bovine serum (*see* item 7), 100 mL horse serum (*see* item 8), and 10 mL Pen/Strep (*see* item 9). Store at 4°C and use within 1 week.
12. Blasticidin S. This nucleoside antibiotic is supplied in solution (*see* **Note 1**).
13. Complete medium with blasticidin: One liter of complete medium (*see* item 11) is mixed with 1 mL of blasticidin S.
14. Freezing medium (50 mL): To 34.5 mL of incomplete medium (*see* item 10) add 5 mL DMSO, 10 mL fetal bovine serum, 50 μL of blasticidin S, and 500 μL of Pen/Strep.
15. Trypsin–EDTA solution: 2.5 g/L of trypsin, 0.38 g/L of EDTA tetrasodium salt in Hanks' balanced salt solution without $CaCl_2$, $MgCl_2$ hexahydrate, and $MgSO_4$ heptahydrate. Store at 4°C.
16. 15- and 50-mL conical polypropylene tubes with screw caps, sterile.

2.3. Construction of hVAChT Transfection Vector

1. Gateway® entry vector pENTR™221/hVAChT (Invitrogen). The product comes as a glycerol stock solution of transformed *Escherichia coli*.
2. Destination vector pcDNA6.2/V5-Dest (Invitrogen).
3. Gateway LR Clonase II enzyme kit (Invitrogen). Components of the kit include separate vials containing

Gateway® LR Clonase™ II enzyme, proteinase K solution (2 μg/μL), and pENTR™-gus-positive control (50 ng/μL).

4. Ampicillin stock solution (100 mg/mL). Place 1 g ampicillin, sodium salt in a 15-mL conical polypropylene tube (*see* **Section 2.4**, item 12). Add about 8 mL of water, vortex to dissolve completely, and bring to 10 mL with more water. Aliquot 1-mL portions into 1.5-mL microcentrifuge tubes and cap. Store at –20°C.
5. LB medium (1 L): Add to 1 L of water to 25 g of powdered LB medium in a media bottle and dissolve. Autoclave at 121°C. Allow to come to RT. To make LB with ampicillin (100 μg ampicillin/mL), ampicillin stock solution (item 4) is diluted 1:1,000 into LB medium. The media can be stored at 4°C for up to a month.
6. LB ampicillin plate. Store at 4°C.
7. Library Efficiency DH5α competent *E. coli* cells (Invitrogen).
8. UV spectrometer.
9. V5-reverse primer.
10. T7-forward primer.
11. Plasmid miniprep kit.
12. TE buffer: 10 mM Tris–HCl, 1 mM EDTA, pH 8.0.
13. Super Optimal broth with Catabolite repression (S.O.C.; Invitrogen). Growth of *E. coli* in S.O.C. medium results in higher transformation efficiencies.
14. LB with kanamycin sulfate (50 μg/mL). Dissolve 5 mg kanamycin sulfate in 100 mL of LB medium (*see* item 5) in a medium bottle. The medium can be stored at 4°C for up to a month.
15. 75% (w/v) glycerol solution in water, sterilized.
16. 17- × 100-mm polypropylene tube.

2.4. Selection, Screening, and Preservation of Stable Transfectants

1. $PC12^{A123.7}$ cells: One 10-cm poly-D-lysine-coated plate containing confluent cells or a freeze of (1.5 – 6) × 10^6 cells.
2. TE buffer: 10 mM Tris–HCl, 1 mM EDTA, pH 8.0.
3. pcDNA6.2/hVAChT (*see* **Section 3.3**, step 16) dissolved in TE buffer at a concentration of 0.2–0.3 mg DNA/mL.
4. Lipofectamine 2000 transfection reagent.
5. Opti-Mem I reduced serum medium, liquid (Invitrogen).
6. 2-, 5-, 10-, 25-, and 50-mL sterile, cotton-plugged serological pipettes.

7. Complete medium with blasticidin (*see* **Section 2.2**, item 13).
8. Transfection medium (50 mL): To 42 mL of incomplete medium (*see* **Section 2.2**, item 10) add 5 mL of horse serum (*see* **Section 2.2**, item 8), 2.5 mL fetal bovine serum (*see* **Section 2.2**, item 7), and 500 μL of nonessential amino acids.
9. Cryovials.
10. Liquid nitrogen.
11. Trypsin–EDTA solution (*see* **Section 2.2**, item 15).
12. 15- and 50-mL conical polypropylene tubes with screw caps, sterile.
13. 10-cm poly-D-lysine-coated cell culture plates.
14. 17- × 100-mm polypropylene tube.
15. S.O.C. medium (Invitrogen).
16. Blasticidin S.
17. 1.5-mL microcentrifuge tubes.
18. 24-well plates, sterile.
19. Phosphate-buffered saline (PBS), pH 7.4: 1.06 mM KH_2PO_4, 155.17 mM NaCl, and 2.96 mM Na_2HPO_4. Store at 4°C.
20. Pen/Strep (10,000 units penicillin/mL and 10,000 μg streptomycin/mL).

2.5. Preparation of Postnuclear Supernatant

1. Protease inhibitor cocktail: Dissolve one tablet (Roche) in 2 mL of distilled water. This makes a 25-fold stock solution. Aliquot in 50-μL portions. Store at –20°C (*see* **Note 2**).
2. Phenylmethylsulfonyl fluoride stock solution (PMSF). Dissolve 17.4 mg of PMSF in 1 mL of isopropanol to prepare 100 mM PMSF. Store at room temperature.
3. Paraoxon stock solution. Paraoxon is a potent inhibitor of acetylcholine esterase. Do not get on skin. Mix 123 μL of isopropanol with 1 g of paraoxon oil to lower the viscosity of the oil and make a 3.3 M solution. Of this, 30.3 μL is further diluted to 1 mL with isopropanol to prepare 100 mM paraoxon (*see* **Note 3**). Store at 4°C.
4. Homogenization buffer (1 mL): Dissolve 10.95 g of sucrose and 0.238 g of *N*-2-hydroxyethylpiperazine-*N*′-2-ethanesulfonic acid (HEPES) in 90 mL of water. Adjust the pH to 7.4 with 1 M KOH and bring to 96.0 mL to prepare a pre-stock solution of 0.333 M sucrose and 10.42 mM HEPES pH 7.4. This can be stored in the dark at 4°C. Two microliters of 100 mM PMSF (*see* item 2)

and 1 μL of 100 mM paraoxon (*see* item 3), each in isopropanol, are evaporated to dryness under gentle nitrogen stream in a 1.5-mL microcentrifuge tube. Pre-stock solution (960 μL) and 40 μL protease inhibitor cocktail (*see* item 1) are added to the microcentrifuge tube and mixed shortly before use. Dissolve 0.2 mg of 1,4-dithio-DL-threitol (DTT) immediately before use to make fresh 0.320 M sucrose and 10 mM HEPES, pH 7.4, containing broad-spectrum protease inhibitors and DTT.

5. Motor-driven stirrer, 5-mL Potter–Elvehjem (Delrin AF-glass) homogenizer, with a pestle clearance of 0.100–0.115 mm.
6. Disposable transfer pipettes.
7. 1.5- and 2-mL microcentrifuge tubes.
8. 0.4% trypan blue solution.
9. Liquid nitrogen and tongs.
10. Hemocytometer.
11. Set of positive displacement micropipettes (e.g., Gilson Microman; 10–250 μL volumes).
12. 15- and 50-mL conical polypropylene tubes with screw caps, sterile.
13. Phosphate-buffered saline (PBS), pH 7.4: 1.06 mM KH_2PO_4, 155.17 mM NaCl, and 2.96 mM Na_2HPO_4. Store at 4°C.

2.6. Western Blot of hVAChT

1. 1° antibody: Goat anti-VAChT N-19 antibody (Santa Cruz Biotech).
2. 2° antibody: Donkey anti-goat IgG HRP (Santa Cruz Biotech).
3. SDS–PAGE precast gel, 8 × 7.3 cm, 12% resolving gel, 4% stacking gel, 15 wells (e.g., Bio-Rad; Ready Gel precast gel).
4. Mini-Protean 3 Cell (Bio-Rad).
5. PVDF membrane (8.3 × 7.3 cm, 0.45 μm pore size) with filter paper.
6. 8- × 11-cm fiber pads.
7. Electrophoretic transfer cell.
8. Supersignal chemiluminescent substrate (Pierce). Combine 1 mL each of the two reagents in the kit immediately before use.
9. Transfer buffer (4 L): 25 mM Tris Base, 192 mM glycine, and 20% w/v methanol in water. Combine 12.1 g Tris Base, 57.6 g glycine, and 800 mL methanol. Add water

to 3.5 L. Adjust to pH 8.3 with HCl. Bring to 4 L with water. Store at room temperature.

10. TBS/Tween 20 (TBST, 10-fold, 1 L): 250 mM Tris base, 1.4 M NaCl, 0.5% w/v Tween 20 in water. Dissolve 30.29 g Tris base, 81.89 g NaCl, and 5 mL Tween 20 in 900 mL water. Adjust to pH 7.4 with HCl. Bring to volume. Mix 100 mL with 900 mL water for use (1X TBST).
11. Sample loading buffer (1 mL): Mix 633 μL homogenization buffer (*see* **Section 2.5**, item 4) with 333 μL 3-fold SDS sample buffer and 34 μL 30-fold DTT (both from New England Biolabs) shortly before use.
12. Blocking solution: 2.5% (w/v) nonfat dry milk is dissolved in 1X TBST (*see* item 10) and made 0.05% (w/v) in thimerosal.
13. Molecular weight markers.
14. Colored molecular weight markers.
15. Chemiluminescence station (e.g., UVP-EPI Chemi Darkroom Apparatus with Lab Works analysis software).

2.7. Preparation of Filters

1. Glass microfiber filter circles: GF/F 1.3-cm diameter.
2. 0.25% (w/v) polyethylenimine: Dissolve 2.5 g of 50% (w/v) polyethylenimine aqueous solution in water and bring to 500 mL.
3. 186-mm diameter Büchner filter funnel and 4 L vacuum flask with side arm.

2.8. [^{3}H]Vesamicol Off-Rate Measurements

1. 10 mM (±)-vesamicol HCl stock solution in ethanol (10 mL): Dissolve 29.5 mg (±)-vesamicol HCl in 10 mL ethanol. Store in an upright, 15 mL capped conical polypropylene tube (*see* item 12) at –20°C.
2. 200 μM (±)-vesamicol HCl in ethanol (5 mL): Transfer 100 μL of (±)-vesamicol HCl vesamicol stock solution in ethanol (*see* item 1) to a 15-mL conical polypropylene tube and mix it with 4.9 mL of ethanol. Cap and store upright at –20°C.
3. Uptake binding buffer (UBB, 1 L): 110 mM L-(+) tartaric acid dipotassium salt, and 20 mM HEPES, pH 7.4. Dissolve 25 g of L-(+) tartaric acid dipotassium salt and 4.8 g of HEPES in 800 mL water. Adjust to pH 7.4 with 1.0 M KOH and dilute to 1 L. Store closed at 4°C. Add 15.4 mg of DTT per 100 mL shortly before each use.
4. 200 μM (±)-vesamicol HCl in UBB (5 mL): Transfer 100 μL of (±)-vesamicol HCl vesamicol stock solution in ethanol to a 15-mL conical polypropylene tube, gently evaporate the ethanol under a stream of nitrogen, add

5 mL of UBB (*see* item 3), and dissolve the (±)-vesamicol HCl. Cap and store at –20°C.

5. UBB with 5 μM (±)-vesamicol HCl (UBBw, 200 mL): Transfer 100 μL of (±)-vesamicol HCl stock solution in ethanol to a clean media bottle, gently evaporate the ethanol under a stream of nitrogen, add 200 mL of UBB, fully dissolve the (±)-vesamicol HCl, and close.
6. [^{3}H]Vesamicol in ethanol (V**), 1 mCi/mL, 15–30 Ci/mmol, 70–30 μM (Perkin-Elmer). Store at –20°C.
7. About 1 μM [^{3}H]vesamicol in UBB (1 μMV*): Transfer 15 μL of [^{3}H]vesamicol in ethanol (V**) to a 1.5-mL microcentrifuge tube. Carefully evaporate the ethanol to dryness under a stream of nitrogen in a chemical fume hood. Add 1 mL of UBB. Allow to equilibrate with occasional mixing for 30 min, after which transfer the solution to a fresh microcentrifuge tube. Cap and place on ice.
8. Disposable transfer pipettes.
9. 1.5-mL microcentrifuge tubes.
10. Set of positive displacement micropipettes (e.g., Gilson Microman; 10–250 μL volumes).
11. Liquid scintillation vials and cocktail.
12. 15- and 50-mL conical polypropylene tubes with screw caps, sterile.

2.9. pH Dependence of ACh Transport

1. 0.5 M dipotassium tartrate (50 mL): Dissolve 5.66 g of L-(+) tartaric acid dipotassium salt in water, bring to 50 mL in a 50 mL conical polypropylene tube (*see* item 15), and cap. Store at 4°C.
2. 0.5 M AMPSO (50 mL): Dissolve 5.69 g 3-[(1,1-dimethyl-2-hydroxyethyl)amino]-2-hydroxy propane sulfonic acid (AMPSO) in water, bring to 50 mL in a 50-mL conical polypropylene tube, and cap. Store at 4°C.
3. 0.5 M HEPES (50 mL): Dissolve 5.96 g of *N*-2-hydroxyethylpiperazine-*N*′-2-ethanesulfonic acid (HEPES) in water, bring to 50 mL in a 50-mL conical polypropylene tube, and cap. Store at 4°C.
4. 0.5 M MES (50 mL): Dissolve 5.33 g of 2-(*N*-morpholino)ethanesulfonic acid in water, bring to 50 mL in a 50-mL conical polypropylene tube, and cap. Store at 4°C.
5. 0.8 mM (±)-vesamicol HCl stock solution in ethanol (10 mL): Dissolve 2.3 mg of (±)-vesamicol HCl in ethanol, bring to 10 mL in a 15-mL conical polypropylene tube (*see* item 15), and cap. Store upright at –20°C.

6. SUBB (specific uptake binding buffer, 10 mL): 119 mM dipotassium tartrate, 24 mM MgATP, 20 mM $MgCl_2$, and 0.1 mM paraoxon. Transfer 10 μL paraoxon stock solution (*see* **Section 2.5**, item 3) to a 15-mL conical polypropylene tube. Evaporate the isopropanol to dryness under gentle nitrogen stream. Add 2.38 mL of 0.5 M dipotassium tartrate (*see* item 1), 6.5 mL water, and 100 μL of 2.0 M magnesium chloride hexahydrate. Dissolve 122 mg of magnesium ATP. Using 1 M KOH, carefully bring pH to 7.0. Make up to 10 mL with water, mix, and cap. Prepare fresh and keep on ice.
7. UBB-low (10 mL, pH 4.5): 99.6 mM each of AMPSO, HEPES, and MES, 4 mM DTT, and 0.1 mM paraoxon. Transfer 10 μL of paraoxon stock solution to a 15-mL conical polypropylene tube. Evaporate isopropanol under gentle nitrogen stream. Mix 6.1 mg of DTT with 2 mL each of 0.5 M AMPSO, 0.5 M HEPES, and 0.5 M MES (*see* items 2–4) in the tube. Bring to 10 mL with water and mix. The pH should be approximately 4.5. Make fresh, cap, and keep on ice.
8. UBB-high (10 mL, pH 10.5): 53.8 mM each of AMPSO, HEPES, and MES, 4 mM DTT, and 0.1 mM paraoxon. Transfer 10 μL paraoxon to a 15-mL conical polypropylene tube. Evaporate isopropanol to dryness under gentle nitrogen stream. Mix 6.1 mg DTT with 1.08 mL each of 0.5 M AMPSO, 0.5 M HEPES, and 0.5 M MES (*see* items 2–4) in the tube. Bring to 8 mL with water. Adjust pH carefully to 10.5 using 1 M KOH. Bring to 10 mL with water and cap. Make immediately before use and keep on ice.
9. UBB with 5 μM (±)-vesamicol HCl (UBBw, 200 mL, *see* **Section 2.8**, item 5). Keep on ice.
10. UBB with 80 μM (±)-vesamicol HCl (80 μMV): Transfer 5 μL of 0.8 mM (±)-vesamicol HCl stock solution in ethanol (*see* item 5) to a 1.5-mL microcentrifuge tube. Evaporate under gentle nitrogen stream. Add 50 μL of UBB (*see* **Section 2.8**, item 3), dissolve the (±)-vesamicol HCl, and cap. Keep on ice.
11. [^{3}H]ACh (ACh**): Dissolve 1 mCi of [^{3}H]ACh iodide powder (50–100 mCi/mmol, Perkin-Elmer) in 1 mL of ethanol, transfer to a 1.5-mL microcentrifuge tube, and cap. Store upright at –20°C.
12. 1% (w/v) SDS solution (1 L): 10 g SDS in 1 L of water. Store capped at room temperature.
13. Scintillation vials and scintillation cocktail (e.g., Fisher Scientific and Research Products Intl).
14. 1.5-mL microcentrifuge tubes.

15. 15- and 50-mL conical polypropylene tubes with screw caps, sterile.
16. Disposable transfer pipettes.
17. Set of positive displacement micropipettes (e.g., Gilson Microman; 10–250 μL volumes).
18. Calomel semimicro pH probe (e.g., Fisher).
19. Homogenization buffer (*see* **Section 2.5**, item 4).

2.10. Binding of ACh at Equilibrium

1. UBB (500 mL): *See* **Section 2.8**, item 3.
2. UBB with paraoxon (UBBp, 50 mL): Transfer 50 μL of 100 mM paraoxon stock solution (*see* **Section 2.5**, item 3) into a 50-mL conical polypropylene tube (*see* item 14). Evaporate the isopropanol to dryness under a stream of gentle nitrogen. Add 50 mL of UBB. Mix thoroughly and cap. Prepare fresh and keep on ice.
3. ACh chloride: 150 mg per vial (premeasured powder, e.g., Sigma-Aldrich). Store at –20°C.
4. 2 M ACh (2 M ACh, 825 μL): Bring two vials of ACh chloride (*see* item 3) to room temperature (*see* **Note 4**). Open one vial. Using 500 μL of room temperature UBBp (*see* item 2), quantitatively rinse the inside of the vial top into the vial bottom and dissolve all ACh. Recover all solution out of the vial bottom using an adjustable pipette, and use the solution to quantitatively rinse the inside of the second vial top into the second vial bottom. Pipette up-and-down until a clear solution forms. Adjust to pH 7.4 slowly with agitation using 0.25 M KOH. Use several 25-μL portions of additional UBBp to rinse residual ACh off surfaces, such as the pH electrode, into the second vial. Mix the combined solutions and determine the total volume by drawing the solution into an adjustable pipette of 1,000-μL capacity. Expel the air residing in the pipette tip by dialing down the pipette plunger until the solution begins to emerge from the tip opening. Dispense the ACh solution into a 1.5-mL microcentrifuge tube (*see* item 9) labeled 2 M ACh. Add a volume of UBBp equal to the difference between 825 μL and the volume showing on the pipette dial. Mix well and cap. Make fresh and keep on ice.
5. 1 M ACh (1 M ACh, 400 μL): To a 1.5-mL microcentrifuge tube labeled 1 M ACh, add 200 μL UBBp and 200 μL 2 M ACh (*see* item 4), mix, and cap. Make fresh and keep on ice.
6. 0.1 M ACh (0.1 M ACh, 150 μL): To a 1.5-mL microcentrifuge tube labeled 0.1 M ACh, add 135 μL UBBp and 15 μL 1 M ACh (*see* item 5), mix, and cap. Make fresh and keep on ice.

7. $[^3H]$Vesamicol in ethanol (V**), 1 mCi/mL, 15–30 Ci/mmol, 70–30 μM (e.g., Perkin-Elmer). Store at –20°C.
8. About 1 μM $[^3H]$vesamicol in UBBp (1 μMV*p): Transfer 1.5 μL of $[^3H]$vesamicol in ethanol (V**) to a 1.5-mL microcentrifuge tube. Carefully evaporate the ethanol to dryness under a stream of nitrogen in a chemical fume hood. Add 100 μL of UBBp. Allow to incubate with occasional mixing for 30 min, after which transfer the solution to a fresh 1.5 mL microcentrifuge tube labeled 1 μMV*p. Cap and place on ice.
9. 1.5-mL microcentrifuge tubes.
10. Disposable transfer pipettes.
11. Scintillation vials and scintillation cocktail (e.g., Fisher Scientific and Research Products Intl.).
12. Set of positive-displacement micropipettes (e.g., Gilson Microman; 10–250 μL volumes).
13. 15- and 50-mL conical polypropylene tubes with screw caps, sterile.

3. Methods

3.1. Twenty-Channel Filter Manifold

A house vacuum line supplies the best vacuum, as large capacity for air handling is required. High vacuum is not desirable, as it will boil the filtrate and aerosolize radioactivity.

3.2. Growth of $PC12^{A123.7}$ Cells and Stable Transfectants

$PC12^{A123.7}$ cells are maintained in 10% CO_2 at 37°C in complete medium without blasticidin (*see* **Section 2.2**, item 11). Cells transfected with plasmid pcDNA6.2/hVAChT (*see* **Section 3.3**, step 15) are maintained in complete medium with blasticidin (*see* **Section 2.2**, item 13). Cultures should be split at confluency every 72–96 h (*see* **Note 5**). All media are brought to 37°C before addition to cells, unless otherwise indicated.

3.2.1. Growth of Cells from a Freeze

1. Place a cryovial containing frozen cells (e.g., *see* **Section 3.2.4**, step 2) in a 37°C water bath until thawed.
2. Using a pipette, transfer the cells to a 15-mL conical polypropylene tube (*see* **Section 2.2**, item 16) containing 9 mL of complete medium and cap. Complete medium should not contain or should contain blasticidin S (*see* **Section 2.2**, items 11 and 13, respectively) as appropriate through all steps. Centrifuge at 800×*g* for 10 min. Aspirate medium to waste and resuspend in 1 mL of appropriate complete medium. This step removes DMSO used to make freezes (*see* **Section 2.2**, item 14).

3. Transfer the washed, resuspended cells to a T-25 flask (*see* **Section 2.2**, item 6) containing 4 mL appropriate complete medium and cap. Incubate until cells are confluent.

3.2.2. Expansion of Cells

1. To split confluent cells in a T-25 flask, use a 2-mL pipette attached to the vacuum tubing leading to a waste trap (*see* **Section 2.2**, item 3) and aspirate the growth medium to waste (*see* **Note 6**). Living cells remain attached to the surface of the flask.
2. Add 1 mL of trypsin/EDTA (*see* **Section 2.2**, item 15). Gently rock the flask by hand to coat adherent cells with trypsin. Lay the flask horizontally with the bevel facing down for 5 min at room temperature (*see* **Note 7**).
3. Add 4 mL of fresh complete medium to the T-25 flask. Resuspend the cells by pipetting the medium up-and-down several times onto the surface of the flask to detach cells still adhering to the plastic.
4. Immediately after resuspending the cells and, before they settle significantly, transfer the suspension to a T-175 flask and add 45 mL of complete medium. Grow the cells until they are confluent.
5. To split confluent cells growing in a T-175 flask, aspirate growth medium to waste (*see* step 1) and add 4 mL of trypsin/EDTA. Gently rock the flask by hand to expose all adherent cells to trypsin. Lay the flask horizontally with the bevel facing down for 5 min at room temperature.
6. Add 16 mL of fresh complete medium to the T-175 flask. Resuspend the cells by pipetting the medium up-and-down several times onto the surface of the flask to detach cells still adhering to the plastic.
7. Before the cells settle significantly, transfer 4 mL of the suspension into each of five new T-175 flasks and add 46 mL of complete medium to each flask. Transfer the flasks to the incubator with the bevel facing downward (*see* **Note 8**).
8. If more cells are required, repeat steps 5–7.

3.2.3. Maintenance of Growing Cells

Not infrequently, it is uncertain how much postnuclear supernatant (PS) will be required to complete investigation of a mutant, yet continued expansion of the clonal line would be wasteful if additional PS proved unnecessary. To facilitate production of additional cells quickly should the need arise, a flask of cells expressing the mutant of interest can be maintained indefinitely at moderate cost. The clone then could be expanded quickly.

1. To maintain cells in a T-25 flask (*see* **Section 2.2**, item 6), proceed according to **Section 3.2.2**, steps 1–3. Transfer 250 μL of a cellular suspension to a new T-25 flask. Add 4.75 mL

of new complete medium (with or without blasticidin S, as appropriate). Incubate at 37°C, refreshing the medium as necessary until the cells are confluent. Repeat.

2. To maintain cells in a T-175 flask, proceed according to **Section 3.2.2**, steps 5 and 6. Transfer 800 μL of cellular suspension to a new T-175 flask. Add 49.2 mL of new complete medium (with or without blasticidin S, as appropriate). Incubate at 37°C, refreshing the medium as necessary until confluent. Repeat.
3. To prepare PS from the cells not used in step 1 or 2, proceed to **Section 3.5**.

3.2.4. Preservation of Cells

1. Release confluent cells from the surface of a T-175 flask by trypsinization (*see* **Section 3.2.2**, steps 5 and 6).
2. Transfer the resuspended cells to a 50-mL conical polypropylene tube and cap. Centrifuge at 800×*g* for 10 min at room temperature. Aspirate the medium to waste and resuspend the cells in 10 mL of fresh freezing medium (*see* **Section 2.2**, item 14). Aliquot 1-mL portions of resuspended cells into each of 10 cryovials, immerse the closed cryovials in liquid nitrogen using tongs, and store in liquid nitrogen.

3.3. Construction of hVAChT Transfection Vector

The open reading frame for hVAChT is flanked by site-specific attachment (*att*) site L in the pENTR221/hVAChT entry vector. The *att*L recognition site matches up with the *att*R recognition site in the pcDNA6.2/V5 destination vector during an LR recombination reaction.

1. Disperse pcDNA6.2/V5-Dest vector (*see* **Section 2.3**, item 2) in 40 μL of TE buffer (*see* **Section 2.3**, item 12), which gives a vector concentration of 150 μg/mL.
2. In 10 mL of LB with kanamycin (*see* **Section 2.3**, item 14), grow *E. coli* transformed with pENTR221/hVAChT (*see* **Section 2.3**, item 1) with vigorous shaking for 15–17 h at 37°C.
3. Extract the plasmid DNA using a standard plasmid miniprep kit (*see* **Section 2.3**, item 11). Follow manufacturer's instructions.
4. Determine the concentration of plasmid DNA by measuring absorbance at 260 nm using an UV spectrometer. The concentration is calculated as DNA (μg/mL) = (A260) × (dilution factor) × (50 μg DNA/mL)/(1 A260 unit) and should be between 200 and 500 μg/mL. Dilute the plasmid with TE to 150 μg/mL.
5. Thaw the components of the LR Clonase II enzyme kit (*see* **Section 2.3**, item 3) on ice and vortex them briefly.

6. Set up the reactions specified in **Table 10.1** by adding the indicated volumes of pENTR221/hVAChT, pcDNA6.2/V5-Dest, pENTR-gus (*see* steps 1–5), and TE buffer to three 1.5-mL microcentrifuge tubes at room temperature. Mix each reaction by pipetting up-and-down.

Table 10.1
Sample preparation for LR recombination reaction[a]

Component	Sample (μL)	Positive control (μL)	Negative control (μL)
pENTR221/hVAChT	1	0	1
pcDNA6.2/V5-Dest	1	1	1
pENTR-gus	0	2	0
TE buffer	6	5	8

[a] *See* **Section 3.3**, step 6. LR Clonase II (2 μL) will be added to the sample and positive control mixtures to bring final volumes to 10 μL in all cases.

7. To each of the sample and positive control reactions, add 2 μL of LR Clonase II enzyme. Add none to the negative control reaction. Mix each reaction by pipetting up-and-down.
8. Incubate reactions at 25°C for 1 h.
9. Add 1 μL of proteinase K solution (*see* **Section 2.3**, item 3) to each reaction and incubate for 10 min at 37°C.
10. Thaw DH5α competent *E. coli* cells (*see* **Section 2.3**, item 7) on ice. Place a 17- × 100-mm polypropylene tube (*see* **Section 2.3**, item 16) on ice. Gently mix the bacteria and aliquot 100 μL of cells into the chilled tube. Refreeze unused bacteria in a dry ice/ethanol bath before returning them to –80°C. Add 1 μL of the LR recombination sample reaction to the bacteria, moving the pipette through them while dispensing. Gently tap the tube to mix contents. Incubate bacteria on ice for 30 min. Heat-shock bacteria 45 s in a 42°C water bath. Place on ice for 2 min. Add 900 μL of room temperature S.O.C. medium (*see* **Section 2.3**, item 13). Shake vigorously for 1 h at 37°C. Spread the bacterial reaction on an LB ampicillin plate (*see* **Section 2.3**, item 6) and incubate at 37°C for 15–17 h.
11. Pick several colonies and grow each in 5 mL of LB with ampicillin (*see* **Section 2.3**, item 5) for 15–17 h at 37°C.
12. Prepare glycerol stocks of all clones by mixing 0.2 mL of 75% (w/v) glycerol (*see* **Section 2.3**, item 15) with 0.8 mL of bacterial culture. The clones are stored in a –80°C freezer.

13. Extract the plasmid DNA from 5 mL cultures of all picked clones (*see* step 11) using a standard plasmid miniprep kit (*see* **Section 2.3**, item 11). Follow manufacturer's instructions. Elute the plasmid in buffer not containing EDTA, which interferes with DNA sequencing protocols.
14. Determine the concentration of plasmid (*see* step 4).
15. Sequence the plasmid from all picked clones with V5-reverse and T7-forward primers (*see* **Section 2.3**, items 9 and 10) using an in-house or a commercial sequencing facility. A clone incorporating hVAChT cDNA (*see* **Note 9**) is called pcDNA6.2/hVAChT.
16. Grow the bacterial clone that contains pcDNA6.2/h-VAChT plasmid from the glycerol stock (*see* step 12). Purify the plasmid (*see* step 13) and quantitate its concentration (*see* step 4). EDTA in TE buffer used to purify plasmid does not interfere with transfection (*see* **Section 3.4**, step 4).
17. The aqueous solution of prepared plasmid can be stored at –20°C.
18. The hVAChT coding sequence in pcDNA6.2/hVAChT can be mutated using the Quik-Change mutagenesis kit (Stratagene; cat. no. 200519) according to the manufacturer's instructions.

3.4. Selection, Screening, and Preservation of Stable Transfectants

1. One day before they are to be transfected, trypsinize confluent $PC12^{A123.7}$ cells grown in complete medium without blasticidin (*see* **Section 2.2**, item 11) in two T-25 flasks (*see* **Section 3.2** for instructions on obtaining confluent $PC12^{A123.7}$ cells in T-25 flasks).
2. Add 4 mL of transfection medium (*see* **Section 2.4**, item 8) to each flask and transfer the now resuspended cells from both T-25 flasks using a 5-mL pipette and a portable hand-held pipettor to one 10-cm poly-D-lysine-coated plate (*see* **Section 2.4**, item 13).
3. Incubate the plate at 37°C in a CO_2 incubator for 24 h, after which the cells should be 75–80% confluent.
4. Dilute 24 μg of pcDNA6.2/hVAChT (*see* **Section 2.4**, item 3) into 1.5 mL of Opti-Mem I (*see* **Section 2.4**, item 5). Mix gently.
5. Mix Lipofectamine 2000 transfection reagent (*see* **Section 2.4**, item 4) gently and dilute 60 μL of it into 1.5 mL of Opti-Mem I. Incubate for 5 min at room temperature.
6. After 5 min, combine the diluted DNA and the diluted Lipofectamine 2000 in a 15-mL conical polypropylene

tube (*see* **Section 2.4**, item 12). Mix gently and incubate for 20 min at room temperature.

7. Add 3 mL of DNA–Lipofectamine mix to the plate containing PC12$^{A123.7}$ cells and transfection medium (*see* step 2). Rock the plate gently back and forth a few times.
8. Incubate the plate at 37°C in a CO_2 incubator for 24 h, during which cells will settle and become adherent.
9. Aspirate the culture medium to waste and add 2 mL of trypsin/EDTA (*see* **Section 2.2**, item 15) to the plate. After 5 min, add 8 mL of fresh transfection medium to the plate.
10. Passage the cells as follows. Using a marking pen, label three 10-cm diameter poly-D-lysine-coated cell culture plates (*see* **Section 2.4**, item 13) 1:20, 1:50, and 1:100. To the plates labeled 1:20, 1:50, and 1:100 add 500, 200, and 100 μL of cells from step 9 and bring total volume in all plates to 10 mL with fresh transfection medium. Gently mix cells and transfection medium by rocking.
11. After 24 h, add 10 μL of blasticidin S (*see* **Section 2.4**, item 16) and 10 μL of Pen/Strep mix (*see* **Section 2.4**, item 20) to each plate and gently mix to begin selection of blasticidin-resistant clones.
12. Replace complete medium with blasticidin (*see* **Section 2.4**, item 7) every 48–72 h.
13. When well-separated colonies appear, usually after 2–3 weeks (*see* **Note 10**), aspirate the growth medium to waste. Gently add 10 mL of PBS (*see* **Section 2.4**, item 19) by pipette so that the PBS covers the colonies without dislodging them, and then aspirate the PBS to waste. Complete the next steps 14–16 quickly to prevent the colonies from becoming dry. If colonies begin to dry, addition and aspiration of PBS can be repeated.
14. Using a fine-tipped marker, mark each chosen colony by putting a dot on the underside of the plate beneath the colony.
15. Clean a microscope with ethanol and move it into the laminar flow hood. Transfer 500 μL of complete medium with blasticidin to each well of a 24-well plate labeled for identification of the VAChT mutant (*see* **Section 2.4**, item 18). Aspirate a colony on one of the 10-cm diameter plates into a 10-μL pipette tip while viewing under the microscope. Transfer the colony by expelling it into one of the wells of the 24-well plate. Label the well number for later identification of the colony.
16. Transfer 23 additional colonies in similar manner.

17. Allow the colonies to grow in the 24-well plate for at least 1 week until cells are close to confluent. Typically, ~ one-fifth of the colonies do not survive this step.
18. Aspirate the growth medium in all wells to waste. Add 500 μL of fresh complete medium with blasticidin to wells containing adherent cells.
19. Resuspend adherent cells in each well by pipetting the medium up-and-down several times onto the surface of the well to detach the cells. Transfer two 240-μL portions of the resuspended cells to two T-25 flasks that each contains 5 mL of fresh complete medium with blasticidin. Add 500 μL of fresh complete medium with blasticidin to the depleted well to expand and maintain the clone.
20. Maintain all clones growing in the 24-well plates until transfectants expressing high levels of hVAChT are identified by Western blot (*see* **Section 3.6**) and preserved. Once each week, aspirate growth medium to waste and add 500 μL of fresh complete medium with blasticidin to each well.
21. When cells in the two T-25 flasks seeded in step 19 are confluent, prepare PS (*see* **Section 3.5**).
22. Analyze each preparation of PS from a new mutant by Western blot (*see* **Section 3.6**).
23. Clones that show an intensely stained band of VAChT in the Western blot are preserved. Cells growing in the 24-well plate (*see* step 20) are resuspended in 500 μL of fresh complete medium with blasticidin by repetitive pipetting up-and-down when they are confluent. Transfer the full volume of resuspended cells to one T-25 flask and grow the cells until confluent. Follow **Section 3.2.2**, steps 1–4, to transfer the cells to a T-175 flask and grow them to confluency. Preserve the cells as directed in **Section 3.2.4**.

3.5. Preparation of Postnuclear Supernatant

The working stocks of hVAChT and its mutants are crude preparations called postnuclear supernatants (PS). Only moderate amounts of PS are required to screen clonal lines by Western blot. Large amounts are required to characterize functional aspects of hVAChT and its mutants. This section describes procedures for preparing moderate and large quantities of PS containing hVAChT.

1. Trypsinize cells from two T-25 flasks (*see* **Section 3.4**, step 21) for Western blot or eight T-175 flasks (*see* **Section 3.2.2**, step 8) for functional characterization.
2. Transfer the cells from two T-25 flasks to a 15-mL conical polypropylene tube (*see* **Section 2.5**, item 12) using a 5-mL pipette and a pipettor, or from eight T-175 flasks to

four 50-mL conical polypropylene tubes (*see* **Section 2.5**, item 12) using a 10-mL pipette and a pipettor. Cap the tubes. From this point onward in this section, except as noted, the steps are conducted at 0–4°C. Centrifuge the tube(s) at 800×*g* for 10 min.

3. Aspirate the culture medium to waste and add 10 mL of ice-cold PBS (*see* **Section 2.5**, item 13) to each cell pellet. Resuspend cells by pipetting up-and-down several times using a 10-mL pipette that has a 1,000-μL disposable tip attached to it (*See* **Note 11**). Combine resuspended cells in four 50-mL conical polypropylene tubes (from eight T-175 flasks) into one 50-mL conical polypropylene tube and cap it.
4. Centrifuge the tube (either 15 or 50 mL) at 800×*g* for 10 min.
5. Aspirate the supernatant to waste and add 10 mL of PBS to the pellet. Resuspend cells as before and cap the tube.
6. Centrifuge the tube at 800×*g* for 10 min.
7. Aspirate the supernatant to waste.
8. Using repetitive pipetting, resuspend the cell pellet in 15-mL conical polypropylene tubes in 100 μL of homogenization buffer (*see* **Section 2.5**, item 4). For 50-mL conical polypropylene tubes, resuspend in 400 μL of homogenization buffer.
9. Transfer the cell suspension from a conical polypropylene tube to a Potter–Elvehjem homogenizer vessel (*see* **Section 2.5**, item 5). Wash the conical polypropylene tube with 100 μL of homogenization buffer and transfer the wash to the homogenizer vessel. With the pestle speed at 800 revolutions per minute, homogenize cells using three up-and-down strokes.
10. Transfer 10 μL of the homogenate to a 1.5-mL microcentrifuge tube. At 23°C, add 90 μL of PBS, and 100 μL of 0.4% trypan blue (*see* **Section 2.5**, item 8). Mix and wait 5 min. Using a hemocytometer, count per square the total number of cells and the number of broken cells, which are stained by trypan blue. About 5–10% of the cells should be unstained (*see* **Notes 12** and **13**). If needed, more homogenization strokes should be done in the cold. When breakage is in the correct range, transfer the homogenate to a 2-mL microcentrifuge tube (*see* **Section 2.5**, item 7) with a pipette and cap.
11. Centrifuge the homogenate at 1,500×*g* for 10 min.
12. Using a 1-mL adjustable pipette, transfer the PS into a 1.5-mL microcentrifuge tube without disturbing the pellet

of cellular debris. Remove a 10-μL portion of PS and determine the concentration of protein using a standard assay at 23°C (*see* **Note 14**).

13. Quick-freeze the PS by immersing the capped microcentrifuge tube in liquid nitrogen using tongs. Store the tube at –80°C until use (*see* **Note 15**).

3.6. Western Blot of hVAChT

The Western blot characterizes the molecular weight and the amount of a particular protein in a complex mixture. Although the technique is common, there are steps specific to hVAChT.

3.6.1. SDS–PAGE

1. Mix 150 μg of PS (*see* **Section 3.5**, step 12) with 500 μL of homogenization buffer (*see* **Section 2.5**, item 4).
2. Centrifuge samples at 171,500 × *g* for 1.5 h.
3. Remove supernatant by aspirating it into a pipette tip without disturbing the membrane pellet (*see* **Note 16**).
4. Dissolve each pellet in 40 μL of sample loading buffer (*see* **Section 2.6**, item 9).
5. To other 40-μL portions of sample loading buffer, add molecular weight markers as instructed by the manufacturer (*see* **Section 2.6**, items 13 and 14).
6. Heat all samples to 65° C for 25 min (*see* **Note 17**).
7. Subject 18 μL of each denatured sample to standard SDS–PAGE using 12% resolving and 4% stacking gels (9) (*see* **Section 2.6**, items 3 and 4).

3.6.2. Protein Transfer

1. Remove the short glass plate from the electrophoresis cell. Wearing clean disposable gloves, remove the gel taking care not to tear it. Place it in a clean 3 × 4 in. plastic tray. Wash gel in transfer buffer (*see* **Section 2.6**, item 9) by gentle agitation on an orbit shaker for 20 min.
2. Meanwhile, in another 3 × 4 in. plastic tray soak PVDF membrane (*see* **Section 2.6**, item 5) in methanol for 30 s. Then soak it in transfer buffer for 5 min. Soak fiber pads and filters (*see* **Section 2.6**, items 5 and 6) in transfer buffer for 5 min.
3. Prepare a transfer sandwich as instructed by the manufacturer, and run the transfer as directed by manufacturer (*see* **Section 2.6**, item 7).

3.6.3. PVDF Membrane Blocking and Antibody Binding

1. Disassemble the transfer sandwich and place the PVDF membrane in blocking solution (*see* **Section 2.6**, item 12) for 60 min.
2. Mix 100 μL of 1° antibody (*see* **Section 2.6**, item 1) with 50 mL of blocking solution to give a 1:500 dilution.

3. Gently agitate the membrane for 120 min or overnight at 4°C. Recover diluted antibodies by pouring them off the membrane into a container and cap. They can be reused up to three times.
4. Wash the membrane twice for 5 min each time in 20 mL of 1X TBST (*see* **Section 2.6**, item 10), then twice for 5 min each time in 20 mL of blocking solution.
5. Mix 16.7 μL of 2° antibody (*see* **Section 2.6**, item 2) with 50 mL of blocking solution to give a 1:3,000 dilution.
6. Gently agitate the membrane in the diluted 2° antibodies for 120 min. Recover diluted antibodies for reuse.
7. Wash the membrane four times for 5 min each in 20 mL of 1X TBST.
8. The PVDF membrane can be transported in the tray in final wash for imaging at another location.

3.6.4. Imaging of Blot

1. The final wash is poured off the PVDF membrane and discarded. The tray containing the membrane is placed inside a chemiluminescence imaging station (*see* **Section 2.6**, item 15).
2. Supersignal chemiluminescent substrate (2 mL, *see* **Section 2.6**, item 8) is poured onto the PVDF membrane, and the imaging station is closed to exclude light.
3. Sequentially collect images over different lengths of time, for example, 15, 20, and 30 s in order to obtain a good-quality image (*see* **Fig. 10.1**).

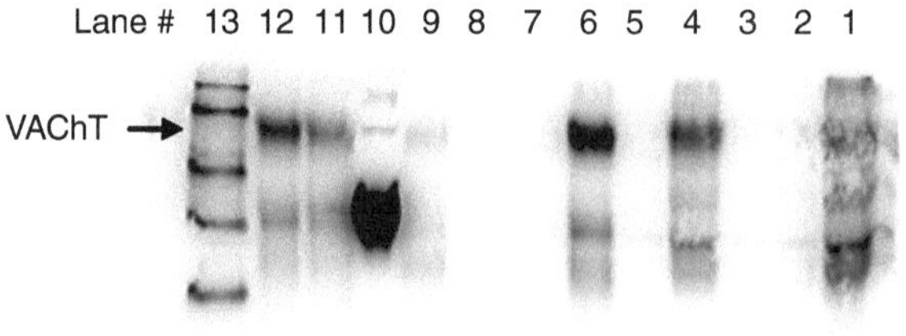

Fig. 10.1. Western blot of wild-type VAChT (~ 45 pmol/mg postnuclear supernatant; lanes 1 and 4), independent clonal lines of the Y432H mutant of VAChT (lanes 2, 3, 5–9, 11, and 12), molecular weight markers (lane 13), and colored molecular weight markers (lane 10). The very intense comet in lane 10 is a staining artifact. Wild-type and mutant VAChTs run as diffuse bands of about 80 kDa due to N-linked glycosylation at two sites in the amino acid sequence. Clone 6 highly expresses the mutant VAChT (more than wild-type), and clones 7 and 8 express almost no mutant VAChT.

3.7. Preparation of Filters

1. Filters (*see* **Section 2.7**, item 1) are prepared several days in advance of need by incubating them in 0.25% (w/v) polyethylenimine (*see* **Section 2.7**, item 2) in water for 2 h, after which they are copiously rinsed on a large Büchner filter funnel (*see* **Section 2.7**, item 3) with water. Dry them overnight at about 80°C and store them at room temperature in a plastic bag in the dark.

3.8. [^{3}H]Vesamicol Off-Rate Measurements

The following procedure determines the rate behavior for dissociation of [^{3}H]vesamicol from hVAChT. A saturating concentration of nonlabeled vesamicol is used to determine nonspecific binding (*see* **Note 18**).

1. Transfer 1 μL of 1 μMV* (*see* **Section 2.8**, item 7) into a scintillation vial containing 3.5 mL of scintillation cocktail and determine the counts per minute (cpm) to <2% error. Calculate the chemical concentration from the specific radioactivity given by the manufacturer and the efficiency of counting (*see* **Note 19**). It should be about 1 μM [^{3}H]vesamicol.
2. Based on its chemical concentration (*see* step 1), calculate the volume of 1 μMV* required to yield 200 μL of 100 nM [^{3}H]vesamicol in UBB (100 nMV*). For example if the concentration is 1 μM, mix 20 μL of 1 μMV* with 180 μL of UBB (*see* **Section 2.8**, item 3) in a 1.5-mL microcentrifuge tube (*see* **Note 20**).
3. Label a 1.5-mL microcentrifuge tube "T" (for total binding) and another tube "NS" (for nonspecific binding). To the tube labeled T, add 75 μL of 100 nMV* (*see* step 2). To the tube labeled NS, add 37.5 μL of 200 μM (±)-vesamicol HCl in ethanol (*see* **Section 2.8**, item 2). Evaporate the ethanol to dryness under a gentle stream of nitrogen in a chemical fume hood and add 75 μL of 100 nMV*.
4. Calculate the volume of PS containing 300 μg of protein. For example, assuming that PS contains 12.5 mg protein/mL (*see* **Section 3.5**, step 12), 24 μL of PS will be required for each of the T and NS reactions.
5. Add (1,387.5 minus the volume of PS calculated) microliter of UBB to the tube labeled T. This is 1,363.5 μL in the example given.
6. Add the required volume of PS (e.g., 24 μL) to tube T for a total volume of 1,462.5 μL.
7. Incubate tube T at room temperature for 20 min. Place 13 filters on the cleaned manifold (*see* **Section 2.1**) with vacuum on and all stopcocks closed.

8. Apply 500 μL of ice-cold UBBw (*see* **Section 2.8**, item 5) to a filter. Vortex tube T briefly and pulse centrifuge to collect. Open the stopcock under the wet filter and pull excess UBBw through the filter. Apply 90 μL from tube T to the center of the filter until the sample has been pulled through. Immediately wash the filter four times with 1-mL portions of ice-cold UBBw delivered at a rate that keeps the full surface of the filter wet, but does not result in overflow. Allow the filter to dry briefly under suction. Close the stopcock under the filter. Using fine-pointed tweezers, remove the filter and immerse it in 3.5 mL of scintillation cocktail contained in an appropriately labeled scintillation vial (*see* **Section 2.8**, item 11). This provides the datum for $t = 0$ (the datum must be adjusted, *see* **Section 3.11.1**).
9. Add 37.5 μL of 200 μM (±)-vesamicol in UBB (*see* **Section 2.8**, item 4) to tube T and simultaneously start a timer ($t = 0$).
10. Repeat step 8 at 12 times ranging from 10 s to 45 min, recording the time at which each sample of 90 μL is filtered. Use a new filter channel for each sample.
11. Repeat steps 5–10, except use tube NS instead of tube T (*see* **Note 21**).
12. When finished with both T and NS reactions, vortex the scintillation tubes thoroughly and count the radioactivity for 5 min or ± 2% error.

3.9. pH Dependence of ACh Transport

In this experiment, a subsaturating concentration of [^{3}H]ACh is used to characterize transport at different values of pH. A saturating concentration of nonlabeled vesamicol is used to determine nonspecific transport (*see* **Note 18**). The buffer over the pH range studied is isotonic with mammalian serum.

1. Transfer 1 μL of [^{3}H]ACh in ethanol (ACh**, *see* **Section 2.9**, item 11) to a scintillation vial containing 3.5 mL of scintillation cocktail (*see* **Section 2.9**, item 13) and determine the cpm to < 2% error. Calculate the chemical concentration from the specific radioactivity given by the manufacturer and the efficiency of counting (*see* **Note 19**). The concentration should be 10–20 mM [^{3}H]ACh.
2. Calculate the volume of ACh** required to give 900 μL of 4.0 mM [^{3}H]ACh. For example, if the concentration of ACh** is 10 mM, 360 μL of it is required. Transfer this volume to a 1.5-mL microcentrifuge tube (*see* **Section 2.9**, item 14) labeled ACh∗. Carefully evaporate the ethanol under nitrogen stream. Add 900 μL SUBB, which contains MgATP to drive ACh transport (*see* **Section 2.9**, item 6). Keep ACh* on ice.

3. Using 1.5-mL microcentrifuge tubes, label one set of 12 tubes 1, 2, 3, and so on to 12. These numbers correspond to the 12 buffers of pH values ranging from 5 to 10 (*see* **Table 10.2**) that will be mixed in these tubes. Label one set of 12 tubes 1-A, 2-A, 3-A, and so on to 12-A for active transport. Label two sets of 12 tubes 1-T, 2-T, 3-T, and so on to 12-T for total transport. Also, label one set of three tubes 1-B, 6-B, and 12-B for blocked transport and two sets of three tubes 1-NS, 6-NS, and 12-NS for nonspecific transport. Label a tube "PS-T" (PS for total transport) and another one "PS-NS" (PS for nonspecific transport).

Table 10.2
Transport of ACh at different pH values[a]

Tube #	UBB-low (μL)	UBB-high (μL)
1	1,000	0
2	900	100
3	800	200
4	700	300
5	600	400
6	500	500
7	450	550
8	400	600
9	300	700
10	200	800
11	100	900
12	0	1,000

[a]pH varies from ~ 5 to ~ 10 from tube 1 to 12. *See* **Section 3.9**, step 4, and **Fig. 10.3**.

4. Using the volumes listed in **Table 10.2**, mix UBB-low and UBB-high (*see* **Section 2.9**, items 7 and 8) buffers in the set of the tubes labeled 1–12 (*see* step 3).
5. Transfer 100 μL from each of tubes 1–12 to the set of tubes 1-A–12-A, placing the buffer from tube 1 into tube 1-A and so on. Cap and place the tubes on ice.
6. Calculate the volume of PS containing 3,250 μg of protein, which will be diluted to 650 μL in the tube labeled PS-T (*see* step 3; *see* **Note 22**). For example, if the PS contains 12.5 mg protein/mL (*see* **Section 3.5**, step 12), mix 260 μL of it with 390 μL of homogenization buffer (*see* **Section 2.5**, item 4) in PS-T. Cap and place the tube

on ice. Adjust volumes of PS and homogenization buffer appropriately for other concentrations of PS.

7. Calculate the volume of PS required to obtain 1,000 μg of protein, which will be diluted to 200 μL with excess non-radioactive vesamicol in the tube labeled PS-NS (*see* step 3). For example, if the PS contains 12.5 mg protein/mL, mix 80 μL of it, 80 μL of homogenization buffer, and 40 μL of 80 μMV (*see* **Section 2.9**, item 10) in PS-NS. Cap and place the tube on ice. Adjust volumes of PS and homogenization buffer appropriately for other concentrations of PS.
8. Distinctively mark one of the two sets (for duplicate data) of 12 empty tubes labeled 1-T–12-T (*see* step 3) to indicate which set will receive the first of duplicate samples. This is done so that the delay in filtration for the duplicates after quench will be approximately the same for all samples. Place both sets of tubes on ice. Transfer 900 μL of ice-cold UBBw (*see* **Section 2.9**, item 9) into each of the 24 tubes. Place 20 filters on the cleaned manifold (*see* **Section 2.1.**) with vacuum on and all stopcocks closed.
9. Place capped tube 1-A containing 100 μL of buffer 1 (which is about pH 4.5) in a 37°C water bath and equilibrate for 4 min.
10. Add 50 μL of ACh*. Incubate for 1 additional minute to equilibrate the temperature.
11. Start transport by adding 50 μL of PS-T (*see* step 6) to tube 1-A, which then contains 200 μL of transporting vesicles, cap the tube, and simultaneously start the timer ($t = 0$). Incubate at 37°C for 4 min.
12. Remove tube 1-A from the 37°C water bath. Pulse vortex. Return to 37°C bath. Read the pH using a calomel semimicro probe (*see* **Section 2.9**, item 18). Make sure the reading is stable and record the value. Drain as much of the reaction off the pH probe as quickly as possible, and cap the tube. Pulse vortex and pulse centrifuge. Return to 37°C bath. Rinse the pH probe with water.
13. When $t = 5$ min, use an adjustable pipette to dispense two 90-μL portions from tube 1-A into the two quench tubes 1-T (which contain ice-cold UBBw), the first portion being delivered into the distinctively marked tube. Close tubes 1-T and vortex briefly to mix. Pulse centrifuge to collect and return tubes 1-T to ice.
14. Wet a filter with 500 μL of ice-cold UBBw. Open the stopcock under the wetted filter and allow the UBBw to drain through. Taking tube 1-T into which the first 90 μL sample was diluted, aspirate the entire 1 mL volume using

an adjustable pipette set to 1,000 μL. Deliver slowly to the center of the wet filter and allow the entire volume to be pulled through. Using a disposable transfer pipette (*see* **Section 2.9**, item 16), transfer 1 mL of ice-cold UBBw to the nearly empty tube 1-T. Transfer the rinse to the center of the same filter and allow it to be pulled through. Wash the filter two more times with 1-mL portions of ice-cold UBBw delivered at a rate that keeps the full surface of the filter wet, but does not result in overflow. Allow the filter to dry briefly under suction.

15. Quickly repeat step 14 for the duplicate tube 1-T.
16. Close both stopcocks of the filter manifold. Using fine-pointed tweezers, remove each filter and immerse it in 350 μL of 1% SDS (*see* **Section 2.9**, item 12; *see* **Note 23**) contained in a scintillation vial labeled 1-T. Cover tightly and vortex intermittently to solubilize the radioactivity.
17. Repeat steps 9–16 for the reactions at other pH values indicated by the numbers 2–15. This process will require loading more filters on the filtration manifold.
18. To obtain estimates of nonspecific transport, transfer 100 μL each of pH buffers 1, 6, and 12 (*see* step 4) to the set of three tubes labeled 1-B, 6-B, and 12-B (*see* step 3), respectively, and cap and place the tubes on ice.
19. Distinctively mark one of two sets (for duplicate data) of three empty tubes labeled 1-NS, 6-NS, and 12-NS (*see* step 3) to indicate which set will receive the first of the duplicate samples. Place both sets of tubes on ice. Transfer 900 μL of ice-cold UBBw into each of the six tubes. Place six filters on the filter manifold with vacuum on and all stopcocks closed.
20. Place capped tube 1-B containing 100 μL of buffer 1 (which is about pH 4.5) in a 37°C water bath and incubate for 4 min.
21. Add 50 μL of ACh*. Incubate for 1 additional minute.
22. Start nonspecific transport by adding 50 μL of PS-NS (*see* step 7) to tube 1-B, which then contains 200 μL of non-transporting vesicles, cap the tube, and simultaneously start the timer ($t = 0$). Incubate at 37°C for 4 min.
23. Continue with steps 12–16, except that tube 1-B is processed instead of 1-A, and the quench tubes and scintillation vials are labeled 1-NS.
24. Repeat steps 20–23, except that tubes 6-B and 12-B are processed instead of 1-B, and the quench tubes and scintillation vials are labeled 6-NS and 12-NS, respectively.

25. After filters are incubated 1 or more hours in SDS, add 3.5 mL of scintillation fluid to each scintillation vial. Vortex scintillation tubes thoroughly before counting for 5 min or 2% error.

3.10. Binding of ACh at Equilibrium

Before it is transported, ACh binds to a saturable site in VAChT. The equilibrium affinity of ACh is too weak to determine using filtration, as bound ACh dissociates during the wash period. Luckily, an indicator method works. When ACh binds, [^{3}H]vesamicol cannot bind (*see* **Note 24**). A saturating concentration of nonlabeled vesamicol is used to determine nonspecific binding (*see* **Note 18**).

1. Transfer 1 μL of 1 μMV*p (*see* **Section 2.10**, item 8) into a scintillation vial containing 3.5 mL of scintillation cocktail and determine the cpm to <2% error. Calculate the chemical concentration from the specific radioactivity given by the manufacturer and the efficiency of counting (*see* **Note 19**). It should be about 1 μM [^{3}H]vesamicol.
2. Label a 1.5-mL microcentrifuge tube 100 nMV*p. Based on the chemical concentration determined in step 1, calculate the volume of 1 μMV*p required to yield 200 μL of 100 nM [^{3}H]vesamicol. For example, if the concentration is 1 μM, mix 20 μL of 1 μMV*p with 180 μL of UBBp (*see* **Section 2.10**, item 2, and **Note 20**). Cap and place on ice.
3. Label a 1.5-mL microcentrifuge tube 100 nMV*pNS. Add 37.5 μL of 200 μM (±)-vesamicol HCl in ethanol (*see* **Section 2.8**, item 2). Evaporate the ethanol to dryness under a gentle stream of nitrogen in a chemical fume hood, add 37.5 μL of 100 nMV*p (*see* item 2), and mix. Cap and place on ice.
4. Label a 1.5-mL microcentrifuge tube "PS-D." Calculate the volume of PS required for 800 μL containing 800 μg of protein. For example, assuming that PS contains 7.8 mg protein/mL (*see* **Section 3.5**, step 12), 102.4 μL of PS is mixed with 697.6 μL of UBBp to make 800 μL of PS-D, which is sufficient volume to determine both total and nonspecific binding. Adjust the volumes of PS and UBBp appropriately to obtain the same volume and concentration of PS-D for other concentrations of PS. Cap and place on ice.
5. Label 11 1.5-mL microcentrifuge tubes with the following numbers corresponding to concentrations of ACh in mM: 0, 6, 10, 20, 50, 100, 200, 300, 400, 700, and 1,000. In addition, label three 1.5-mL microcentrifuge tubes NS-0, NS-300, and NS-1000 for three nonspecific incubations at

the concentrations of ACh indicated by the labels. Place these reaction tubes on ice.

6. Using **Table 10.3** as a guide, transfer the indicated volume of UBBp into each tube and add 50 μL of PS-D (*see* step 4) to each one. Cap the tubes.

Table 10.3
Equilibrium displacement of [^{3}H]vesamicol by ACh[a]

ACh (mM)	UBBp (μL)	PS-D (μL)	0.1 M ACh (μL)	1 M ACh (μL)	2 M ACh (μL)	100 nMV*p (μL)	100 nMV*pNS (μL)
			Total binding				
0	140	50	0	0	0	10	0
6	128	50	12	0	0	10	0
10	120	50	20	0	0	10	0
20	100	50	40	0	0	10	0
50	130	50	0	10	0	10	0
100	120	50	0	20	0	10	0
200	100	50	0	40	0	10	0
300	80	50	0	60	0	10	0
400	60	50	0	80	0	10	0
700	70	50	0	0	70	10	0
1,000	40	50	0	0	100	10	0
			Nonspecific binding				
NS-0	140	50	0	0	0	0	10
NS-300	80	50	0	60	0	0	10
NS-1000	40	50	0	0	100	0	10

[a] *See* **Section 3.10**, step 6, and **Fig. 10.4**.

7. Place 20 filters on the clean filter manifold (*see* **Section 2.1.**) with vacuum on and all stopcocks closed.

8. Transfer one of the reaction tubes (i.e., 0, 6, 10, ... to NS-1000) to a 37°C bath and incubate it for 5 min. Add the designated volume of concentrated ACh (2 M ACh, 1 M ACh, or 0.1 M ACh, *see* **Table 10.3**) to the warm reaction tube, cap the tube, pulse vortex it, and start a timer. Wet two filters with 500 μL of ice-cold UBB (*see* **Section 2.10**, item 1).

9. At $t = 10$, add 10 μL of 100 nMV*p (*see* item 2) or 100 nMV*pNS (*see* item 3) as appropriate, for a total volume of 200 μL in the warm reaction tube. Cap the tube and pulse vortex it. Return to 37°C for 10 min.
10. When $t = 19{:}40$, remove the tube from a 37°C bath, pulse vortex, and pulse centrifuge it.
11. On the filter manifold, open the stopcock under a wet filter and allow the UBB (*see* step 8) to drain through. Using an adjustable pipette, transfer 90 μL from the reaction tube to the center of the wet filter where it will be pulled through. Using a disposable transfer pipette (*see* **Section 2.10**, item 10), immediately apply 1 mL of ice-cold UBB slowly to the center of the filter and allow it to be pulled through. Immediately wash the filter three more times with 1-mL portions of ice-cold UBB at a rate that keeps the full surface of the filter wet but does not result in overflow. Allow the filter to dry briefly under suction.
12. Repeat step 11 for a second 90-μL portion of the contents in the same tube.
13. Close the stopcocks under both filters. Using fine-pointed tweezers, remove the filters and immerse them in separate 3.5-mL portions of scintillation cocktail in appropriately labeled scintillation vials.
14. Repeat steps 8–13 for the other tubes. The manifold will have to be reloaded with more filters. Instead of waiting 20 min for each equilibration, subsequent reactions can be started at times staggered to avoid overlapping actions. A table dictating which action is to be taken at what time is recommended. For example, the first three tubes will have the following schedule. ACh will be added to tube 0 at $t = 0$, tube 6 at $t = 4$, and tube 10 at $t = 8$. The 100 nMV*p will be added to tube 0 at $t = 10$, tube 6 at $t = 14$, and tube 10 at $t = 18$. Filtration will occur for tube 0 at $t = 20$, tube 6 at $t = 24$, and tube 10 at $t = 28$. To avoid conflict between addition of ACh and filtration of an incubating reaction, add ACh 10 s before the time indicated in the table to avoid overlap with filtration (*see* **Note 25**).
15. Vortex scintillation tubes thoroughly before counting for 3 min or 2% error.

3.11. Data Reduction and Model Fitting

The analysis detailed here uses simultaneous fitting of equations for total and nonspecific binding or transport to data. This approach yields better-determined values for the parameters than when nonspecific data are subtracted from total data before a fit to specific binding or nonspecific binding is inferred from the slope of data at high concentrations of radiolabeled ligand.

3.11.1. [^{3}H] Vesamicol Off-Rate Measurements

Addition of excess of nonradioactive vesamicol to the preformed complex of [^{3}H]vesamicol and hVAChT results in nonradioactive vesamicol replacing the specifically bound [^{3}H]vesamicol after it spontaneously dissociates (*see* **Fig. 10.2**).

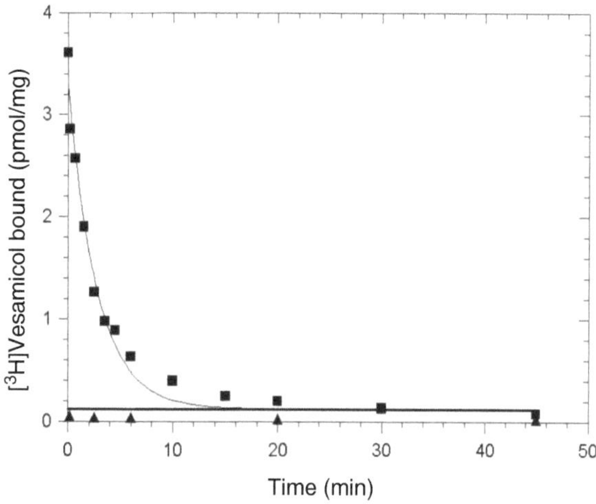

Fig. 10.2. [^{3}H]Vesamicol off-rate determination. Total binding (■) and nonspecific binding (▲) to postnuclear supernatant from cells expressing wild-type hVAChT are shown. The regression lines are defined by B0 = 3.16 ± 0.10 pmol/mg, k = 0.36 ± 0.03 min^{-1}, ns = 0.12 ± 0.04 pmol/mg (one standard error is indicated, *see* **Section 3.11.1**).

1. In a spreadsheet (e.g., Microsoft Excel), the data in cpm must be transformed into the amounts of [^{3}H]vesamicol bound/mg PS given by

$$\text{cpm} \times \frac{1\text{Ci}}{2.2 \times 10^{12}\text{dpm}} \frac{1}{\text{efficiency}} \frac{1}{\text{PSper datum}} \frac{10^3\mu\text{g}}{\text{mg}} \frac{1}{\text{specific activity}},$$

   where efficiency is in cpm/dpm, specific radioactivity is in Ci/pmol given by the manufacturer, and PS per datum is the amount of PS filtered per datum, which is 45 μg in the example given (*see* **Note 26**).

2. Open Scientist. Select under the *File* menu, *New*, and *Model* and press enter. A model template appears that you overwrite with your own model file. Replicate the following model file.

   IndVars: t

   DepVars: Btot, Bns

   Params: B0, *k*, ns

Bns= ns

Btot = Bns + B0*exp(−k*t)

***

where t is the time after addition of nonradioactive vesamicol, Bns is the amount of nonspecifically bound [^{3}H]vesamicol, Btot is the amount of total bound [^{3}H]vesamicol, B0 is the initial specific binding at $t = 0$, and k is the dissociation rate constant for specific binding per minute. Select under the *Model* menu, *Compile*, and press enter. A window appears that says "Compilation done." Save the file.

3. Select under the *File* menu, *New*, and *Spreadsheet* and press enter. The columns in the spreadsheet window are labeled with the names of the independent and dependent variables. Under the t column, type in the times of sampling in minutes after addition of the nonradioactive vesamicol (including 0). Copy the data for [^{3}H]vesamicol bound/mg PS (*see* step 1) into the columns for Bns and Btot, being careful to associate data with the correct times. Save the file.
4. Select under the *File* menu, *New*, and *Parameter Set* and press enter. The rows in the parameter spreadsheet are labeled with the names of the parameters. Scan the picomole per milligram data in the spreadsheet file (*.mmd) to estimate values for the parameters. The k will be about the reciprocal of the time required for one-half of the initial specifically bound [^{3}H]vesamicol to dissociate, and B0 will be about the difference between the total binding and the nonspecific binding at time 0. The value for ns will be about the average of all nonspecific binding values. Enter these values in the Value column.
5. Select under the *Calculate* menu, the *Least Squares Fit*, and press enter. The Spreadsheet file appears containing new columns filled with calculated values for Btot and Bns. Select under the *Calculate* menu, *Statistics*, and press enter. Scroll down the Statistics Report to find the optimized values for the parameters, their standard errors, and 95% confidence intervals. Save the file.
6. Select under the *Plot* menu, *Plot Template*, and press enter. In the window that opens there are two pairs of curves (four in total). Each pair contains a file for a set of observations and a file for the calculated best-fit line to those observations. The default assignment of Curve Styles associates a data file with a Curve Style having an odd number (e.g., 1) and a best-fit line file with a Curve Style having an even number (e.g., 2). The default assignment of Curve Styles also associates odd-numbered styles with data symbols and

even-numbered styles with types of lines (e.g., solid or broken). Click on the Plot tab and change the Plot Title if desired. Click on the X-Axis tab and change the label to Time (min). Click on the Y-Axis tab and change the label to [^{3}H]vesamicol bound (pmol/mg). The Default range and int. (which means intervals for ticks) should be checked for both X- and Y-Axes. Select under the *Plot* menu, *Save Template*, assign a file name, and save the file.

3.11.2. pH Dependence of ACh Transport

VAChT, like essentially all proteins, exhibits strong pH dependence for function (*see* **Fig. 10.3**). Fortunately, the pH dependence observed for VAChT does not include the pH dependence of the vacuolar ATPase that supplies protons used for transport. Vacuolar ATPase does have a pH dependence, but not over the range of pH that affects VAChT activity (1).

1. In a spreadsheet (e.g., Microsoft Excel), average the cpm for duplicates at each concentration of [^{3}H]ACh in each of the data sets for total and nonspecific transport. The averaged data in cpm must be transformed into picomole [^{3}H]ACh transported/mg PS, as was done in step 1 of **Section 3.11.1**, except that the PS per datum is 112.5 μg in the example given and the specific activity is about 1,000-fold lower. Make this calculation for all cpm data.
2. Open Scientist. Select under the *File* menu, *New*, and *Model* and press enter. Replicate the following model file:

   ```
   IndVars: pH
   DepVars: AChtot, AChns
   Params: ns, AChmax, pKa1, pKa2
   AChns = ns
   AChtot = AChns+AChmax/(1+(10^(-pH)/10^(-pKa1))+
     (10^(-pKa2)/10^(-pH)))
   ***
   ```

   where pH is the recorded value of the transporting solution, AChns is the amount of nonspecific transport, and AChtot is the amount of total transport. pKa1 describes a site in VAChT that must be deprotonated for transport, pKa2 describes a site in VAChT that must be protonated for transport, and AChmax is the pH-independent amount of transport (by the optimally protonated form of VAChT). Select under the *Model* menu, *Compile*, and press enter. A window appears that says "Compilation done." Save the file.
3. Select under the *File* menu, *New*, and *Spreadsheet* and press enter. Under the pH column, type in the measured pH for each solution 1–12 (*see* **Table 10.2**). Copy the data for picomole [^{3}H]ACh transported/mg PS into the columns for

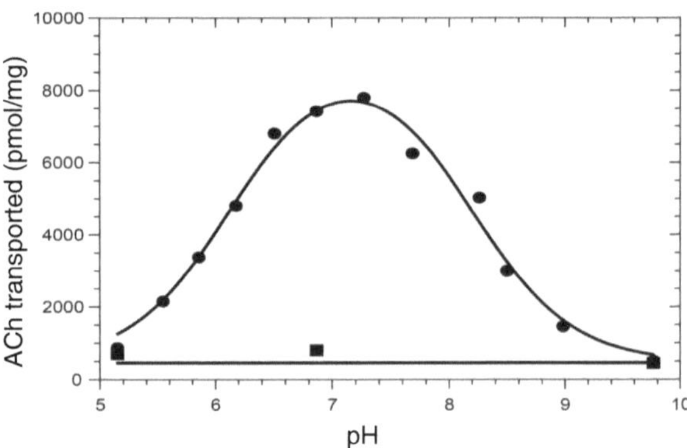

Fig. 10.3. pH dependence for transport of 1 mM ACh in 5 min. Total (•) and nonspecific (■) transport at pH values from 5.1 to 9.8 are shown for wild-type hVAChT. The regression lines are defined by ACh$_{max}$ = 8,700 ± 400 pmol/mg, p*K*a1 = 6.14 ± 0.08 and p*K*a2 = 8.17 ± 0.09, and ns = 460 ± 170 pmol/mg (one standard error is indicated, *see* **Section 3.11.2**).

AChns and AChtot, being careful to associate data with the correct pH values. Save the file.

4. Select under the *File* menu, *New*, and *Parameter Set* and press enter. Scan the picomole [^{3}H]ACh transported/mg PS data in the spreadsheet file (*.mmd) to estimate values for the parameters. For example, ns will be the averaged value of the three AChns data, pKa1 will be approximately the pH at the midpoint of the ascending curve (from low to high pH), and pKa2 will be approximately the pH at the midpoint of the descending curve (from low to high pH). AChmax will be about the difference between the total binding and the nonspecific binding at pH 7.0. Enter these values into Value column. Save the file.
5. Select under the *Calculate* menu, *Least Squares Fit*, and press enter. Select under the Calculate menu, Statistics, and press enter. Scroll down the Statistics Report to find the optimized values for the parameters, their standard errors, and 95% confidence intervals. Save the file.
6. Select under the *Plot* menu, *Plot Template*, and press enter. Click on the Plot tab and change the Plot Title if desired. Click on the X-Axis tab and change the label to pH. Click on the Y-Axis tab and change the label to ACh transported (pmol/mg). Select under the *Plot* menu, *Save Template*, assign a file name, and save the file.
7. Select under the *File* menu, *New*, *Plot*, and press enter. The plot appears. Save the file.

3.11.3. Binding of ACh at Equilibrium

ACh binds to high- and low-affinity sites in VAChT. Mutational studies (10) have indicated that a given molecule of VAChT

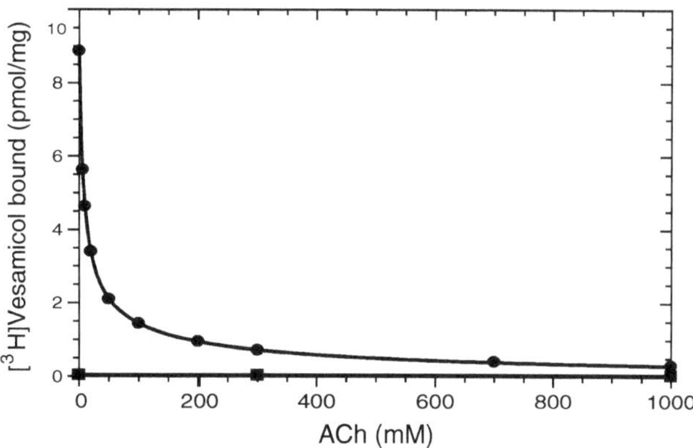

Fig. 10.4. Binding of ACh at equilibrium. Total (•) and nonspecific (■) binding of subsaturating [^{3}H]vesamicol in the presence of zero to high concentrations of ACh are shown for wild-type hVAChT. The regression lines are defined by B0 = 8.83 ± 0.01 pmol/mg, perct_ha = 85.3 ± 0.5, Kach_ha = 8.1 ± 0.1 mM, Kach_la = 180 ± 10 mM, and ns = 0.037 ± 0.007 pmol/mg (one standard error is indicated, *see* **Section 3.11.3**).

contains either a high- or a low-affinity binding site for ACh but not both. The cause of high and low affinity is not known (*see* **Fig. 10.4**).

1. In a spreadsheet (e.g., Microsoft Excel), average the cpm for duplicates at each concentration of ACh in the data sets for total and nonspecific binding of [^{3}H]vesamicol. The averaged data in cpm must be transformed into picomole [^{3}H]vesamicol bound/mg PS, as was done in step 1 of **Section 3.11.1**, except that the PS per datum is 22.5 μg in the example given.
2. Open Scientist. Select under the *File* menu, *New*, and *Model* and press enter. Replicate the following model file:

```
IndVars: ACh
DepVars: Btot, Bns
Params: ns, B0, perct_ha, Kach_ha, Kach_la
Bns = ns
Btot = B0*perct_ha*Kach_ha/(100*(Kach_ha + ACh)) +
  B0*(100 - perct_ha)*Kach_la/(100*(Kach_la +
  ACh)) + Bns
***
```

ACh is in millimolar, Btot is total bound vesamicol, Bns is nonspecifically bound vesamicol, B0 is specifically bound vesamicol in the absence of ACh, perct_ha is the percent of high-affinity site for ACh binding, Kach_ha is the dissociation constant in millimolar for high-affinity binding of ACh, and Kach_la is the dissociation constant for low-affinity binding of ACh. Select under the *Model* menu, *Compile*, and press enter. A window appears that says "Compilation done." Save the file.

3. Select under the *File* menu, *New*, and *Spreadsheet* and press enter. Under the ACh column, type in the concentrations of ACh, including 0 mM. Copy the data for picomole [^{3}H]vesamicol bound/mg PS into the columns for Btot and Bns, being careful to associate data with the correct concentrations of ACh. Save the file.
4. Select under the *File* menu, *New*, and *Parameter Set* and press enter. Enter the values for initial estimates of parameters and their limits shown in **Table 10.4**. Save the file.

Table 10.4
Initial parameter estimates for displacement of [^{3}H]vesamicol by ACh

Parameter	Lower limit	Value	Upper limit
ns	0	0.03	Infinity
B0	0	8	Infinity
Perct_ha	0	80	Infinity
Kach_ha	0	5	Infinity
Kach_la	0	300	Infinity

5. Select under the *Calculate* menu, *Least Squares Fit*, and press enter. Select under the *Calculate* menu, *Statistics*, and press enter. Scroll down the Statistics Report to find the optimized values for the parameters, their standard errors, and 95% confidence intervals. Save the file.
6. Select under the *Plot* menu, *Plot Template*, and press enter. Click on the Plot tab and change the Plot Title if desired. Click on the X-Axis tab and change the label to ACh (mM). Click on the Y-Axis tab and change the label to [^{3}H]vesamicol bound (pmol/mg). Select under the *Plot* menu, *Save Template*, assign a file name, and save the file.
7. Select under the *File* menu, *New*, *Plot*, and press enter. The plot appears. Save the file.

3.12. Results and Discussion

PC12$^{A123.7}$ cells are desirable for neurosecretory research, as they have a relatively short doubling time and are not difficult to culture. They can be induced to assume neuronal-like morphology, although that process is not described here. Stable transfectants of these cells yield several-fold higher levels of expression of a protein than transient transfectants do (8). The result is significantly higher ratios of specific to nonspecific binding and transport when studying VAChT. For characterization of mutants, the greater signal provided by stable transfectants can be important when the mutant has decreased function. Different clonal lines of the same transfection reaction express hVAChT at very different levels. The variation probably is due to integration of pcDNA6.2/hVAChT into different regions of the genome. For this reason, we typically screen 10–15 clonal lines by Western blot, choose two highly expressing lines for preservation, and grow biochemical quantities of the most highly expressing line.

In off-rate measurements (*see* **Fig. 10.2**), [^{3}H]vesamicol at 5 nM concentration is used to form the initial complex with hVAChT. Addition of excess nonradioactive vesamicol immediately occupies all unoccupied binding sites for vesamicol and prevents newly unoccupied sites formed by dissociation of [^{3}H]vesamicol from rebinding [^{3}H]vesamicol. The amount of specific binding at time zero (B0) found by the regression has little importance. It is arbitrary because subsaturating [^{3}H]vesamicol was used and the amount of nonspecific binding (ns) is low. The rate constant for dissociation of the specifically bound [^{3}H]vesamicol (k) contains the useful information. The rate at room temperature is slow enough that the filtration assay is reliable, especially as the assay uses ice-cold washes that slow the dissociation further. Dissociation was assumed to be a single first-order process in the regression analysis. However, the data show systematic deviation from the fit. The result might indicate that fast and slow dissociation occurs. More comprehensive study would be required to test this possibility.

In the measurement characterizing the pH dependence of ACh transport (*see* **Fig. 10.3**), subsaturating [^{3}H]ACh is used to avoid shifting the intrinsic pH dependence of hVAChT. Nonspecific transport is determined in the presence of saturating vesamicol (1). The value obtained by regression for the pH-independent transport rate has little importance. It is arbitrary because subsaturating ACh was used. The amount of nonspecific binding (ns) is low. The pKa values contain the useful information. They are well determined, which implies that the simple protonation model used here describes hVAChT well. The pKa values are conserved in VAChT and vesicular monoamine transporters 1 and 2, but their origins are not understood (1).

In the measurement characterizing the binding of ACh at equilibrium (*see* **Fig. 10.4**), subsaturating [^{3}H]vesamicol is utilized to avoid shifting the apparent affinity of hVAChT for ACh. The amount of specific binding by [^{3}H]vesamicol at time zero (B0) determined by regression has little importance. It is arbitrary because subsaturating [^{3}H]vesamicol was used, and the amount of nonspecific binding (ns) is low. The useful information is in the curve generated by inhibition of [^{3}H]vesamicol binding by different concentrations of ACh. Because the amount of bound (but not transported) ACh per se cannot be quantitated due to weak binding, the B_{max} value for ACh binding is not available. Previous research consistently has shown single-affinity binding for vesamicol and high- and low-affinity binding for ACh (10). In other words, the deviation from simple competition between ACh and vesamicol occurs because of heterogeneity in the ACh binding site and not in the vesamicol binding site. In the current experiment, the percent of high-affinity binding is 85.3 ± 0.5, the high-affinity dissociation constant for ACh is 8.1 ± 0.1 mM, and the low-affinity dissociation constant for ACh is 180 ± 10 mM.

4. Notes

1. Blasticidin S yields more stable transfectants than gentamicin does. It inhibits mammalian protein synthesis. Resistance is conferred by the blasticidin S deaminase gene (*bsr*).
2. Some of the reagents used here are moderately unstable toward spontaneous hydrolysis or oxidation when in dilute aqueous solutions. In particular, admonitions to use only freshly prepared aqueous solutions of PMSF, paraoxon, and DTT must be heeded. The protease inhibitor cocktail can be stored in concentrated, frozen aqueous solution, but it too is unstable after thawing and dilution.
3. Transfer microliter volumes of isopropanol or ethanol with a positive-displacement pipette. Standard pipettes that use aspiration to fill the pipette, such as the Gilson Pipetmans, yield very poor accuracy with organic solvents because organics wet plastics and have relatively high vapor pressures.
4. Each molecule of ACh chloride occupies the volume of about 10 water molecules. When preparing a concentrated solution of ACh, one must allow for an increase in solution volume. Because ACh chloride is hygroscopic, it is difficult to weigh accurately. Luckily, pre-weighed ACh chloride is available commercially. However, the experimenter must dissolve the pre-weighed ACh chloride in an unknown volume of buffer to obtain a convenient, known concentration of ACh. For two vials containing a total of 300 mg of ACh chloride, the final volume required to make 2 M ACh chloride is

$$\frac{300\text{mgAChCl}}{2\text{vials}} \times \frac{1\text{g}}{1000\text{ mg}} \times \frac{1\text{ mol AChCl}}{181\text{ g AChCl}} \times \frac{1\text{ L}}{2\text{ mol}} \times \frac{10^{6}\mu\text{L}}{1\text{ L}} =$$

825μL of solution.

The text provides a method to bring 300 mg of ACh chloride powder to pH 7.4 in a final volume of 825 μL with an estimated accuracy of ±1%.

5. Cells normally have a round appearance. Only a few cells should appear flat and polygonal because of differentiation toward neuronal morphology. An adherent cell culture is confluent when cells have formed a single layer over the entire area available for growth. Cultures must be split every 72–96 h before some cells begin to grow on top of adherent cells.
6. Remove the cotton plug from the pipette to avoid pulling it into the vacuum line.
7. Most cells will lift from the plastic. Not all surfaces require exposure to trypsin to cause release of the cells. For example, cells are released from some brands of 24-well plates without trypsin merely by pipetting the growth medium up-and-down several times onto the surface of the well. Unfortunately, some manufacturers treat the surface chemistry of their products as proprietary, and a researcher must take an empirical approach to finding a satisfactory, economical method to release cells from the surfaces of different products.
8. Flasks are upright during pipetting. Up to eight upright flasks can be placed side by side with the bevels in the same orientation, clamped together with both hands, and tilted as a group to orient the bevels downward. Move flasks slowly so that the culture medium does not splash into the flask necks.
9. The sequence is at http://www.ncbi.nlm.nih.gov/entrez/viewer.fcgi?db = nuccore&id = 118582256
10. If the plate is overgrown and colonies are not well isolated, then a new plate with higher dilution may be prepared from the overgrown plate. Blasticidin S should be in all media used to select and maintain stable transfectants.
11. Movement of the cells through the narrow opening of a disposable pipette tip ensures they are dispersed.
12. The number of cells per square in the visual field of the hemocytometer should be between 5 and 35. If it is outside of these limits, an appropriate dilution should be prepared.
13. Trypan blue does not interact with cells unless the plasma membrane is broken. Approximately 90–95% breakage of the plasma membrane optimizes the release of synaptic-like microvesicles with minimal damage to vacuolar ATPase, which is broken off vesicular membranes by excessive shear during homogenization. Microvesicles damaged in

this manner will bind vesamicol, but they will not transport ACh.

14. The protein concentration should be 10–15 mg/mL.
15. Be careful not to allow liquid nitrogen to reach the top of the tube. There is no difference in the ligand binding and transport properties of fresh and frozen PS.
16. Removal of supernatant greatly lowers nonspecific staining.
17. The standard method of heating to 100°C causes VAChT to aggregate and not enter the gel.
18. Two types of interaction occur when radiolabeled ligand interacts with a biological membrane. Specific interaction mediated by the protein of interest occurs, and nonspecific interaction mediated by unknown components occurs. To disentangle specific and nonspecific, total interaction, which is the sum of specific and nonspecific interactions, is determined in the presence of radiolabeled ligand alone. Nonspecific interaction is determined in the presence of a saturating concentration of a nonlabeled ligand that blocks specific interaction. The nonlabeled ligand generally does not affect nonspecific interaction, which behaves as if it occurs with very low affinity but a large number of sites. Specific interaction is the difference between total and nonspecific interactions in a computer regression model. The specific interactions monitored here encompass both binding and transport.
19. The chemical concentration [M] is given by

$$\frac{\text{cpm}}{\mu\text{L}}\,\frac{1\text{ Ci}}{2.2{*}10^{12}\text{ dpm}}\,\frac{1}{\text{efficiency}}\,\frac{10^{6}\mu L}{1\text{ L}}\,\frac{1}{\text{specific radioactivity}},$$

   where efficiency is given in cpm/dpm (typically about 0.45) and specific radioactivity is given in Ci/mol.

20. The concentration need not be exactly 100 nM. The goal is to bind subsaturating [^{3}H]vesamicol to hVAChT (zero bound [^{3}H]vesamicol would be ideal) but to also have sufficient specifically bound [^{3}H]vesamicol to generate reliable data. The 100 nM of [^{3}H]vesamicol in this step results in 5 nM of [^{3}H]vesamicol in the reaction, which is a compromise concentration relative to the vesamicol dissociation constant of about 20 nM (8).
21. At $t > 0$, the NS reaction will have twice the concentration of (±)-vesamicol in it as the T reaction does, but this difference does not affect the data.
22. The [^{3}H]ACh transport assay requires more PS than the [^{3}H]vesamicol binding assay does. Transporting vesicles

are diluted into UBBw not only to quench transport at a well-defined time but also to avoid the clogging of filters that occurs when concentrated PS is applied to a small fraction of the filter surface. Such clogging means that wash buffer flows poorly through the part of the filter containing the highest amount of unbound radioactivity, thus inadequately washing the most important the part of the filter. The result would be greatly increased data scatter.

23. Tritium produces such weak radioactivity that it is poorly detected if it is not fully solubilized. 1% SDS solubilizes [^{3}H]ACh bound to the filters. [^{3}H]vesamicol is hydrophobic enough not to require SDS to solubilize it.
24. The order of addition of ACh and [^{3}H]vesamicol does not matter to the outcome, but equilibrium is reached faster when the ligands are added in the order indicated here.
25. As this is an equilibrium experiment, exact timing is not critical.
26. The 0-time samples are taken before addition of nonradioactive vesamicol. Thus, the cpm data for these samples are multiplied by 0.947 to correct for the more concentrated PS relative to samples taken after addition of nonradioactive vesamicol.

References

1. Parsons, S.M. (2000) Transport mechanisms in acetylcholine and monoamine storage. *FASEB J.* **14**, 2423–2434.
2. Alfonso, A., Grundahl, K., Duerr, J.S., Han, H.P., and Rand, J.B. (1993) The *Caenorhabditis elegans* unc-17 gene: a putative vesicular acetylcholine transporter. *Science* **261**, 617–619.
3. Hirai, T., Heymann, J.A.W., Shi, D., Sarker, R., Maloney, P.C., and Subramaniam, S. (2002) Projection structure of the bacterial oxalate transporter OxlT at 3.4 Å resolution. *Nat. Struct. Biol.* **9**, 597–600.
4. Abramson, J., Smirnova, I., Kasho, V., Verner, G., Kaback, H.R., and Iwata, S. (2003) Structure and mechanism of the lactose permease of *Escherichia coli*. *Science* **301**, 610–615.
5. Huang, Y., Lemieux, M.J., Song, J., Auer, M., and Wang, D.-N. (2003) Structure and mechanism of the glycerol-3-phosphate transporter from *Escherichia coli*. *Science* **301**, 616–620.
6. Yin, Y., He, X., Szewczyk, P., Nguyen, T., and Chang, G. (2006) Structure of the multidrug transporter EmrD from *Escherichia coli*. *Science* **312**, 741–744.
7. Inoue, H., Li, Y.P., Wagner, J.A., and Hersh, L.B. (1995) Expression of the choline acetyltransferase gene depends on protein kinase A activity. *J. Neurochem.* **64**, 985–990.
8. Ojeda, A.M., Bravo, D.T., Hart, T.L., and Parsons, S.M. (2003) Equilibrium binding and transport studies in the mathematical modeling of acetylcholine transport by transporter VAChT. *Meth. Mol. Biol.* **227** *Membrane Transporters* (Yan, Q., ed.). Humana Press, Totowa, NJ, USA, pp. 155–177.
9. Weber, K. and Osborn, M. (1969) The reliability of molecular weight determinations by dodecyl sulfate-polyacrylamide gel electrophoresis. *J. Biol. Chem.* **244**, 4406–4412.
10. Bravo, D.T., Kolmakova N.G., and Parsons S.M. (2005) Mutational and pH analysis of ionic residues in transmembrane domains of vesicular acetylcholine transporter. *Biochemistry* **44**, 7955–7966.

Chapter 11

ABC Transporters in Ophthalmic Disease

Corey Westerfeld

Abstract

ABC transporters have been implicated in a variety of human diseases. The *ABCR* gene and its protein have been linked to Stargardt's disease, fundus flavimaculatus, cone–rod dystrophy, retinitis pigmentosa, and age-related macular degeneration. The genetic and molecular pathways involved in the pathogenesis of ABCR-related ophthalmic conditions will be explored. Future diagnostic and therapeutic objectives for these diseases will also be discussed.

Key words: ABC transporters, ABCR gene, ABCA4 gene, Stargardt's disease, fundus flavimaculatus, cone–rod dystrophy, retinitis pigmentosa, age-related macular degeneration.

1. Introduction

ATP-binding cassette (ABC) transporters are found across a wide spectrum of species from bacteria and yeast to humans (1–4). They constitute one of the largest protein families known to date. Among the human ABC transporter family, seven subfamilies have been classified based on a common molecular architecture. The subfamilies are classified as ABC A–G. Among the subfamilies, 48 specific subtypes of ABC transporters have been identified (5). ABC transporters are a family of evolutionarily highly conserved transmembrane proteins that use the energy of ATP to translocate a variety of molecules across extracellular and intracellular membranes.

ABC transporters show structural similarity. Full-structured transporters have two symmetric halves each consisting of one nucleotide-binding domain (NBD), the ATP-binding cassette, and one transmembrane domain (TMD). They may also form

Q. Yan (ed.), *Membrane Transporters in Drug Discovery and Development*, Methods in Molecular Biology 637,
DOI 10.1007/978-1-60761-700-6_11, © Springer Science+Business Media, LLC 2010

homo- or heterodimers which consist of one NBD and one TMD. The NBDs encompass characteristic Walker A and B motifs separated by a center region and an ABC "signature" C motif (6). The transmembrane domains contain 6–12 membrane-spanning α-helices and are important in substrate specificity. Each ABC transporter has a unique substrate for which it is specific. Most eukaryotic ABC transporters have two tandem sets of six transmembrane helices followed by a Walker type A and B nucleotide-binding motif (7).

In eukaryotes, ABC transporters may be found in the plasma membrane and the membranes of intracellular compartments such as the Golgi, endoplasmic reticulum, endosomes, peroxisomes, and mitochondria. ABC transporters function by using ATP to transport substrates across cellular membranes. As a result, they are referred to as "traffic ATPases" due to the fact that they regulate the transport of various substrates and hydrolyze ATP into ADP and P_i (8). Many substrates may be transported by ABC transporters including peptides, lipids, amino acids, carbohydrates, vitamins, ions, glucuronide and glutathione conjugates, and xenobiotics (9). In their role as transporters, they may also function as channel proteins, receptors, and/or conductance regulators for other channel proteins (10).

ABC transporters have been implicated in a wide variety of human-inherited disorders, all of which are characterized by defects in the transport of specific substances. These include Tangier disease or familial HDL deficiency (*ABCA1*), neonatal surfactant deficiency (*ABCA3*), Stargardt's disease, cone–rod dystrophy, retinitis pigmentosa (*ABCA4*), ichthyosis (*ABCA12*), familial intrahepatic cholestasis (*ABCB4*, *ABCB11*), Dubin–Johnson syndrome (*ABCC2*), pseudoxanthoma elasticum (*ABCC6*), cystic fibrosis (*ABCC7*), familial hyperinsulinemic hypoglycemia of infancy (*ABCC8*), adrenoleukodystrophy (*ABCD1*), and β-sitosterolemia (*ABCG5*, *ABCG8*) (11). As such, improved understanding of the structure and pathophysiology of these proteins has implications in a variety of clinical arenas.

2. ABCR (ABCA4) Gene

Linkage studies initially mapped the Stargardt/fundus flavimaculatus (*STGD/FFM*) gene to 1p13–p21 with little evidence of genetic heterogeneity (12–14). Further linkage studies refined the location of the gene to a 2-cM interval between polymorphic markers D_IS406 and D_IS236 (15). Positional cloning was then performed by Allikmets and coworkers to identify the underlying gene (16). Fluorescence in situ hybridization has since localized the gene to 1p22.1–p21 (17). The *ABCA4* gene consists of 50 exons and is estimated to span 150 kb. Exon sizes range from 33 to 266 base pairs (18). The gene encodes a 2,273 amino acid protein, the Rim Protein (RmP) (19). The ABCR protein is homologous to the bovine and *Xenopus* Rim proteins previously identified in the rims of rod outer segment discs (20). The ABCR protein has been localized to the disc membrane of rod outer segments, but it may be expressed in cones as well (18, 21).

3. ABCR (ABCA4) Mutations

Mutation in the *ABCA4* gene may be in the form of missense, splice site, and frameshift mutations (16, 17, 22). To date, over 400 sequence variations in the *ABCA4* gene have been described. Certain mutant alleles such as G863A, A1038V, and G1961E cause Stargardt's disease and appear to be more common and may have altered frequencies in different populations, presumably because of founder effect (23, 24). However, these most common variants have only been found in about 10% of Stargardt patients demonstrating the tremendous heterogeneity in the disease genotype. Genotype microarray chips currently have the capability of detecting more than 98% of the existing known mutations in the *ABCA4* gene (25). *ABCA4* mutations have been linked to Stargardt's disease, fundus flavimaculatus, autosomal recessive cone–rod dystrophy type 3 (arCRD), retinitis pigmentosa type 19 (RP19), and age-related macular degeneration (AMD) (26, 27).

4. Stargardt's Disease and Fundus Flavimaculatus

Stargardt's disease is an autosomal recessive form of juvenile macular degeneration first described in 1909 by Karl Stargardt (28). Patients with Stargardt's disease generally report a bilateral, gradual decline in vision beginning between the ages of 6 and 20 years. Initially, there may be no visible ophthalmoscopic changes, and the earliest sign is disappearance of the foveolar reflex. Yellowish round or pisciform flecks at the level of the retinal pigment epithelium (RPE) may then appear in the macula and extend into the mid-periphery. Later stages are characterized by a zone of atrophic RPE in the macula that gives the appearance of a "bulls-eye" maculopathy. Fluorescein angiography (FA) may be useful at the time of initial presentation. Early cases often show an ovoid zone of faint hyperfluorescent flecks around the fovea due to RPE atrophy. The FA reveals a "dark choroid" in approximately 86% of patients, resulting from blockage of normal choriocapillaris fluorescence by RPE cells engorged with lipofuscin (29).

Fundus flavimaculatus is clinically characterized by similar yellow-white flecks which may extend further into the periphery. This term is used in cases in which the disease begins later in life, the flecks are scattered throughout the fundus, and there is relative preservation of visual acuity (30). Stargardt's disease and fundus flavimaculatus most likely represent different phenotypic spectrums of the same disease genotype. This has been supported by genetic studies demonstrating that Stargardt's disease and fundus flavimaculatus are caused by mutations in the same gene (14).

5. ABCA4 Mutations in Cone–Rod Dystrophy and Retinitis Pigmentosa

Cone–rod dystrophy (CRD) and retinitis pigmentosa (RP) are progressive hereditary retinal degenerations. Both conditions exhibit significant overlap and result in progressive deterioration of cone and rod photoreceptor cells often leading to blindness. CRD may be inherited in an autosomal recessive or dominant pattern and usually presents in young adulthood with loss of central acuity. The rate of rod and cone degeneration is equal, and as such, patients rarely report loss of night vision specifically. Progressive decline in central acuity and eventual loss of peripheral vision are characteristic. RP generally presents in young adulthood with early loss of night vision and peripheral vision. Numerous genes and mutations have been implicated in RP which shows autosomal recessive, autosomal dominant, and X-linked inheritance patterns. The classic mutation in RP affects genes encoding rhodopsin, a protein in the photoreceptor outer segments necessary for visual transduction. In RP, the rods are often primarily affected leading to early loss of night vision. Cones are generally affected late in the disease process (31). Both CRD and RP show characteristic electroretinographic (ERG) abnormalities which reveal reduced amplitudes and prolonged implicit times. CRD often affects both cones and rods equally early in the disease while RP may affect only the rods. Late in the disease, the ERG findings are similar with uniform reductions in amplitudes and prolonged implicit times.

Studies examining consanguineous families with RP19, an autosomal recessively inherited form of RP, and cone–rod dystrophy have revealed mutations in the *ABCA4* gene. A severe homozygous 5' splice site mutation was found in patients with RP19. The cone–rod dystrophy patients were heterozygous for the same mutation and a different 5' splice site mutation (26). In fact, one study concluded that *ABCA4* mutations are the major cause of autosomal recessive cone–rod dystrophy with 65% (13/20) of patients having mutations (32). In concert with these findings, Ducroq et al. examined 55 patients with CRD. They identified *ABCA4* mutations in 20.7% (6/29) of autosomal recessive CRD and in 26.9% (7/26) of sporadic cases of CRD. The study screening was estimated to have detected approximately 80% of mutations in the families studied, giving a corrected implication of the *ABCA4* gene at 30% of all cases (33). Although this is lower than that seen in the prior study, both studies confirm that *ABCA4* is a major gene responsible for CRD.

6. ABCA4 Mutations in Age-Related Macular Degeneration

Age-related macular degeneration (AMD) is a leading cause of vision loss in the elderly. The development of macular degeneration is multifactorial, likely resulting from a combination of genetic and environmental factors. Clinically, the condition may be subdivided into dry (non-neovascular) and wet (neovascular) forms. Patients with dry AMD generally present with slowly progressive loss of central acuity. Clinically, the disease is characterized by the presence of drusen, the hallmark finding of AMD. Drusen are yellowish deposits at the level of the RPE and Bruch's membrane and are composed of lipofuscin and other phototransduction end-products. Other features include atrophy and pigmentary changes of the RPE. Wet AMD may result in a sudden loss of central acuity and/or metamorphopsia. The key clinical finding is choroidal neovascularization, often demonstrated using fluorescein angiography, and it may result in subretinal fluid or hemorrhage. The end stage of wet AMD is characterized by subretinal fibrosis and scarring, often referred to as a disciform scar.

The *ABCA4* gene has been implicated in the pathogenesis of AMD as well. Allikmets et al. examined 167 patients with AMD and found heterozygous mutations in one *ABCA4* allele in 26 patients (16%) (16). However, other groups have been unable to replicate these findings (34–36). It has been hypothesized that the higher number of *ABCA4* mutations in AMD patients versus controls in this study may have been due to differences in the intensity of the screening process (37). The estimated carrier frequency of *ABCA4* mutations in the general population is 2–3%, making it difficult to infer pathogenicity from heterozygous mutations in the *ABCA4* gene (16). Furthermore, AMD is very common with over 30% of individuals over the age of 70 showing signs of the disease, the development of which is likely confounded by multiple genetic and environmental factors (38). In sum, the combination of variable expressivity and high disease prevalence make it difficult to achieve statistical significance in studies on the subject. A 15-center meta-analysis of the published data on the two most common *ABCA4* variants, the D2177N and G1961E alleles, found the two variants to be present in 3.4% of patients with AMD in comparison to 0.95% of controls (39). The results were statistically significant, although showed only a small correlation with AMD. Presently, the role of *ABCA4* mutations in AMD remains unclear. However, there is evidence to suggest that heterozygous mutations in the *ABCA4* gene may predispose to the development of AMD in a subset of patients.

7. The Residual Activity Model

Given that *ABCA4* has been implicated in a variety of retinal degenerations, a model has been proposed to correlate the genotype–phenotype findings in *ABCA4*-associated diseases. The presumption is that the severity of disease can be explained by the amount of residual RmP activity produced by mutant *ABCA4* genotypes. The inference is that severe mutations resulting in little or no RmP activity would produce a more severe disease phenotype such as RP. Moderate–severe mutations would result in a phenotype such as cone–rod dystrophy. Mild–moderate mutations would produce either Stargardt's disease or fundus flavimaculatus. Finally, it is hypothesized that mutations resulting in only a mild decline in RmP activity would produce a disease that manifests at an older age such as AMD (40). This model, referred to as the "residual activity model," has been supported in part by recent investigations. For instance, the homozygous, frameshift mutation 5917delG is associated with a relatively severe Stargardt phenotype. Alternatively, the truncating mutations Y362X and R1300X are associated with milder clinical symptoms (26, 37). Also consistent with the residual activity model are the aforementioned studies revealing homozygous splice site mutations in patients with RP19 and heterozygous distinct splice site mutations in patients with cone–rod dystrophy (26, 41, 42). Finally, heterozygous mutations in the *ABCA4* gene have been implicated in AMD.

8. Rim Protein

The *ABCR* gene encodes a retina-specific ATP-binding cassette transporter protein known as the rim protein (RmP). RmP shows the characteristic structure of an ABC transporter consisting of two transmembrane domains, each with six membrane-spanning hydrophobic segments, and two highly conserved

ATP-binding domains (18). RmP was initially believed to be exclusively expressed in the rims of rod outer segment discs (27, 43), but it has more recently been shown to be expressed within cones as well (21). The protein is essential in facilitating retinoid cycling between the photoreceptors and the retinal pigment epithelium (27). In the phototransduction cascade, a photon of light is captured by rhodopsin in the photoreceptor outer segment leading to the isomerization of 11-*cis* retinal to all-*trans* retinal. All-*trans* retinal then leaves the activated opsin molecule and enters the intradiscal space where it combines with phosphatidylethanolamine (PE) to form *N*-retinylidene-PE (N-RPE). RmP functions as an outwardly directing ATP-dependent flippase that transports N-RPE out of the intradiscal space into the cytoplasm of the photoreceptor outer segment. While doing so, it cleaves the bond between all-*trans* retinal and PE. All-*trans* retinal is then reduced to all-*trans* retinol and is transported to the RPE where it is reconverted back to 11-*cis* retinal. RmP has been demonstrated in experimental studies to bind N-RPE with high affinity (44). Without RmP, N-RPE accumulates in the intradiscal space. High levels of N-RPE favor the secondary condensation of N-RPE with all-*trans* retinal to form *N*-retinylidene-*N*-retinyl-PE (A2PE). A2PE is hydrolyzed to form *N*-retinylidene-*N*-retinyl-ethanolamine (A2E) (45).

N-retinylidene-*N*-retinyl-ethanolamine (A2E) is one of the major components of lipofuscin in the RPE. A2E is toxic to RPE cells (46), and in cell culture studies, A2E delays clearance of photoreceptor outer segments by the RPE (47). A2E is thought to exert this inhibitory effect by acting on either lipid hydrolases or lipid trafficking, thus slowing the ability of the RPE to perform its phagocytic role. Over time, A2E results in a decline in RPE support functions and ultimately photoreceptor cell death.

The primary involvement of ABCR-related conditions on the macula is likely due to the high ratio of photoreceptors to RPE cells in this region of the retina. The density of RPE cells is greatest at the fovea and gradually declines with foveal eccentricity (48). The density of photoreceptors (rods + cones) is highest in a ring centered at the fovea and approximately 3–5 mm in radius (49, 50), meaning that the perifoveal and macular areas have the highest photo-oxidative production of lipofuscin per RPE cell in the retina. This explains the predisposition for involvement of the macula in Stargardt's disease, fundus flavimaculatus, and age-related macular degeneration (51). CRD and RP generally affect the entire retina, although CRD may primarily affect the macula early in the condition. Additionally, most RP cases are due to mutations in genes other than *ABCA4*.

In summary, the pathophysiology of ABCR-related diseases is explained by a genetic mutation in the *ABCA4* gene resulting in insufficient or defective RmP in photoreceptor outer segments leading to the accumulation of N-RPE. N-RPE is converted into A2E that impairs RPE cell function and leads to the formation and accumulation of lipofuscin. RPE cells in the macula are affected earliest due to the high concentration of photoreceptors with respect to RPE cells in this region. RPE cell dysfunction and loss of RPE support function ultimately results in photoreceptor cell death.

9. Conclusions

An understanding of the molecular biology of ABC transporters has provided invaluable insight into the etiology and pathogenesis of numerous diseases. Ongoing research may lead to novel diagnostic and therapeutic advances. Further understanding of the rim protein, its function, and interaction with other proteins involved in the visual cycle will allow a more complete identification of the pathogenesis of *ABCA4* gene mutations. Continued advancement from a research perspective will lead to improved genetic testing and identification of all mutations involved. From a clinical standpoint, such advances will enable physicians to give individualized prognoses for patients and may ultimately lead to new, targeted therapies.

Gene therapy may ultimately produce a cure for ABC transporter-related diseases. With respect to ophthalmic conditions, it is conceivable that intravitreal injections of a vector containing wild-type *ABCR* gene could be used to produce functional RmP and treat patients with *ABCR* mutations. For now, ongoing research continues to reveal new information regarding the genetic and molecular basis of these conditions and will hopefully lead to new screening techniques and ultimately the development of novel therapies for these and other conditions associated with the ABC transporter family.

References

1. Nikaido, H. (1994) Maltose transport system of Escherichia coli: an ABC-type transporter. *FEBS Lett.***346**, 55–58.
2. Goffeau, A., Park, J., Paulsen, I.T., et al. (1997) Multidrug-resistant transport proteins in yeast: complete inventory and phylogenetic characterization of yeast open reading frames with the major facilitator superfamily. *Yeast* **13**, 43–54.
3. Gottesman, M.M. and Pastan, I. (1993) Biochemistry of multidrug resistance mediated by the multidrug transporter. *Annu. Rev. Biochem.* **62**, 385–427.
4. Allen, S.S., Cutting, G.R., Guggino, W.B., et al. (1995) CFTR regulates outwardly rectifying chloride channels through an autocrine mechanism involving ATP. *Cell* **81**, 1063–1073.
5. Efferth, T. (2001) The Human ATP-binding cassette transporter genes: from the bench to the bedside. *Curr. Mol. Med.* **1**, 45–65.
6. Michaelis, S. and Berkower, C. (1995) Sequence comparison of yeast ATP-binding cassette proteins. *Cold Spring Harb. Symp. Quant. Biol.* **60**, 291–307.

7. Walker, J.E., Saraste, M., Runswick, M.J., and Gray, N. (1982) Distantly related sequences in the alpha and beta subunits of ATP synthase, myosin, kinases, and other ATP requiring enzymes and a common nucleotide binding fold. *EMBO J.* **1**, 945–951.
8. Hwang, T.C., Nagel, G., Nairn, A.C., and Gadsby, D.C. (1994) Regulation of the gating of cystic fibrosis transmembrane conductance regulator Cl channels by phosphorylation and ATP hydrolysis. *Proc. Natl. Acad. Sci. USA* **91**, 4698–4702.
9. Dean, M. and Annilo, T. (2005) Evolution of the ATP-binding cassette (ABC) transporter superfamily in vertebrates. *Annu. Rev. Genomics Hum. Genet.* **6**, 123–142.
10. Higgins, C.F. (1995) P-glycoprotein and cell volume-activated chloride channels. *J. Bioenerg. Biomembr.* **27**, 63–70.
11. Kaminski, W.E., Piehler, A., and Wenzel, J.J. (2006) ABC A-subfamily transporters: structure, function, and disease. *Biochim. Biophys. Acta* **1762**, 510–524.
12. Gerber, S., Rozet, J.M., et al. (1995) A gene for late-onset fundus flavimaculatus with macular dystrophy maps to chromosome 1p13. *Am. J. Hum. Genet.* **56**, 396–399.
13. Anderson, K.L., Baird, L., Lewis, R.A., et al. (1995) A YAC contig encompassing the recessive Stargardt disease gene (STGD) on chromosome 1p. *Am. J. Hum. Genet.* **57**, 1351–1363.
14. Kaplan, J., Gerber, S., et al. (1993) A gene for Stargardt's disease (fundus flavimaculatus) maps to the short arm of chromosome 1. *Nat. Genet.* **5**, 308–311.
15. Hoyng, C.B., Poppelaars, F., van de Pol, T.J.R., et al. (1996) Genetic fine mapping of the gene for recessive Stargardt disease. *Hum. Genet.* **98**, 500–504.
16. Allikmets, R., Shroyer, N.F., Singh, N., et al. (1997) Mutation of the Stargardt disease gene (ABCR) in age-related macular degeneration. *Science* **277**, 1805–1807.
17. Nasonkin, I., Illing, M., Koehler, M.R., et al. (1998) Mapping of the rod photoreceptor ABC transporter (ABCR) to 1p21-p22.1 and identification of novel mutations in Stargardt's disease. *Hum. Genet.* **102**, 21–26.
18. Illing, M., Molday, L.L., and Molday, R.S. (1997) The 220-kDa rim protein of retinal rod outer segments is a member of the ABC transporter superfamily. *J. Biol. Chem.* **272**, 10303–10310.
19. Broccardo, C., Luciani M.F., and Chimini, G. (1999) The ABCA class of mammalian transporters. *Biochim. Biophys. Acta.* **1461**, 395–404.
20. Papermaster, D.S., Schneider, B.G., Zorn, M.A., and Kraehenbuhl, J.P. (1978) Immunocytochemical localization of a large intrinsic membrane protein to the incisures and margins of frog rod outer segment disks. *J. Cell Biol.* **78**, 415–425.
21. Molday, L.L., Rabin, A.R., and Molday, R.S. (2000) ABCR expression in foveal cone photoreceptors and its role in Stargardt macular dystrophy. *Am. J. Ophthalmol.* **130**, 689.
22. Gerber, S., Rozet, J.M., van de Pol, T.J.R., et al. (1998) Complete exon-intron structure of the retina specific ATP binding transporter gene (ABCR) allows the identification of novel mutations underlying Stargardt disease. *Genomics* **48**, 139–142.
23. Maugeri, A., van Driel, M.A., van de Pol, D.J., et al. (1999) The 2588GàC mutation in the ABCR gene is a mild frequent founder mutation in the Western European population and allows the classification of ABCR mutations in patients with Stargardt disease. *Am. J. Hum. Genet.* **64**, 1024–1035.
24. Simonelli, F., testa, F., de Crecchio, G., et al. (2000) New ABCR mutations and clinical phenotype in Italian patients with Stargardt disease. *Invest. Ophthalmol. Vis. Sci.* **41**, 892–897.
25. Jaakson, K., Zernant, J., Kulm, M., et al. (2003) Genotyping microarray (gene chip) for the ABCR (*ABCA4*) gene. *Hum. Mutat.* **22**, 395–403.
26. Cremers, F.P., van de Pol, D.J., van Driel, M. et al. (1998) Autosomal recessive retinitis pigmentosa and cone-rod dystrophy caused by splice site mutations in the Stargardt's disease gene ABCR. *Hum. Mol. Genet.* 7, 355–362.
27. Allikmets, R., Singh, N., Sun, H., et al. (1997) A photoreceptor cell-specific ATP-binding transporter gene (ABCR) is mutated in recessive Stargardt macular dystrophy. *Nat. Genet.* **15**, 236–246.
28. Stargardt, K. (1909) Uber familiare, progressive degeneration in der makulagegend des auges. *Graefes Arch. Clin. Exp. Ophthalmol.* **71**, 534–550.
29. Bonin, P., Passot, M., and Triolaire-Colton, M-Th. (1976) Le signe du silence choroidien dans les degenerscences tapetoretiniennes posterieures. In: De Laey JJ, ed. International Symposium on Fluorescein Angiography. *Doc. Ophthalmol. Proc. Ser.* **9**, 461–463.
30. Weleber, R.G. (1994) Stargardt's macular dystrophy. *Arch. Ophthalmol.* **112**, 752–754.

31. Hartong, D.T., Berson, E.L., and Dryja, T.P. (2006) Retinitis pigmentosa. *Lancet* **368**, 1795–1809.
32. Maugeri, A., Klevering, B.J., Rohrschneider, K., et al. (2000) Mutations in the ABCA4 (ABCR) gene are the major cause of autosomal recessive cone-rod dystrophy. *Am. J. Hum. Genet.* **67**, 960–966.
33. Ducroq, D., Rozet, J.M., Gerber, S., et al. (2002) The ABCA4 Gene in autosomal recessive cone-rod dystrophies. *Am. J. Hum. Genet.* **71**, 1480–1482.
34. Kuroiwa, S., Kojima, H., Kikuchi, T., and Yoshimura, N. (1999) ATP binding cassette transporter retina genotypes and age-related macular degeneration: an analysis on exudative non-familial Japanese patients. *Br. J. Ophthalmol.* **83**, 613–615.
35. Stone, E.M., Brain, E., Kimura, A.E., et al. (1994) Clinical features of a Stargardt-like dominant progressive macular dystrophy with genetic linkage to chromosome 6q. *Arch. Ophthalmol.* **112**, 765–772.
36. Rivera, A., White, K., Stohr, H., et al. (2000) A comprehensive survey of sequence variation in the ABCA4 (ABCR) gene in Stargardt disease and age-related macular degeneration. *Am. J. Hum. Genet.* **67**, 800–813.
37. Dryja, T.P., Briggs, C.E., Berson, E.L., Rosenfeld, P., et al. (1998) ABCR gene and age-related macular degeneration. *Science* **20**, 1107.
38. Seddon, J.M. (1999) Epidemiology of age-related macular degeneration. In *The Principles and Practice of Ophthalmology* (Albert, D.M., Jakobiec, F.A., eds.). WB Saunders, Philadelphia, **1,** 521–531.
39. Allikmets, R. and the International ABCR Screening Consortium. (2000) Further evidence for an association of ABCR alleles with age-related macular degeneration. *Am. J. Hum. Genet.* **67**, 487–491.
40. van Driel, M.A., Maugeri, A., Klevering, B.J., Hoyng, C.B., et al. (1998) ABCR unites what ophthalmologists divide. *Ophthalmic Genet.* **19**, 117–122.
41. Martinez-Mir, A., Palmoa, E., Allikmets, R., et al. (1998) Retinitis pigmentosa caused by a homozygous mutation in the Stargardt disease gene ABCR. *Nat. Genet.* **18**, 11–12.
42. Klevering, B.J., Yzer, S., Rohrschneider, K., et al. (2004) Microarray-based mutation analysis of the ABCA4 (ABCR) gene in autosomal recessive cone-rod dystrophy and retinitis pigmentosa. *Eur. J. Hum. Genet.* **12**, 1024–1032.
43. Sun, H. and Nathans, J. (1997) Stargardt's ABCR is localized to the disc membrane of retinal rod outer segments. *Nat. Genet.* **17**, 15–16.
44. Beharry, S., Zhong, M., and Molday, R.S. (2004) N-retinylidene-phosphatidylethanolamine is the preferred retinoid substrate for the photoreceptor-specific ABC transporter ABCA4 (ABCR). *J. Biol. Chem.* **279**, 53972–53979.
45. Mata, N.L., Weng, J., and Travis, G.H. (2000) Biosynthesis of a major lipofuscin fluorophore in mice and humans with ABCR-mediated retinal and macular degeneration. *Proc. Natl. Acad. Sci. USA* **97**, 7154–7159.
46. Eldred, G.E. and Laskey, M.R. (1993) Retinal age pigments generated by self-assembling lysosomotropic detergents. *Nature* **361**, 724–726.
47. Finnemann, S.C., Leung, L.W., and Rodriguez-Boulan, E. (2002) The lipofuscin component A2E selectively inhibits phagolysosomal degradation of photoreceptor phospholipid by the retinal pigment epithelium. *PNAS* **99**, 3842–3847.
48. Del Priore, L.V., Kuo, T.H., and Tezel, T.H. (2002) Age-related changes in human RPE cell density and apoptosis proportion in situ. *Invest. Ophthalmol. Vis. Sci.* **43**, 3312–3318.
49. Curcio, C.E., Sloan, K.R., and Kalina, R.E. (1990) Human photoreceptor topography. *J. Comp. Neurol.* **292**, 497–523.
50. Jonas, J.B., Schneider, U., and Naumann, G.O. (1992) Count and density of human photoreceptors. *Graefes Arch. Clin. Exp. Ophthalmol.* **230**, 505–510.
51. Weiter, J.J., Delori, F., and Dorey, C.K. (1988) Central sparing in annular macular degeneration. *Am. J. Ophthalmol.* **106**, 286–292.

Chapter 12

Imaging of Protein Translocation In Situ in Skeletal Muscle of Living Mice

Hans P.M.M. Lauritzen

Abstract

Skeletal muscle plays a key role in regulating whole body glucose homeostasis and severe dysfunction in insulin-mediated glucose uptake is the hallmark of insulin-resistant states and type II diabetes. Therefore it is highly pathophysiologically relevant to perform detailed studies of insulin signaling inside skeletal muscle cells in order to elucidate the specific molecular events during both normal and insulin-resistant conditions. So far, cell biology imaging techniques have been limited to in vitro cultured muscle originating from primary or cell line-based myoblasts. However, these types of cultured muscle lack many characteristics of fully differentiated muscle cells. By performing intravital protein translocation analysis directly in situ in living animals, we have been able to give a high-resolution account of the spatial and temporal details during insulin signaling in vivo in muscle that does not have the limitations of in vitro cultures. We have shown that after i.v. insulin injection, PI3-kinase activation and, in turn, GLUT4 translocation are initiated at the plasma membrane proper, the sarcolemma. Then insulin signaling progresses into the t-tubules with a velocity corresponding to the diffusion of sulforhodamine B-conjugated insulin molecules. By using intravital confocal time-lapse analysis we have revealed that the t-tubules are the membrane surface where the majority of the insulin signaling is located.

Key words: Glucose transporter translocation, intravital imaging, GFP fusion.

1. Introduction

Despite the important role of skeletal muscle as a metabolic sink in whole body glucose uptake, the majority of studies on signaling and protein translocation in skeletal muscle cells have been carried out during in vitro culture conditions (1–5). Under these conditions muscle cells are not able to fully differentiate from the primary or cell line-based myoblast stage to mature muscle, but the muscle cells stop in

Q. Yan (ed.), *Membrane Transporters in Drug Discovery and Development*, Methods in Molecular Biology 637,
DOI 10.1007/978-1-60761-700-6_12, © Springer Science+Business Media, LLC 2010

the myotube stage (for review see (6)). This is due to several in vitro conditions, lack of adequate growth factors, nervous innervation, and stretching, all factors that cannot be simulated sufficiently in vitro. Furthermore, in vitro cultivation of fully developed muscle fibers leads to significant dedifferentiation within a short time span and also due to the inadequacy of in vitro conditions. The differentiation of cultured muscle results in a lack of important characteristics, e.g., lateralized nuclei, fully developed t-tubule networks, neuromotor end plates, and insulin sensitivity. Thus, in order to study protein translocation in muscle fibers that are fully differentiated and accordingly represent mature muscle, the protein translocation assay has to be prepared, performed, and analyzed directly in situ in the living animal, without ever removing the muscle fibers from their physiological context of skeletal muscle within the animal.

2. Materials

2.1. Plasmid Preparation for In Vivo Transfection

1. Sterile 1 M $CaCl_2$ in Milli-Q H_2O (dH_2O) (*see* **Note 1**).
2. Freshly made sterile 0.05 M spermidine (*see* **Note 2**) in dH_2O.
3. 8% (w/v) polyvinylpyrrolidone in freshly opened absolute alcohol (*see* **Note 3**). Store aliquots at –20°C.
4. 0.6 μm gold particles.
5. Endotoxin-free plasmid DNA (Qiagen Endofree mega kit) at a concentration of 1 μg/μl in dH_2O. CMV-driven plasmids coding for either ARNO-EGFP (6) or GLUT4-EGFP (7) were used.
6. Tefzel Tubing.
7. Tubing prep station.
8. Tubing cutter.
9. 50 ml centrifuge tubes.

2.2. In Situ Transfection of Muscle Fibers

1. Bio-Rad Helios gene gun.
2. Bio-Rad helium regulator.
3. 100-μm nylon filter (*see* **Note 4**).
4. Vicryl sutures 5-0.
5. Pentobarbital sodium (50 mg/ml).
6. Autoclaved physiological saline with 2% penicillin–streptomycin.
7. Autoclaved surgical blades.
8. Autoclaved forceps.

9. Autoclaved scissor.
10. Hemostat.
11. Sterile desiccator.

2.3. Immune Staining for Co-localization with Endogenous Protein

1. Picric acid solution.
2. EM grade paraformaldehyde.
3. 1 M NaOH solution.
4. Magnetic stirrer and heater.
5. 500 ml Corning 0.22 μm filter system.
6. 0.1 M Sorensen buffer: 115 ml 0.2 M NaH_2PO_4 X H_2O+385 ml 0.2 M Na_2HPO_4 X 7 H_2O + 500 ml H_2O.
7. Freshly 95% O_2/5% CO_2-gassed Krebs–Ringer buffer.
8. Procaine.
9. Fine forceps.
10. Leica MZ16 FA stereomicroscope or equivalent stereo microscope for fluorescence.
11. 24-well plates.
12. Pasteur pipettes.
13. Non-mouse primary antibody and non-mouse secondary antibody.
14. Immunobuffer (IB): 0.25% BSA (5% stock) + 50 mM Glycine + 0.033% Saponin + 0.05% azide (10% stock in PBS) in PBS.
15. Microscope slides and cover glasses.
16. VECTASHIELD® mounting medium and nail polish.

2.4. Intravital Imaging

1. Coltene President mounting medium.
2. Circular cover glass.
3. Cover glass holder and mouse mounting platform (*see* **Note 5**).
4. Pentobarbital sodium (50 mg/ml).
5. Objective heater and controller or equivalent.
6. Zeiss C-Apochromat ×63 1.2 NA water immersion objective or equivalent objective. Objectives with long working distances are preferred.
7. Zeiss Achrostigmat ×10 0.25 NA air immersion objective or equivalent objective.
8. A Zeiss LSM-410 confocal microscope with ArKr 488 nm laser line and 515–525 band pass emission filter for EGFP detection or an equivalent microscope. For sulforhodamine B-conjugated insulin detection sulforhodamine 543 nm

laser line was used and emission light collected between 570 and 700 nm.

9. Sterile physiological saline.
10. Portex fine bore polyethylene tubing 0.28 mm inner diameter or equivalent (*see* **Note 6**).
11. 30 G needles (*see* **Note** 7).
12. CMA 400 microdialysis syringe pump.
13. Sulforhodamine B-conjugated insulin solution (17 μg/μl). Insulin solutions are dissolved together with 0.01% BSA to prevent insulin binding to glassware. To ensure that the insulin–SRB is fully dissolved, the solution is sonicated 3 × 40 s.
14. Metamorf imaging software V.7.6 (Universal imaging, USA).

3. Methods

In order to produce healthy transfected muscle fibers it is important to proceed as gently and "sterile" as possible. Furthermore, confocal imaging of fluorescent-tagged proteins is restricted to cells near the surface of the muscle. Thus, it is important to transfect fibers in upper fiber layers of the muscle. By using gene gun transfection with low pressure (100–240 psi) only a few fibers will be transfected, but these will be located close to the muscle surface and only a few will be damaged by the transfection procedure (7).

After successful in vivo transfection co-localization of the transfected EGFP fusion protein with the endogenous non-tagged counterpart should be analyzed before applying the technique to time-lapse experiments. By fixing the transfected muscle and isolating the individual transfected fibers, these fibers can be immuno-staining (a modified protocol after (8)) to determine the degree of overlap between EGFP fusion and endogenous protein (*see* **Fig. 12.1a–c**). Once correct localization has been verified then confocal microscopy time-lapse experiments of GFP fusion localization and translocation can be initiated. By injecting insulin i.v. in a tail vein, events like PI3-kinase and GLUT4 translocation can be followed (*see* **Fig. 12.2a,d**). The timing of intracellar signaling (GFP linked) events can be linked to the actual extracellular arrival of insulin (by injecting fluorescent tagged insulin (Sulforhodamine B coupled)) (*see* **Fig. 12.2b,c**) by capturing confocal images in two channels simultaneously (10).

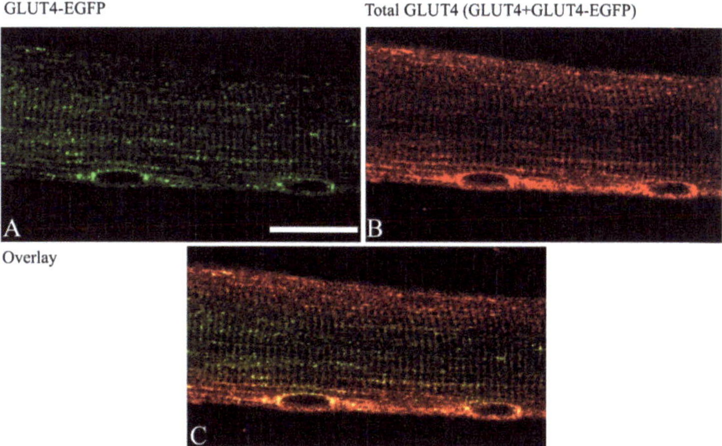

Fig. 12.1. Imaging of isolated gene gun GLUT4-EGFP-transfected mouse flexor digitorum brevis (FDB) muscle fiber immunostained for co-localization of GLUT4-EGFP with endogenous GLUT4. (**a**) In the GFP channel the isolated fibers show *green* fluorescence corresponding to GLUT4-EGFP. (**b**) Imaging of total immunostained (*red*) GLUT4 (GLUT4-EGFP+ endogenous GLUT4). (**c**) Overlay of (**a**) and (**b**) showing that that the *green* EGFP signal co-localized with the *red* GLUT4 staining giving rise to a *yellow* color in picture. This indicates that GLUT4-EGFP and endogenous GLUT4 co-locates. Bar is 20 μm. (Reproduced with permission from (9)).

3.1. Plasmid/Gold Particle Preparation

1. Frozen aliquots of sterile 50 μl 1 M $CaCl_2$, 3 ml absolute alcohol with 8% polyvinylpyrrolidone, and 50 μl plasmid solution are defrosted at RT.
2. 25 mg of 0.6 μm gold beads are added to an Eppendorf tube and mixed with 50 μl freshly mixed 0.05 M spermidine. Add 50 μl plasmid solution and vortex and mix for 10 s.
3. 50 μl 1 M $CaCl_2$ is added dropwise to the gold/DNA/spermidine solution at moderate vortexing. The gold beads should precipitate on the bottom of the tube. The mixture is incubated 5 min at RT with a brief vortexing every 30 s. The ratio 50 μg plasmid (1 μg/μl) to 25 mg gold from previous experience has been set to achieve high expression in the muscle fibers (*see* **Note 8**).
4. The gold/DNA mix is spun for 15 s at 3,000×*g* in an Eppendorf centrifuge to precipitate all gold particles.
5. The supernatant is discharged and the gold pellet is resuspended in 3 ml absolute alcohol with 8% polyvinylpyrrolidone by pipetting up and down several times.
6. The gene gun tubing is installed in the tubing prep station, gassed with nitrogen at approximately 0.4 liquid flow rate

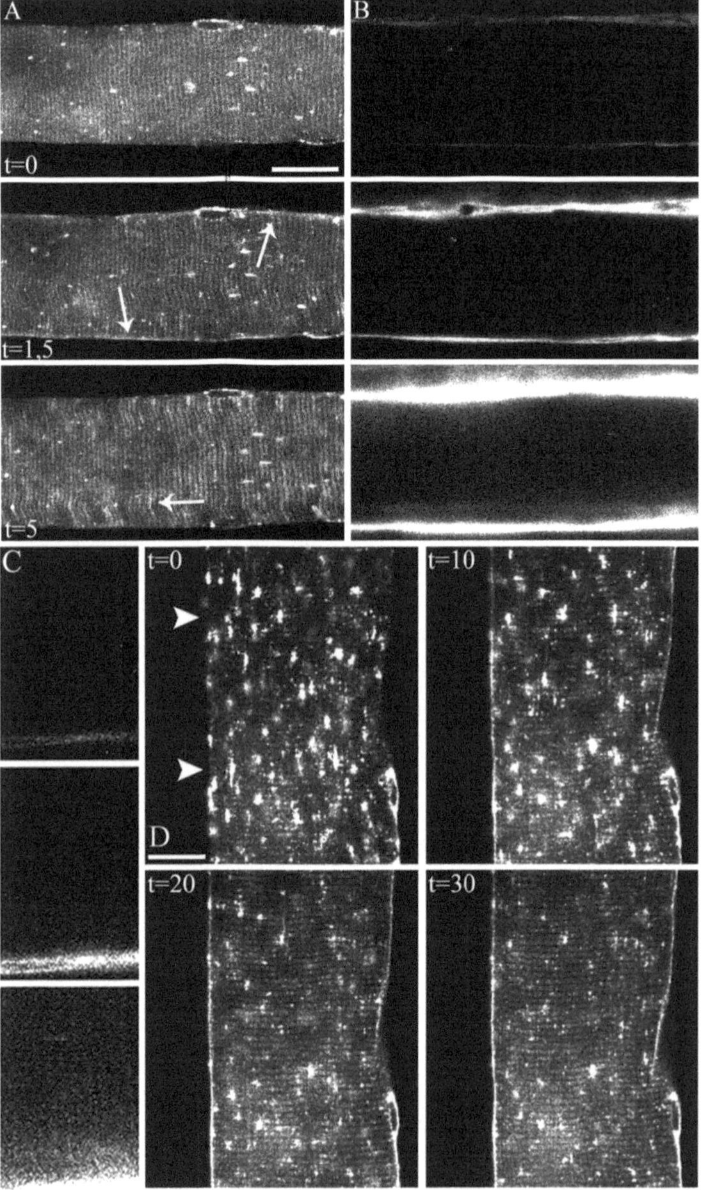

Fig. 12.2. Insulin-mediated PIP3 production and, in turn, GLUT4 translocation are delayed in the t-tubules compared to the sarcolemma corresponding to a delay in insulin diffusion. (**a**) *t*=0 shows a confocal image of a basal EGFP-ARNO expressing muscle fiber imaged in the EGFP channel just after i.v. injection (12 s after i.v. bolus) of sulforhodamine B-labeled insulin. Images of EGFP-ARNO are shown for the time points indicated (minute). Bar = 20 μm. At *t*=0 no PI3K is activated; thus the PIP3 receptor EGFP-ARNO is evenly distributed throughout the cell. At *t*=1.5 min after insulin injection, PI3K activity and, in turn, PIP3 production are first initiated at the sarcolemma to which EGFP-ARNO translocates (*arrows*). Later at *t*=5 min, EGFP-ARNO translocates to the t-tubules (striated white lines, arrows). (**b**) *t*=0 shows the same fiber as in (**a**) imaged for sulforhodamine B-labeled insulin in the sulforhodamine B channel. Insulin is seen to arrive at the sarcolemma (12 s after injection) at *t*=0. At *t*=1.5 min no insulin is present in the t-tubules. Insulin is seen only at the sarcolemma corresponding to the PI3K activity seen at *t*=1.5 min in (**a**). However, at *t*=5 min the tagged insulin gives rise to a striated pattern in the t-tubule region, indicating diffusion into the t-tubules. (**c**) An enlarged part of (**b**) is shown.

(LPM) for 20 min, and then removed from the tubing prep station. Nitrogen flow is turned off.

7. DNA/gold mix is vortexed vigorously to ensure full resuspension. Then the 3 ml DNA/gold mix is immediately sucked into the nitrogen-gassed gene gun tubing with a syringe.
8. The gene gun tubing containing the DNA/gold mix is immediately placed back in the tubing prep station. The ethanol solution (still containing some gold mix) is immediately removed by a peristaltic pump (0.25 ml/sec) (*see* **Note 9**).
9. Start spinning the tubing prep station to ensure even distribution of gold mix on the interior of the gene gun tubing.
10. When even distribution has been achieved the nitrogen gas flow is turned on and kept at a flow rate of 0.4 LPM for 15 min in order to dry the DNA/gold smear on the inside of the tubing.
11. The gene gun tubing is removed from the prep station and visual inspection ensures that the gold/plasmid mix has been distributed evenly. If not evenly distributed, then discard and start over. The tubing is cut in 0.5″ pieces by use of the tubing cutter, and the pieces are stored in a 50 ml centrifuge tube. The centrifuge tube is stored in a desiccator.

3.2. Transfection of Mice

1. The mice of choice are anesthetized. We have successfully used C57/Black (Charles River, Germany (7; 9–11)), ICR and SJL (Taconic, USA, H.P.M.M. Lauritzen, unpublished results), and FVB (Charles River, Germany (10, 11)) mice. Mice are first fully anesthetized and placed under a heating lamp or on a heating pad to prevent hypothermia. On each mouse the hair on the upper right leg is trimmed down by a hair trimmer. After trimming the upper leg is gently washed in 70% ethanol to remove any remaining hair and clean the area.
2. The gene gun is gently washed in 70% ethanol and the autoclaved barrel liner is installed. The autoclaved cartridge holder is filled with one empty cartridge in position 1. In the remaining positions are placed cartridges with DNA/gold

Fig. 12.2. (continued) (**d**) t=0 shows a confocal image of a basal GLUT4-EGFP expressing muscle fiber just before i.v. insulin injection. Images of GLUT4-EGFP are shown for t=10, 20, and 30 min. *Arrowheads* indicate sarcolemma. Maximal sarcolemma staining of GLUT4-EGFP is reached at t=20 min. In contrast, maximal GLUT4-EGFP staining of t-tubules is reached later at t=30 min. Bar = 20 μm. (Reproduced with permission from (8)).

coating of choice. A sterile 100-μm nylon filter is cut to fit the tip of the barrel liner. Pressure is adjusted to the desired psi (*see* **Note 10**).

3. For one mouse at a time, the skin covering the hip side of the upper right leg is gently washed in ethanol. Then the mouse is placed on an ethanol-washed flamingo box, and the skin is opened by a sterile surgical blade to expose the quadriceps muscle. The quadriceps muscle can subsequently be gene gun bombarded (*see* **Note 11**).
4. Immediately after bombardment, some drops of autoclaved saline with 2% penicillin/streptomycin are added to the exposed muscle surface to prevent drying. The skin is closed using vicryl sutures, and the animals are allowed to wake up under heated conditions.

3.3. Immunostaining of Transfected Fibers

1. 500 ml Zamboni's fixative is prepared by adding 75 ml picric acid to 11 g of paraformaldehyde in 500-ml glass flasks. In the hood the mixture is magnetically stirred and heated to 65°C for 15 min with foil covering the flask opening to avoid evaporation. Two to three drops of 1 M NaOH are added until the liquid changes from yellow to golden clear. The solution is diluted to 500 ml in 0.1 M Sorensen buffer and filtered with a Corning filtration system.
2. The mice are anesthetized, the skin opened and the transfected quadriceps muscle is removed by cutting around the kneecap and at the hip region.
3. Immediately after removal the muscle is incubated for 1 min in freshly gassed Krebs–Ringer buffer with procaine to avoid muscle fiber contraction during fixation (*see* **Note 12**).
4. The muscle is transferred to freshly made Zamboni fixative and left for 30 min at RT, followed by 3 h and 30 min incubation at 4°C.
5. After incubation in Zamboni, the muscles are transferred to PBS and left overnight at 4°C. The next day the muscles are transferred to 50% glycerol in PBS and can be stored at –20°C.
6. The muscles are taken out to equilibrate to RT for 3 min. The white superficial part of the vastus lateralis is removed from the bulk of the muscle using fine forceps under the stereo microscope.
7. Using the fluorescence aggregate and the EGFP filter of the stereo microscope, muscle segments of the white quadriceps are scanned for EGFP-fusion proteins (appearing as fine green fibers). The transfected fibers are sub-

sequently isolated from the other fibers by use of fine forceps and transferred to wells containing PBS in a 24-well plate.

8. Wells containing muscle fibers are incubated with immunobuffer (IB) for 30 min and then the buffer is removed gently by use of glass Pasteur pipettes (*see* **Note 13**). Immediately after buffer removal, IB with primary antibody is added (1 ml/well) and left to incubate overnight at RT under low shaking.
9. The next day wells are washed three times 30 min in IB followed by incubation for 1.5–3 h with secondary antibody.
10. After incubation with secondary antibody the fibers are washed three times 30 min with IB. If nuclei are to be stained, Hoechst dye (1:2,000) can be included in the first wash.
11. Finally, fibers are washed in PBS and mounted in Vector-shield mounting medium on glass slides. Cover glasses are sealed with use of nail polish.
12. Slides can be analyzed for the degree of color overlap (green + red= yellow) between exogenous EGFP fusion protein and endogenous protein (*see* **Fig. 12.1**).
13. In case the antibody cannot distinguish between endogenous protein and the EGFP-tagged version, the degree of overexpression can also be determined by measuring antibody staining within the transfected fibers. Since EGFP fusion protein expression is often distributed in a gradient from the site of gold bead entrance, nontransfected areas of the fiber can be compared with areas of increasing EGFP fusion protein expression (1, 3).

3.4. Mounting of Transfected Mice for Intravital Imaging

1. To mix the dental cement, a few drops of the president base is mixed with president catalyst in a dish and immediately distributed along the rim of the cover glass holder (*see* **Fig. 12.3a**). The president mix forms a thin gasket sealing possible water leak through the cover glass as well as keeping the cover glass in place.
2. A transfected mouse (5 days after transfection) is anesthetized and the skin covering the quadriceps is re-opened by a simple gentle pull of the skin, due to the vicryl sutures being partly absorbed. The exposed muscle is immediately covered by few drops of physiological saline to ensure that it does not dry out.
3. 20 ml of president base is mixed with 20 ml of president catalyst in a dish and distributed on top of the mounting

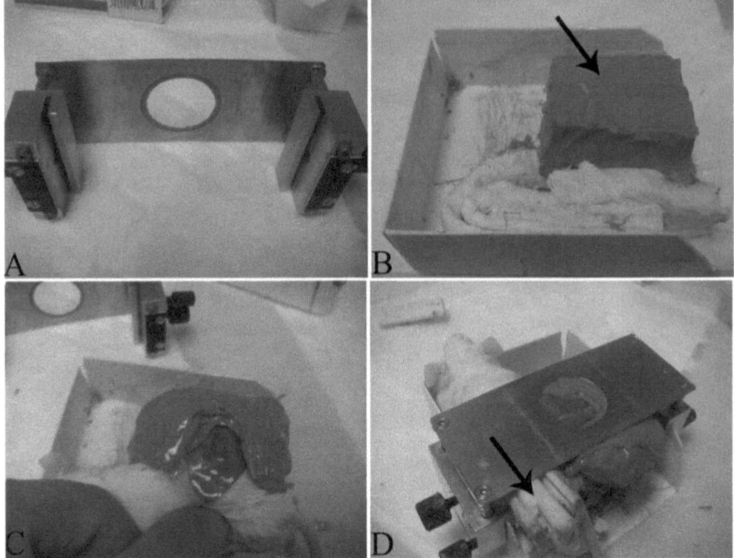

Fig. 12.3. Schematic representation of the setup for mounting mice for intravital confocal microscopy. (**a**) The cover glass holder with cover glass and the sealing dental cement gasket are shown. (**b**) The mounting platform consisting of a plastic box, a raised platform made of hardened dental cement (*arrow*), and a pillow made of paper are shown. (**c**) The open mouse leg imbedded into freshly mixed dental cement on top of the raised platform is shown. (**d**) The finished mounting of the animal, with the cover glass holder gently pressed into the dental cement and the cover glass covering the exposed quadriceps muscle, is shown. Note the additional paper pillow (*arrow*) on the left side used to reduce respiratory movements.

platform (*see* **Fig. 12.3b**). As fast as possible the right leg of the mouse is mounted and embedded into the president mixture (*see* **Fig. 12.3c**). A few drops of saline are added to the muscle and the cover glass holder is gently pressed down into the dental cement, gently touching the muscle surface (*see* **Fig. 12.3d**). A paper pillow is added (arrow) in the left side to stabilize and dampen movements.

4. While the dental cement hardens, an i.v. catheter can be gently placed in a tail vein if needed. After placement of the catheter it is important to keep a steady flow of saline (~1 μl/min) (microdialysis pump regulated) to prevent blood clots blocking the catheter needle. The animal setup is transferred to the confocal stage (*see* **Fig. 12.4a**). At low magnification, using an air immersion objective, the muscle surface is scanned for transfected fibers (*see* **Fig. 12.4b**).
5. When an appropriate transfected fiber is found, Zeiss immersol medium is added to the ×63 objective and this objective is used for the confocal recordings (*see* **Fig. 12.4c**).

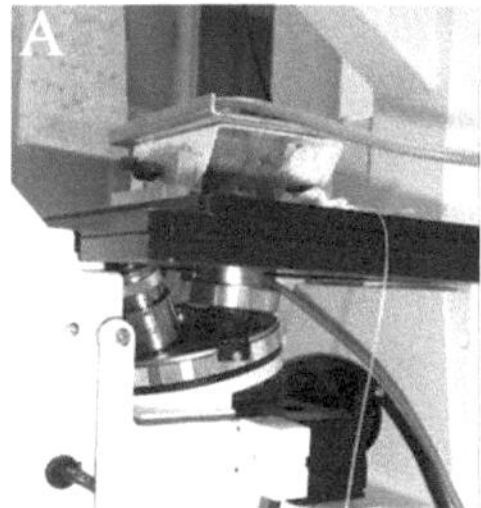

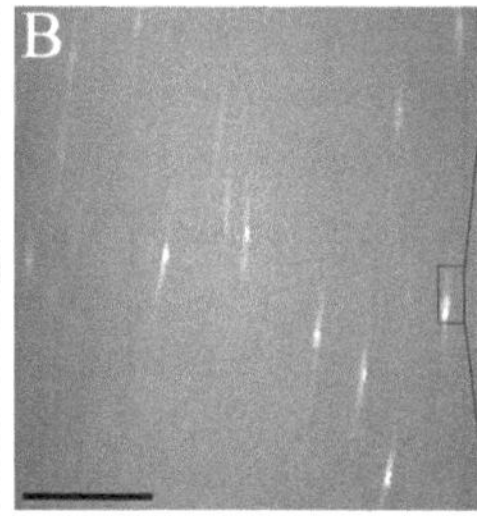

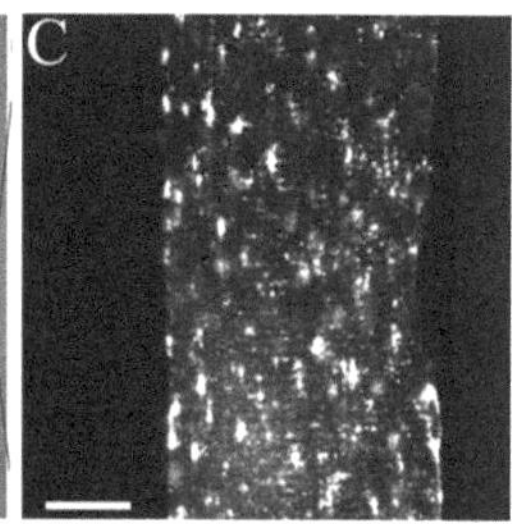

Fig. 12.4. The experimental setup for intravital confocal microscopy and images showing localization of and recording from GLUT4-EGFP transfected fibers in quadriceps muscle in situ in anesthetized mice. (**a**) The animal setup on the confocal stage with heating plate on top (*arrow*) and objective heater on the bottom. Note the catheter tubing crossing the stage into the tail vein. (**b**) A low-magnification image of the surface of the quadriceps muscle with several transfected fibers (*white lines*) is shown. Bar = 1 mm. (**c**) At high magnification an image shows GLUT4-EGFP in vesicle structures inside the muscle fiber depicted in (**b**). Bar is 20 μm.

6. During the anesthesia and mounting of the animal some cooling of the muscle has taken place. Thus, it is important to leave the animal on the confocal stage in contact with (via the immersion medium) the heated objective (37.5°C) and the stage heater plate for at least 30 min before starting time-lapse imaging (*see* **Note 14**).
7. After the initial setting period, confocal image acquisition can start at the time resolution permitted by the microscope at hand.
8. Infusion of dye-tagged insulin (*see* **Note 15**) or ordinary insulin solution can be performed by connecting the insulin solution to the catheter line feeding the continually saline flow. The rate of insulin solution flow can be adjusted precisely using the microdialysis pump. Following the insulin bolus an infusion of glucose can be given to prevent hypoglycemia. However, the appropriate glucose flow rates have to be established by monitoring glucose concentration, e.g., in tail vein blood.

3.5. Image Analysis

1. The TIF images obtained with the confocal software are imported into Metamorf Software (V. 7.6, Universal Imaging Corp.). In metamorf, time-lapse image stacks can be created and pixel values can be corrected for noncellular fluorescence and further processed in order to increase signal to noise ratio.
2. Image quantification of protein translocation, e.g., GLUT4-EGFP translocation, to membrane surfaces can be done by drawing a thin region of interest (ROI) including, e.g., the plasma membrane proper, the sarcolemma, or region of t-

tubules and excluding vesicular structures. Since the expression level of EGFP fusion protein varies between fibers and sometimes within the individual fibers (1), image quantification ROI values should be expressed relative to ROI values at $t = 0$ to compensate for such variation. Thus, arbitrary units illustrate the actual gray value of ROI fluorescence divided by ROI fluorescence value at $t = 0$.

4. Notes

1. Aliquots of 1 M $CaCl_2$ can be stored at –20°C for up to 6 weeks.
2. The spermidine solution has to be made fresh for each day to optimize transfection efficiency.
3. Aliquots of absolute alcohol with 8% polyvinylpyrrolidone can be stored at –20°C sealed with parafilm for up to 6 weeks.
4. Smaller diameter-sized filters, e.g., 40 μm, can be used. However, the number of transfected fibers are decreasing with decreasing filter size.
5. The cover glass holder has been made and customized by our local machine shop to fit our confocal microscope stage.
6. To be used as catheter together with the 30G needle.
7. The tip of the needle is gently cut with a fine file. The needle tip is then gently squeezed into the fine bore polyethylene tubing and used as catheter.
8. The DNA/gold ratio can be lowered with highly expressing plasmids.
9. Only a part of the DNA/gold will settle in the tubing, some will remain in solution and be removed with the ethanol by the pump.
10. Generally, the helium pressure has to be increased with decreasing filter size.
11. Alternatively if transfection efficiency is very low, the fascia covering the muscle can be removed before gene gun bombardment (the bombardment pressure may have to be reduced), ensuring a greater penetration through the surface of the muscle.
12. When removing the quadriceps muscle, it can be complicated to isolate the muscle at the tendons, thus leaving some of the muscle fibers open (as with biopsies). In case of open fibers the use Krebs–Ringer buffer without calcium

is advocated in order to avoid contraction of fibers before fixation.

13. It is important to carefully monitoring that no muscle fibers are removed together with the medium during washes when using the Pasteur pipette.
14. This period ensures temperature equilibrium of the animal and allows the whole setup to settle, thereby reducing out of focus drift during time-lapse recordings.
15. Dye-tagged insulin tends to precipitate at the high concentration (up to 60–70 μg/μl) used for imaging in vivo in skeletal muscle. Thus, immediately before i.v. injection the solution should be sonicated.

Acknowledgments

This work was supported by the Weimann Foundation, the Beckett Foundation, and the Danish National Research Foundation and the DERC P30DK036836 grant at the Joslin Diabetes Center.

References

1. Antonescu, C.N., Diaz, M., Femia, G., Planas, J.V., and Klip, A. (2008) Clathrin-dependent and independent endocytosis of glucose transporter 4 (GLUT4) in myoblasts: regulation by mitochondrial uncoupling. *Traffic* **9**, 1173–1190.
2. Schertzer, J.D., Antonescu, C.N., Bilan, P.J., Jain, S., Huang, X., Liu, Z., Bonen, A., and Klip, A. (2008) A transgenic mouse model to study GLUT4myc regulation in skeletal muscle. *Endocrinology*.
3. Karlsson, H.K., Chibalin, A.V., Koistinen, H.A., Yang, J., Koumanov, F., Wallberg-Henriksson, H., Zierath, J.R., and Holman, G.D. (2009) Kinetics of GLUT4 trafficking in rat and human skeletal muscle. *Diabetes*.
4. Fazakerley, D.J., Lawrence, S.P., Lizunov, V.A., Cushman, S.W., and Holman, G.D. (2009) A common trafficking route for GLUT4 in cardiomyocytes in response to insulin, contraction and energy-status signalling. *J. Cell Sci.* **122**, 727–734.
5. Lauritzen, H.P. (2009) In vivo imaging of GLUT4 translocation. *Appl. Physiol. Nutr. Metab.* **34**, 420–423.
6. Lauritzen, H.P. and Schertzer, J.D. (2010) Measuring GLUT4 translocation in mature muscle fibers. In press. *Am. J. Physiol. Endocrinol. Metab.*
7. Lauritzen, H.P.M.M., Reynet, C., Schjerling, P., Ralston, E., Thomas, S., Galbo, H., and Ploug, T. (2002) Gene gun bombardment-mediated expression and translocation of EGFP-tagged GLUT4 in skeletal muscle fibers in vivo. *Pflugers Arch. Eur. J. Physiol.* **444**, 710–721.
8. Ploug, T., van Deurs, B., Ai, H., Cushman, S.W., and Ralston, E. (1998) Analysis of GLUT4 distribution in whole skeletal muscle fibers: identification of distinct storage compartments that are recruited by insulin and muscle contractions. *J. Cell Biol.* **142**, 1429–1446.
9. Lauritzen, H.P., Galbo, H., Brandauer, J., Goodyear, L.J., and Ploug, T. (2008) Large GLUT4 vesicles are stationary while locally and reversibly depleted during transient insulin stimulation of skeletal muscle of living mice: imaging analysis of GLUT4-enhanced green fluorescent protein vesicle dynamics. *Diabetes* **57**, 315–324.

10. Lauritzen, H.P., Ploug, T., Prats, C., Tavare, J.M., and Galbo, H. (2006) Imaging of insulin signaling in skeletal muscle of living mice shows major role of T-tubules. *Diabetes* **55**, 1300–1306.

11. Lauritzen, H.P., Ploug, T., Ai, H., Donsmark, M., Prats, C., and Galbo, H. (2008) Denervation and high-fat diet reduce insulin signaling in T-tubules in skeletal muscle of living mice. *Diabetes* **57**, 13–23.

Chapter 13

Glucose Transporters in Parasitic Protozoa

Scott M. Landfear

Abstract

Glucose and related hexoses play central roles in the biochemistry and metabolism of single-cell parasites such as *Leishmania*, *Trypanosoma*, and *Plasmodium* that are the causative agents of leishmaniasis, African sleeping sickness, and malaria. Glucose transporters and the genes that encode them have been identified in each of these parasites and their functional properties have been scrutinized. These transporters are related in sequence and structure to mammalian facilitative glucose transporters of the SLC2 family, but they are nonetheless quite divergent in sequence. Hexose transporters have been shown to be essential for the viability of the infectious stage of each of these parasites and thus may represent targets for development of novel anti-parasitic drugs. The study of these transporters also illuminates many aspects of the basic biology of *Leishmania*, trypanosomes, and malaria parasites.

Key words: Glucose transporters, parasitic protozoa, nutrient uptake, inhibitors of glucose transport, drug development.

1. Introduction

The objective of this chapter is to provide a brief review of glucose transporters in three parasitic protozoa that cause important diseases in humans: *Leishmania mexicana*, *Trypanosoma brucei*, and *Plasmodium falciparum*. These parasites are unicellular eukaryotes that are the causative agents of cutaneous leishmaniasis, African sleeping sickness, and malaria, respectively, diseases that cause major health burdens worldwide (http://www.who.int/mediacentre/factsheets/fs094/en/). The review focuses on these parasites because (i) glucose transporters play central roles in the nutrition and metabolism of each organism and appear to be essential for their viability and

Q. Yan (ed.), *Membrane Transporters in Drug Discovery and Development*, Methods in Molecular Biology 637,
DOI 10.1007/978-1-60761-700-6_13,

(ii) substantial work has been accomplished at the molecular level on the glucose transporters from each species. Given the central role of glucose uptake in these parasites, an understanding of this process at the molecular, cellular, and genetic levels will promote a better understanding of the fundamental biochemistry and physiology of each pathogen. In addition, the molecular dissection of parasite glucose transporters may provide a target for development of novel therapeutics that selectively inhibit the essential parasite permeases without altering uptake of glucose in the mammalian host.

2. Leishmania mexicana

The genus *Leishmania* encompasses a variety of species that are pathogenic to humans and which causes distinct diseases collectively designated as leishmaniasis (1). These parasites exhibit two principal stages in their life cycles. Promastigotes are slender, flagellated protozoa (**Fig. 13.1a**) that live in the gut of the insect vector, any of several species of hematophagous sand flies, but can be cultured easily in a variety of tissue culture media. When infected female sand flies feed upon a vertebrate host (to obtain protein from the blood meal which is required for egg formation), they inject promastigotes into the skin, where they are taken up by host macrophages. Inside the macrophage, promastigotes transform into amastigotes, smaller, oval, non-flagellated forms (**Fig. 13.1b**), that are adapted for survival and growth within the acidified phagolysosomal vesicles of the macrophage. The physiological environments of extracellular promastigotes and intracellular amastigotes are quite distinct, with promastigotes being exposed to high levels of sugars from the plant nectar meals taken by the sand flies for general nutrition (2) and amastigotes, probably being exposed to much lower levels of sugars inside the macrophage phagolysosome (3–5).

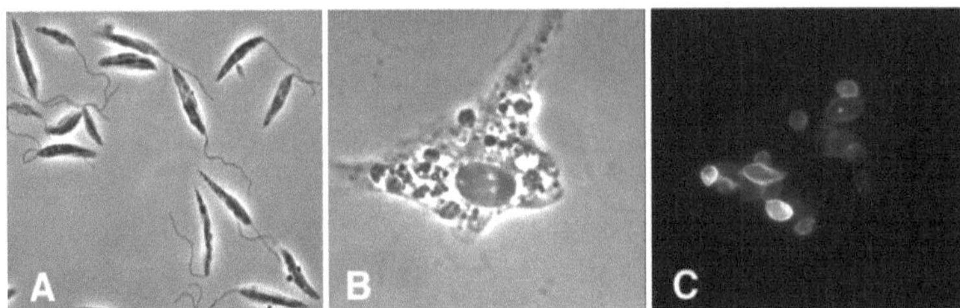

Fig. 13.1. Microscopic images of promastigotes and amastigotes of *L. mexicana*. Panels represent a phase contrast image of promastigotes (**a**), a phase contrast image of a murine peritoneal macrophage infected with amastigotes (**b**), and immunofluorescence (**c**) of the same image as in (**b**) stained with an antibody directed against the amastigote-specific surface protein HASPB (62) to visualize intracellular parasites.

L. mexicana, one agent of cutaneous leishmaniasis that induces ulcerating skin lesions, has been employed by my laboratory to examine the function of glucose transporters in these parasites. Initial studies revealed that the genome of *L. mexicana* encompasses a cluster of three linked glucose transporter genes designated *LmGT1*, *LmGT2*, and *LmGT3* (6) (**Fig. 13.2**). The proteins encoded by these genes are related in sequence and predicted secondary structure to mammalian facilitative glucose transporters (SLC2 family, http://www.bioparadigms.org/slc/intro.asp), which have 12 transmembrane domains (**Fig. 13.3**), but were only ~20% identical in sequence to the closest homologue, human GLUT1. The three transporters are isoforms that are similar but not identical to each other over the region encompassing the transmembrane domains and connecting loops, but they differ significantly in sequence in the hydrophilic domains at the NH_2 and COOH termini.

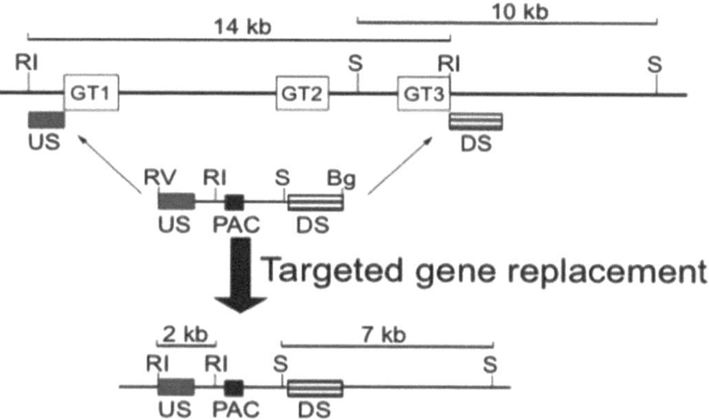

Fig. 13.2. Diagram of the glucose transporter gene cluster in *L. mexicana* and the strategy for generation of the Δ*lmgt* null mutant by targeted gene replacement. *Open rectangles* designate the open reading frames for the *GT1*, *GT2*, and *GT3* genes. The construct below the gene cluster contains a puromycin acetyl transferase gene (PAC) employed as a marker to confer resistance to puromycin and cloned segments from upstream (US) and downstream (DS) of the *GT* cluster to target gene replacement by homologous recombination. Integration of this construct (*heavy vertical arrow*) leads to replacement of the *GT* cluster with the drug resistance marker. Since *Leishmania* are diploid, generation of the homozygous null mutant required a subsequent gene replacement with a second drug resistance marker (*22*). This figure was reproduced from reference (22) with permission.

Expression of each isoform, initially in *Xenopus* oocytes (6) but later in a glucose transporter null mutant of *L. mexicana* (7), revealed that they were hexose transporters with the ability to take up glucose, fructose, mannose, and galactose. The affinity for glucose was high relative to mammalian homologues, with K_m values ranging from ~100 μM to ~1 mM.

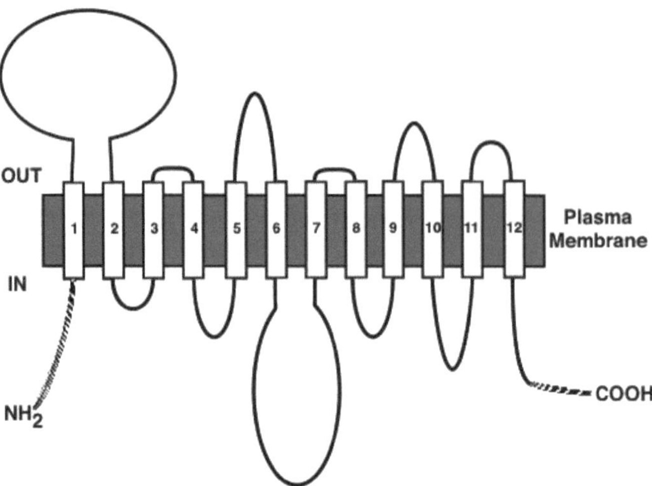

Fig. 13.3. Topology of *Leishmania* glucose transporters and regions of sequence divergence. They *gray horizontal rectangle* represents the plasma membrane, the *numbered white vertical rectangles* are predicted hydrophobic transmembrane segments, and the *black lines* indicate hydrophilic segments that connect transmembrane domains. The *stippled lines* at the NH_2 and COOH termini indicate sequences that are highly divergent between the LmGT1, LmGT2, and LmGT3 proteins. The figure is modified from Fig. 2 in reference (63).

There has been some uncertainty about whether the *Leishmania* glucose transporters are facilitative permeases, similar to their homologues in mammals (8) referred to as GLUTs, or concentrative proton symporters (9, 10). However, glucose transport studies carried out on intact parasites reveal that glucose uptake is inhibited by protonophores such as FCCP and respiratory inhibitors such as sodium azide (11–13), suggesting a role for proton symport. In addition, glucose-induced deacidification of the medium was observed when D-glucose, but not L-glucose, was added to de-energized parasites, also suggesting a role for proton symport. Furthermore, studies with sealed plasma membrane ghosts from *Leishmania donovani* (14) showed that an imposed pH gradient could drive uptake of labeled 2-deoxyglucose into the ghosts, as would be expected for a proton symporter. Addition of Mg^{+2}-ATP to everted membrane vesicles resulted in extrusion of preloaded [^{14}C]D-glucose from the vesicles by activating the proton-ATPase of the plasma membrane to concentrate protons in the interior of the vesicle. This observation that acidification of the everted vesicle lumen promoted extrusion of labeled glucose also supports an active proton symport mechanism. Although studies with *Leishmania* glucose transporters expressed in *Xenopus* oocytes did not reveal substrate-induced currents (15), this may have been due to loss of proton coupling in the heterologous expression system. It is also noteworthy that the structurally related *myo*-inositol transporter from *L. donovani* did generate substrate-induced currents with a charge to substrate stoichiometry of 1:1 when expressed in oocytes (16), establishing that members of the SLC2 family in *Leishmania* can function as protons symporters.

Despite the apparent similarity of LmGT1, LmGT2, and LmGT3 to each other regarding biochemical properties, some important biological distinctions emerged. First, *LmGT2* mRNA levels were strongly developmentally regulated being ~15-fold more abundant in promastigotes compared to amastigotes (6), consistent with previous biochemical studies demonstrating that promastigotes both imported (13) and metabolized (17, 18) glucose at much higher rates compared to amastigotes. In contrast, *LmGT1* and *LmGT3* mRNAs were constitutively expressed in both life cycle stages. Localization of functionally active fusion proteins between each permease and green fluorescent protein revealed that LmGT2 and LmGT3 were targeted to the pellicular plasma membrane surrounding the cell body of the amastigote, but they were excluded from the flagellar membrane. In striking contrast, LmGT1 was targeted preferentially to the flagellar membrane (**Fig. 13.4**). Studies performed with the GT1 isoform from

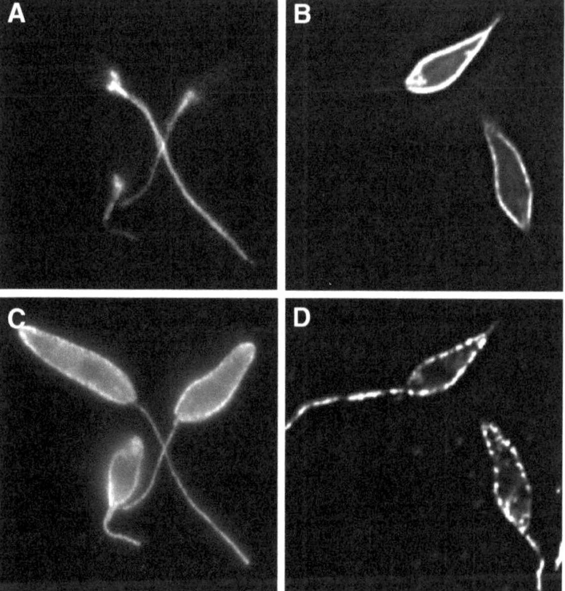

Fig. 13.4. Fluorescence images of *L. mexicana* promastigotes expressing LmGT1-GFP (**a**, **c**) and LmGT2-GFP (**b**, **d**) fusion proteins. The *top* images (**a**, **b**) represent GFP fluorescence and indicate that LmGT1 is localized to the flagellum and LmGT2 to the pellicular plasma membrane surrounding the cell body. The *bottom* images (**c**, **d**) represent the same cells as in (**a**, **b**) stained with anti-α-tubulin antibody to localize the flagellar axoneme and the pellicular microtubules that are attached to the cytosolic side of the pellicular plasma membrane. (**a**, **c**) represent conventional fluorescence micrographs, whereas (**b**, **d**) represent confocal micrographs representing a 0.5 μm section through the image.

the related *Leishmania* species *Leishmania enriettii* (19) revealed that two distinct regions of the divergent amino terminus of the flagellar isoform are required for flagellar targeting. These distinctions in developmental regulation and subcellular location of each isoform suggest that each permease subserves distinct functions for the parasite. The function of the flagellar isoform is not clear at present, but it is possible that it could serve some role in sensing glucose. This speculation is consistent with the observation that membrane proteins that are selectively localized to flagellar or ciliary membranes in a variety of eukaryotes are often involved in sensing the environment (20) and with the well-established role of glucose transporter-like proteins as nutrient sensors in yeast (21).

A genetic approach to defining glucose transporter function in *L. mexicana* was applied by generating a null mutant deficient in all three linked glucose transporter genes. Targeted gene replacement was employed to replace the *LmGT1–LmGT2–LmGT3* cluster on both homologous chromosomes with drug resistance markers (22) (**Fig. 13.2**). This Δ*lmgt* null mutant was deficient in uptake of glucose, fructose, mannose, and galactose, and complementation of the null mutant with each open reading frame (ORF) on an episomal expression vector restored uptake of all four sugars (7). These results confirmed that each permease is a hexose transporter with broad substrate specificity. The Δ*lmgt* promastigotes, the life cycle stage in which the null mutant was made, grew more slowly and to a lower density than wild-type parasites, and complementation with each *LmGT* ORF was able restore growth to near wild-type levels.

Infection of murine primary peritoneal macrophages with the Δ*lmgt* null mutant revealed an unexpected result. While the null mutant was able to enter macrophages with equal efficiency as wild-type promastigotes (22) and to transform into intracellular amastigotes (23), the mutant parasites did not replicate but died over the course of 6 days, whereas wild-type *L. mexicana* amastigotes replicated. Complementation with *LmGT2* or *LmGT3* was able to restore, at least partially, growth of Δ*lmgt* amastigotes in macrophages. These results suggest that glucose transporters are essential for the viability of intracellular amastigotes, despite the fact that glucose transport (13) and metabolism (17, 18) are downregulated in amastigotes compared to promastigotes. Consistent with the above observations, Δ*lmgt* promastigotes did not transform into axenic culture form amastigotes, and wild-type axenic amastigotes lost viability when glucose was withdrawn from the medium (22). Finally, while either wild-type or hemizygous knockouts in the glucose transporter gene cluster were able to form cutaneous lesions in BALB/c mice, the Δ*lmgt* promastigotes were unable to establish lesions (Richard Burchmore and Graham Coombs, unpublished data).

All the above observations point to the essential nature of glucose transporters in the infectious stage of the parasite life cycle. These results are unanticipated for several reasons. First, these permeases are required for viability in the life cycle stage in which their expression, as well as glucose metabolism, is downregulated, but they are not essential in promastigotes where the wild-type parasites express higher levels of the permease and metabolize glucose robustly. Second, while glucose and fructose reach high levels in the sandfly gut at least temporarily (2), hexoses are thought to be relatively less abundant within the parasitized phagolysosome (3–5), although no direct measurements of metabolites exist for these organelles. Finally, gluconeogenesis is active and essential in amastigotes of *L. major*, as demonstrated by the nonviability of null mutants in the gluconeogenic enzyme fructose 1,6-bisphosphatase in the amastigote stage of the life cycle (24). Curiously, it appears that both glucose uptake and gluconeogenesis are essential for survival of these intracellular parasites.

Several metabolic and cellular phenotypes of the Δ*lmgt* promastigotes suggest reasons why they may be impaired in viability inside the macrophage (7, 25). First, these null mutants are more susceptible to oxidative killing than wild-type promastigotes or null mutants that have been complemented with the *LmGT2* ORF. Second, the null mutants have reduced viability at elevated temperatures (e.g., 33°C compared to the usual temperature for cultivation of promastigotes, 27°C) compared to either wild-type parasites or null mutants complemented with the *LmGT2* gene (23). Third, null mutants are more susceptible to nutrient limitation than are wild-type or complemented mutants (23). All three of these 'environmental insults' are conditions that amastigotes are exposed to when they reside within macrophage phagolysosomes, and the increased susceptibility of the null mutants to these stresses likely contributes to their non-viability inside the host cell. In addition, the major storage carbohydrate of *Leishmania* promastigotes and amastigotes, the mannose polymer β-mannan, is reduced ~3-fold in level in Δ*lmgt* promastigotes compared to wild-type promastigotes (7), and the reduction of this established virulence factor (26) is likely to impair amastigote viability and possibly contribute to the increased susceptibility to nutrient limitation, increased oxidative stress, and elevated temperature noted above.

Another confirmation of the essential role of glucose uptake in amastigotes emerged from the detection and analysis of spontaneous suppressor mutants that arose when the promastigotes were passaged for several months in glucose-replete medium (23). These variants were initially observed to survive within primary murine macrophages significantly better than the original Δ*lmgt* mutants, although they did not replicate as robustly as wild-type amastigotes. In addition, the suppressors grew with similar kinetics compared to wild-type promastigotes rather than at the lower rate and density characteristic of Δ*lmgt* promastigotes. Furthermore, the suppressors had re-established the ability to transport glucose and fructose, although not to the same degree as wild-type promastigotes. Molecular analysis of these suppressors revealed that they had amplified another permease gene, *LmGT4*, on a circular amplicon and that they overexpressed *LmGT4* mRNA and protein compared to either wild-type or null mutant parasites. These suppressors had also at least partially restored relative resistance to oxidative stress, nutrient limitation, and elevated temperature, and they had re-established wild-type levels of β-mannan. The *LmGT4* gene (previously called the *D2* gene) had been characterized previously as a single-copy hexose transporter gene unlinked to the *LmGT1–LmGT2–LmGT3* locus (15). Curiously, Δ*lmgt* mutants do not exhibit any detectable hexose uptake even though the *LmGT4* gene is intact, and it is the amplification by ~50–100-fold of the subchromosomal region encompassing this gene that partially restores hexose transport. It is likely that the suppressors arose initially by the rare spontaneous generation of a circular element containing the *LmGT4* gene. Generation of circular extrachromosal elements that subsequently undergo amplification upon genetic selection for a gene encompassed within that element is an event that has been well documented for other loci in *Leishmania* (27). Since null mutants were passaged in glucose-replete medium, a genetic selection was being applied inadvertently for any variant that amplified an alternative hexose transporter gene such as *LmGT4*. Such genetic variants would be able to grow more rapidly in glucose-replete medium than the null mutants and hence would overtake the culture, leading to a uniform population of suppressor mutants.

3. Use of the Δ *lmgt* Null Mutant to Express Heterologous Glucose Transporters

In addition to studying the biological roles of glucose transporters in *Leishmania* parasites, the Δ*lmgt* mutant provides a useful glucose transport-deficient background in which to express heterologous glucose transporters. The THT1 hexose transporter from *T. brucei* (28), the PfHT1 hexose transporter from the malaria parasite *P. falciparum* (29), and the human glucose transporter GLUT1 (30) were all able to restore hexose transport to mutants when expressed in the null mutant using episomal expression vectors (31). These heterologous expressors provide a potential assay system for screening chemical libraries for selective inhibitors of those permeases. Indeed the Δ*lmgt* null mutants require amino acids such as proline as an alternate source of metabolic energy

(9, 22), since they cannot take up and metabolize hexoses. Consequently, growth of the Δ*lmgt* null mutant in proline-deficient medium is dependent upon both glucose in the medium and a functional complementing hexose transporter gene (31). Thus compounds that inhibit the heterologous transporter, such as the glucose analogue 3-*O*-(undec-10-en)yl-D-glucose that is a selective inhibitor of PfHT1 (32, 33), will inhibit growth of null mutants expressing that heterologous transporter. In principal, this growth assay could be employed to identify compounds that inhibit growth of null mutants expressing the THT1, PfHT1, or any of the *Leishmania* glucose transporters but not inhibiting growth of null mutant expressing the human glucose transporter GLUT1. Such selective inhibitors could be both useful reagents for probing structure and function of each permease and potential leads for development of drugs that inhibit essential parasite transporters.

4. Trypanosoma brucei

The African trypanosome *T. brucei* and its subspecies *T. b. gambiense* and *T. b. rhodesiense* are the causative agents of the disease nagana in cattle and African sleeping sickness in humans (1). Procyclic forms (PFs) are flagellated parasites (**Fig. 13.5a**) that live and divide within the midgut of the tsetse fly, whereas bloodstream forms (BFs) are flagellated parasites (**Fig. 13.5b**) that live extracellularly within the bloodstream of the mammalian host. Since BF trypanosomes live in a high-glucose environment, they are able to generate energy exclusively by glycolysis, and they do not possess a fully functional Krebs cycle or oxidative phosphorylation (34). As a result, BF parasites utilize large amounts of glucose to generate a small yield of ATP, and these life cycle forms are exquisitely dependent upon glucose uptake for viability. As such, trypanosome hexose transporters have been considered as

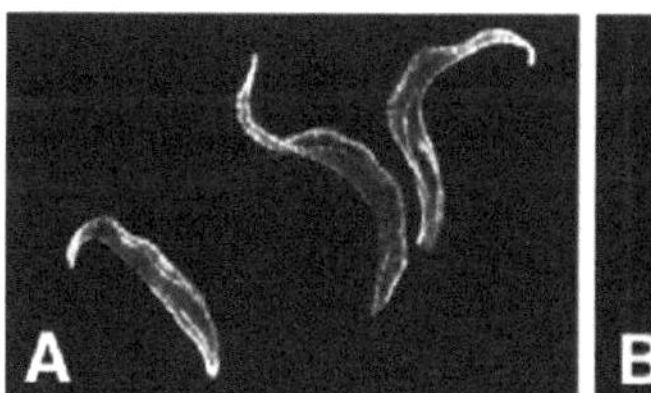

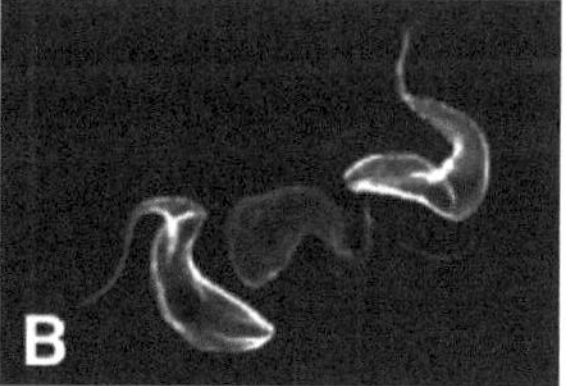

Fig. 13.5. Immunofluorescence micrographs of procyclic (**a**) and bloodstream (**b**) form *T. brucei* parasites stained with an anti-α-tubulin antibody. This antiserum reacts with the subpellicular microtubules attached to the cytosolic face of the plasma membrane and provides an image of the cell body.

potential targets for development of drugs against these important pathogens (35–38).

Early studies on transport of glucose or its analogues in BF trypanosomes (34, 39–41) revealed robust substrate saturable uptake activity with K_m values in the low millimolar range. This uptake could be inhibited by other hexoses such as fructose and mannose, suggesting that the transport system was probably a broad specificity hexose permease. Classical inhibitors of the human facilitative glucose transporter GLUT1, such as cytochalasin B and phloretin, could inhibit uptake of glucose by BF trypanosomes, but cytochalasin B inhibition required ~1,000-fold higher concentrations compared to the concentrations effective in inhibition of glucose uptake in human red blood cells that express GLUT1. Studies on the uptake of 2-deoxyglucose by PF trypanosomes (42) revealed a distinct, high-affinity transport activity (K_m for uptake of ~38 μM compared to ~1 mM for BF trypanosomes) that was sensitive to respiratory inhibitors and ionophores, suggesting that it represented proton-coupled, active transport. These distinctions between hexose transport in BF and PF trypanosomes would be consistent with the high levels of glucose available in mammalian blood compared to the much lower levels present in the gut of the tsetse fly.

Molecular cloning demonstrated that the *T. brucei* genome contains a cluster of hexose transporter genes (43). In the EATRO-164 strain first studied, this cluster encompasses six copies of the *THT1* gene followed by five copies of the *THT2* gene (43). The THT1 and THT2 proteins are 82% identical in sequence. THT1 mRNA is expressed almost exclusively in BF parasites (28), whereas *THT2* mRNA is expressed primarily in PF parasites but is present at a lower but detectable level on BFs (44). Kinetic studies performed on BF trypanosomes revealed that THT1 is a low-affinity, high-capacity transporter that takes advantage of the high concentration of glucose in blood, whereas THT2 is a high-affinity, low-capacity transporter adapted for life in the relatively low-glucose environment of the tsetse midgut. THT1 appears to be a facilitative transporter (44), but it has been suggested that THT2 may function as a proton symporter at low glucose concentration but as a facilitative transporter at high glucose concentration (45), i.e., that it undergoes 'slippage' regarding coupling of protons to glucose transport. To date, no subcellular localization studies have been performed for THT1 or THT2.

Two studies have sought to identify high-affinity inhibitors of glucose transport in BF trypanosomes (37, 38). Triazine compounds, especially Cibacron blue, were found to be moderately good inhibitors of glucose transport (37). Examination of glucose and fructose analogues revealed that adding a bromo-acetamide group onto the 2-*O* position of D-glucose or the 6-*O* position of

D-fructose produced analogues that were competitive inhibitors of D-glucose uptake by BF parasites (38), and it was suggested that the alkylated side chain may react covalently with a nucleophilic group on the trypanosome hexose transporter. These two compounds had trypanocidal effects with LD_{50} values in the low micromolar range. While these compounds by themselves are unlikely to provide novel trypanocides in their own right, their ability to inhibit the parasite hexose transporter and kill trypanosomes in culture suggests that this permease is a potential drug target.

5. Plasmodium falciparum

Malaria in humans is caused by several species of *Plasmodium*, with the most severe, life-threatening disease resulting from infection by *P. falciparum* (46). These parasites display a complex life cycle with various developmentally distinct stages occurring in both the mosquito vector and the human host, but the major disease-causing form of the parasite invades and replicates within the erythrocyte. These life cycle forms occupy an internal membrane-bound parasitophorous vacuole (PV) (**Fig. 13.6**). As such, nutrients must pass across three membranes: the red blood

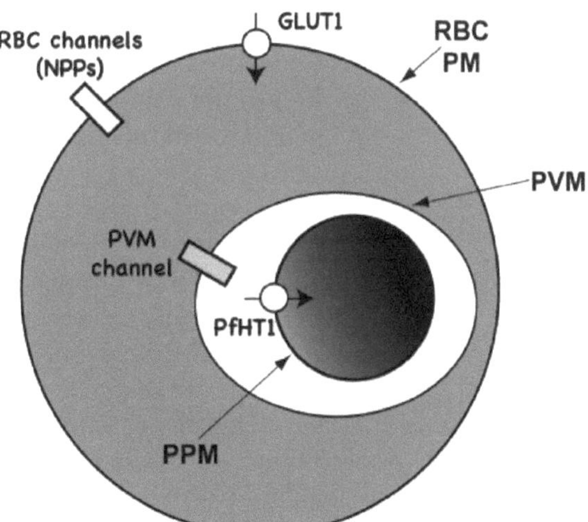

Fig. 13.6. Diagram of an erythrocyte infected with the malaria parasite *P. falciparum*. The *gray circle* represents the host red blood cell outlined by the red blood cell plasma membrane (RBCPM), the *white oval* is the parasitophorous vacuole surrounded by the parasitophorous vacuole membrane (PVM), and the inner *shaded circle* represents the malaria parasite delimited by the parasite plasma membrane (PPM). The red cell glucose transporter GLUT1 and the parasite glucose transporter PfHT1 are indicated by *circles* with *arrows* indicating the direction of glucose uptake. In addition channels with broad permeant specificity are depicted by *rectangles* in the red blood cell membrane (RBC channels or NPPs) and the PVM (PVM channel).

cell plasma membrane (RBCPM), the PV membrane (PVM), and the parasite plasma membrane (PPM).

Intraerythrocytic parasites acquire glucose from host blood and are completely dependent upon anaerobic glycolysis for viability (47). Consequently, parasitized red blood cells utilize glucose at ~25–50-fold higher level than non-parasitized erythrocytes (48–50) and export increased levels of lactate, an end product of glycolysis. The host cell glucose transporter GLUT1 is the major mediator of glucose transport across the RBCPM of the parasitized red blood cell (**Fig. 13.6**) (51). Various parasite-induced channel activities have been proposed to play roles in transport of multiple nutrients across the RBCPM (designated New Permeation Pathways or NPPs) (52, 53), but these NPPs do not play a significant role in glucose uptake compared to GLUT1 (51). There is a high-capacity, low-specificity channel in the PVM (54) that is likely to be the route for passage of glucose across that intracellular membrane. Early studies on uptake of glucose or its analogues by malaria-infected erythrocytes (50, 55–57) suggested that the parasite expressed a facilitative glucose carrier in its plasma membrane that mediated the uptake of this essential nutrient into the parasite cytoplasm. Subsequent studies in which polyadenylated RNA isolated from intraerythrocytic stage parasites was injected into *Xenopus* oocytes (58) revealed a significant stimulation of uptake of radiolabeled 2-deoxy-D-glucose compared to water-injected oocytes, and this uptake was stereospecific and not inhibited by L-glucose. These results implied that the parasite genome encodes endogenous glucose transporters likely to be involved in uptake of this nutrient across the PPM.

A major breakthrough in understanding glucose transport in malaria parasites emerged from identification, among sequences from the then ongoing *P. falciparum* genome project (http://www.sanger.ac.uk/Projects/P_falciparum/), of an open reading frame (ORF) designated *PfHT1* that exhibited significant (20–30%) amino acid sequence identity to human GLUTs (29). Expression of the *PfHT1* ORF in *Xenopus* oocytes induced saturable uptake of 2-deoxyglucose and glucose with a K_m for the latter substrate of 0.48 mM. Transport of both substrates was inhibited by fructose, mannose, galactose, and the glucose analogues 6-deoxyglucose and 3-*O*-methylglucose, indicating that the permease had broad specificity for hexoses. In addition, the classical inhibitors of mammalian GLUTs cytochalasin B, phloretin, and phloridzin exhibited partial inhibition of substrate uptake when applied in the concentration range of 50–500 μM. In synchronized infections of erythrocytes, *PfHT1* mRNA levels were strongly developmentally regulated. Transporter mRNA peaked in early ring stage parasites about 8 h post-invasion, fell about 10-fold in level by 20 h, and then accumulated to intermediate levels in more mature intraerythrocytic parasites between

30 and 40 h. In addition, the mRNA was expressed at very low levels in gametocytes, sexual stages of the parasite present in the blood of the infected host. A polyclonal antibody directed against a PfHT1 peptide recognized a single band on Western blots of lysates from infected erythrocytes and stained the plasma membrane of the intracellular parasites, confirming that this permease is localized to the surface membrane of the parasite rather than being inserted into host cell membranes. In summary, the first molecular studies on glucose transport in malaria parasites identified the gene encoding a high-affinity hexose transporter that mediates uptake of sugars across the parasite plasma membrane.

Further studies by the same group (59) demonstrated that PfHT1 transports both radiolabeled glucose and fructose, either of which can serve as an energy source for intraerythrocytic parasites. Furthermore mutation of the Q169 residue, located within predicted transmembrane helix 5, to N resulted in a permease that still transported glucose but was deficient in fructose uptake, indicating that this amino acid is involved in determining the substrate specificity of PfHT1.

The likely role of PfHT1 in providing essential nutrients to the intraerythrocytic parasites suggested that it played a vital role in survival of the infecting microorganism and that selective inhibitors of PfHT1 might provide promising leads for development of new antimalarial drugs. To validate PfHT1 as a potential drug target, the Krishna group examined the ability of various glucose analogues to selectively inhibit PfHT1 (32). The observation that 3-*O*-methylglucose strongly inhibited uptake of glucose by PfHT1 suggested that other 3-*O* derivatives might provide high-affinity inhibitors. Indeed the glucose derivative 3-*O*-((undec-10-en)-1-yl)-D-glucose, also referred to as compound 3361, inhibited uptake of labeled glucose by PfHT1 with a K_i of ~50 μM but was a very poor inhibitor of human GLUT1 (K_i ~ 3 mM). Notably compound 3361 killed *P. falciparum* intraerythrocytic parasites in culture with an IC_{50} value of 16 μM. Furthermore, treatment of mice infected with the murine malaria parasite *P. yoelii* with 25 mg/kg compound 3361 twice a day for 4 days resulted in significant suppression of parasitemia. This work provides proof of principle that the parasite glucose transporter could serve as a potential target for development of drugs that selectively inhibit its transport activity.

It is noteworthy that the completed genome of *P. falciparum* encodes two other potential sugar transporters in addition to *PfHT1* (60). These other ORFs are more divergent in sequence from known glucose transporters than PfHT1 and thus might represent transporters for other sugar or non-sugar ligands, but they have not been functionally characterized to date. These permeases might be important in the insect stages of the life cycle, where sugars other than glucose are likely to be present from the plant diet of the mosquito. It thus seems likely that PfHT1 is the glucose transporter that plays an essential nutritional role for the intraerythrocytic parasites and that inhibition of this permease by compound 3361 undermines the viability of the parasite within the red blood cell.

6. Summary and Perspectives

The work summarized here underscores the essential role of glucose transporters in the infectious stages of three parasites of importance to global health. For *T. brucei* and *P. falciparum* that live in glucose-rich blood or in erythrocytes and rely upon glucose metabolism exclusively, the essential role of hexose permeases is a logical expectation. The reason for the essential nature of glucose transporters in the intracellular stage of *Leishmania* parasites, where hexoses are thought to be relatively non-abundant and where glucose transport and metabolism are downregulated compared to the insect stage of the life cycle, is more obscure. Nonetheless, the essential nature of hexose permeases in each parasite, demonstrated either genetically in *L. mexicana* or by the inhibition of parasite growth by selective inhibitors for *T. brucei* and *P. falciparum*, indicates that these permeases are central to the metabolism of each parasite and present the potential for exploitation as therapeutic targets. The low degree of sequence identity between the microbial transporters and their GLUT homologues in mammals suggests that many inhibitors may be selective for the parasite over the human permeases.

The identification of hexose analogues that inhibit trypanosome or malaria hexose transporters provides an experimental demonstration that selective inhibitors do exist, and their ability to inhibit growth of parasites in vitro and in animal models of infection confirms the therapeutic potential of inhibiting these nutrient carriers. Nonetheless, the inhibitory substrate analogues identified to date are not likely to provide compounds useful as anti-parasitic drugs. An alternative approach suggested by multiple workers would be to screen large libraries of small compounds for those that selectively inhibit the parasite transporters and then to examine these compounds for those with the most likely properties as drug leads. For parasites that live in the bloodstream and are exposed to high concentrations of glucose, non-substrate analogues that inhibit noncompetitively at sites distinct from the substrate may offer an advantage. The efficacy of such inhibitors would not be compromised by competition with the natural substrate for access to the permease.

In addition to their potential role in drug development, hexose transporters provide opportunities to study many basic problems central to the biochemistry, physiology, and cell biology of these parasites. Where multiple members of the family exist, it is imperative to understand the distinct functions of each permease. For the THT1 and THT2 permeases of trypanosomes, the preferential expression in BF and PF trypanosomes, respectively, correlates with transport properties required to support life in high versus low glucose concentrations. However, for the three glucose transporters expressed by *Leishmania* parasites, the distinct functions of each isoform are less clear. Differential subcellular targeting of distinct glucose transporters provides a model for understanding how membrane proteins are trafficked to specific organelles in parasites. One striking example of discrete organellar targeting is provided by the LmGT1 flagellar glucose transporter in *L. mexicana*. Since other proteins are targeted to the flagellar membranes of kinetoplastid parasites such as *Leishmania* and *Trypanosoma* (20), LmGT1 may provide insights regarding how an entire class of membrane proteins are trafficked to the flagellum. While we know some information about sequences required for flagellar targeting, we do not know what cellular machinery directs such proteins to the flagellar membrane. As for various nutrients in kinetoplastid parasites, growth under conditions of glucose limitation induces higher glucose uptake capacity (61), and at least some of this increase may be due to enhanced expression of glucose transporters. In parallel, deletion of the glucose transporter genes or starvation for glucose in *Leishmania* leads to changes in the levels of many other proteins (Xiuhong Feng, Scott Landfear, and Richard Burchmore, unpublished data), suggesting that the parasites respond to the inability to take up a metabolically important nutrient by altering expression of other proteins in ways that may promote parasite viability or alter metabolism. Thus getting to know the lives of glucose transporters may provide a rich body of information on regulation of gene and protein expression, differential targeting of membrane proteins, and physiological requirements for nutrient uptake and metabolism among parasitic protozoa.

Acknowledgments

The author would like to thank Dr. Richard Burchmore for communicating unpublished data regarding murine infections with *L. mexicana* glucose transporter null mutants and protein expression in these null mutants and for providing a critical reading of the chapter. Images for figures included in this chapter were provided by Cosmo Buffalo, Dayana Rodriguez-Contreras, and Marco Sanchez in the author's laboratory. This work was supported by grant number AI25920 from the National Institutes of Health.

References

1. Stuart, K., Brun, R., Croft, S., Fairlamb, A., Gurtler, R.E., McKerrow, J., Reed, S., and Tarleton, R. (2008) Kinetoplastids: related protozoan pathogens, different diseases. *J. Clin. Invest.* **118**, 1301–1310.
2. Schlein, Y. (1986) Sandfly diet and *Leishmania. Parasitol. Today* **2**, 175–177.
3. Burchmore, R.J. and Barrett, M.P. (2001) Life in vacuoles – nutrient acquisition by Leishmania amastigotes. *Int. J. Parasitol.* **31**, 1311–1320.
4. McConville, M.J., de Souza, D., Saunders, E., Likic, V.A., and Naderer, T. (2007) Living in a phagolysosome; metabolism of Leishmania amastigotes. *Trends Parasitol.* **23**, 368–375.
5. Naderer, T. and McConville, M.J. (2008) The Leishmania–macrophage interaction: a metabolic perspective. *Cell Microbiol.* **10**, 301–308.
6. Burchmore, R.J.S. and Landfear, S.M. (1998) Differential regulation of multiple glucose transporter genes in the parasitic protozoan *Leishmania mexicana. J. Biol. Chem.* **273**, 29118–29126.
7. Rodriguez-Contreras, D., Feng, X., Keeney, K.M., Bouwer, H.G., and Landfear, S.M. (2007) Phenotypic characterization of a glucose transporter null mutant in Leishmania mexicana. *Mol. Biochem. Parasitol.* **153**, 9–18.
8. Manolescu, A.R., Witkowska, K., Kinnaird, A., Cessford, T., and Cheeseman, C. (2007) Facilitated hexose transporters: new perspectives on form and function. *Physiology (Bethesda)* **22**, 234–240.
9. Ter Kuile, B.H. (1993) Glucose and proline transport in kinetoplastids. *Parasitol. Today* **9**, 206–210.
10. Ter Kuile, B.H. (1994) Membrane-related processes and overall energy metabolism in *Trypanosoma brucei* and other kinetoplastid species. *J. Bioenerg. Biomembr.* **26**, 167–172.
11. Zilberstein, D. and Dwyer, D. (1984) Glucose transport in *Leishmania donovani* promastigotes. *Mol. Biochem. Parasitol.* **12**, 327–336.
12. Zilberstein, D. and Dwyer, D. (1985) Proton force-driven active transport of D-glucose and L-proline in the protozoan parasite *Leishmania donovani. Proc. Natl. Acad. Sci. USA* **82**, 1716–1720.
13. Burchmore, R.J.S. and Hart, D.T. (1995) Glucose transport in promastigotes and amastigotes of *Leishmania mexicana*: characterization and comparison with host glucose transporters. *Mol. Biochem. Parasitol.* **74**, 77–86.
14. Mukherjee, T., Mandal, D., and Bhaduri, A. (2001) *Leishmania* plasma membrane Mg^{+2} ATPase is a H+/K+-antiporter involved in glucose symport. *J. Biol. Chem.* **276**, 5563–5569.
15. Langford, C.K., Kavanaugh, M.P., Stenberg, P.E., Drew, M.E., Zhang, W., and Landfear, S.M. (1995) Functional expression and subcellular localization of a high-K_m hexose transporter from *Leishmania donovani. Biochemistry* **34**, 11814–11821.
16. Drew, M.E., Langford, C.K., Klamo, E.M., Russell, D.G., Kavanaugh, M.P., and Landfear, S.M. (1995) Functional expression of a *myo*-inositol/H^+ symporter from *Leishmania donovani. Mol. Cell. Biol.* **15**, 5508–5515.
17. Hart, D.T. and Coombs, G.H. (1982) *Leishmania mexicana*: energy metabolism of amastigotes and promastigotes. *Exp. Parasitol.* **54**, 397–409.
18. Rainey, P.M. and MacKenzie, N.E. (1991) A carbon-13 nuclear magnetic resonance analysis of the products of glucose metabolism in *Leishmania pifano* amastigotes and promastigotes. *Mol. Biochem. Parasitol.* **45**, 307–316.
19. Nasser, M.I.A. and Landfear, S.M. (2004) Sequences required for the flagellar targeting of an integral membrane protein. *Mol. Biochem. Parasitol.* **135**, 89–100.
20. Landfear, S.M. and Ignatushchenko, M. (2001) The flagellum and flagellar pocket of trypanosomatids. *Mol. Biochem. Parasitol.* **115**, 1–17.
21. Johnston, M. and Kim, J.H. (2005) Glucose as a hormone: receptor-mediated glucose sensing in the yeast Saccharomyces cerevisiae. *Biochem. Soc. Trans.* **33**, 247–252.
22. Burchmore, R.J.S., Rodriguez-Contreras, D., McBride, K., Merkel, P., Barrett, M.P., Modi, G., Sacks, D.L., and Landfear, S.M. (2003) Genetic characterization of glucose

transporter function in *Leishmania mexicana*. *Proc. Natl. Acad. Sci. USA* **100**, 3901–3906.

23. Feng, X., Rodriguez-Contreras, D., Buffalo, C., Bouwer, A., Kruvand, E., Beverley, S.M., and Landfear, S.M. (2009) Amplification of an alternate transporter gene suppresses the avirulent phenotype of glucose transporter null mutants in Leishmania mexicana. *Mol. Microbiol.* 71: 369–381.
24. Naderer, T., Ellis, M.I., Sernee, M.F., De Souza, D.P., Curtis, J., Handman, E., and McConville, M.J. (2006) Virulence of Leishmania major in macrophages and mice requires the gluconeogenic enzyme fructose-1,6-bisphosphatase. *Proc. Natl. Acad. Sci. U.S.A.* **103,** 5502–5507.
25. Rodriguez-Contreras, D. and Landfear, S.M. (2006) Metabolic changes in glucose transporter-deficient Leishmania mexicana and parasite virulence. *J. Biol. Chem.* **281**, 20068–20076.
26. Ralton, J.E., Naderer, T., Piraino, H.L., Bashtannyk, T.A., Callaghan, J.M., and McConville, M.J. (2003) Evidence that intracellular {beta}1-2 mannan is a virulence vector in *Leishmania* parasites. *J. Biol. Chem.* **278**, 40757–40763.
27. Beverley, S.M. (1991) Gene amplification in *Leishmania*. *Annu. Rev. Microbiol.* **45**, 417–444.
28. Bringaud, F. and Baltz, T. (1992) A potential hexose transporter gene expressed predominantly in the bloodstream form of *Trypanosoma brucei*. *Mol. Biochem. Parasitol.* **52**, 111–122.
29. Woodrow, C.J., Penny, J.I., and Krishna, S. (1999) Intraerythrocytic Plasmodium falciparum expresses a high affinity facilitative hexose transporter. *J. Biol. Chem.* **274**, 7272–7277.
30. Mueckler, M., Caruso, C., Baldwin, S.A., Panico, M., Blench, I., Morris, H.R., Allard, W.J., Lienhard, G.E., and Lodish, H.F. (1985) Sequence and structure of a human glucose transporter. *Science* **229**, 941–945.
31. Feistel, T., Hodson, C.A., Peyton, D.H., and Landfear, S.M. (2008) An expression system to screen for inhibitors of parasite glucose transporters. *Mol. Biochem. Parasitol.* **162**, 71–76.
32. Joet, T., Eckstein-Ludwig, U., Morin, C., and Krishna, S. (2003) Validation of the hexose transporter of Plasmodium falciparum as a novel drug target. *Proc. Natl. Acad. Sci. USA* **100**, 7476–7479.
33. Ionita, M., Krishna, S., Leo, P.M., Morin, C., and Patel, A.P. (2007) Interaction of O-(undec-10-en)-yl-d-glucose derivatives with the Plasmodium falciparum hexose transporter (PfHT). *Bioorg. Med. Chem. Lett.* **17**, 4934–4937.
34. Ter Kuile, B.H. and Opperdoes, F.R. (1991) Glucose uptake by *Trypanosoma brucei*. *J. Biol. Chem.* **266**, 857–862.
35. Ter Kuile, B.H. (1991) Glucose uptake mechanisms as potential targets for drugs against trypanosomatids. In *Biochemical Protozoology* (Coombs, G.H. and North, M., eds.). Taylor and Francis, London and Washington, pp 635.
36. Bakker, B.M., Westerhoff, H.V., Opperdoes, F.R., and Michels, P.A. (2000) Metabolic control analysis of glycolysis in trypanosomes as an approach to improve selectivity and effectiveness of drugs. *Mol. Biochem. Parasitol.* **106**, 1–10.
37. Bayele, H.K. (2001) Triazinyl derivatives that are potent inhibitors of glucose transport in Trypanosoma brucei. *Parasitol. Res.* **87**, 911–914.
38. Azema, L., Claustre, S., Alric, I., Blonski, C., Willson, M., Perie, J., Baltz, T., Tetaud, E., Bringaud, F., Cottem, D., Opperdoes, F.R., and Barrett, M.P. (2004) Interaction of substituted hexose analogues with the Trypanosoma brucei hexose transporter. *Biochem. Pharmacol.* **67**, 459–467.
39. Eisenthal, R., Game, S., and Holman, G.D. (1988) Specificity of hexose transport in *Trypanosoma brucei*. *Biochim. Biophys. Acta.* **985**, 81–89.
40. Munoz-Antonia, T., Richards, F.F., and Ullu, E. (1991) Differences in glucose transport between bloodstream and procyclic forms of *Trypanosoma brucei rhodesiense*. *Mol. Biochem. Parasitol.* **47**, 73–82.
41. Seyfang, A. and Duszenko, M. (1991) Specificity of glucose transport in *Trypanosoma brucei*: effective inhibition by phloretin and cytochalasin B. *Eur. J. Biochem.* **202**, 191–196.
42. Parsons, M. and Nielsen, B. (1990) Active transport of 2-deoxy-D-glucose in *Trypanosoma brucei*. *Mol. Biochem. Parasitol.* **42**, 197–204.
43. Bringaud, F. and Baltz, T. (1993) Differential regulation of two distinct families of glucose transporter genes in *Trypanosoma brucei*. *Mol. Cell. Biol.* **13**, 1146–1154.
44. Barrett, M.P., Tetaud, E., Seyfang, A., Brignaud, F., and Baltz, T. (1998) Trypanosome glucose transporters. *Mol. Biochem. Parasitol.* **91**, 195–205.
45. Barrett, M.P., Tetaud, E., Seyfang, A., Bringaud, F., and Baltz, T. (1995) Functional expression and characterization of the

Trypanosoma brucei procyclic glucose transporter, THT2. *Biochem. J.* **312**, 687–691.
46. Winzeler, E.A. (2008) Malaria research in the post-genomic era. *Nature* **455**, 751–756.
47. Sherman, I.W. (1979) Biochemistry of Plasmodium (malarial parasites), *Microbiol. Rev.* **43**, 453–495.
48. Roth, E.F., Jr. (1987) Malarial parasite hexokinase and hexokinase-dependent glutathione reduction in the Plasmodium falciparum-infected human erythrocyte. *J. Biol. Chem.* **262**, 15678–15682.
49. Sherman, I.W. (1988) The Wellcome Trust lecture. Mechanisms of molecular trafficking in malaria. *Parasitology* (**96** Suppl), S57–S81.
50. Tanabe, K. (1990) Glucose transport in malaria infected erythrocytes. *Parasitol. Today* **6**, 225–229.
51. Kirk, K. and Saliba, K.J. (2007) Targeting nutrient uptake mechanisms in Plasmodium. *Curr. Drug. Targets* **8**, 75–88.
52. Becker, K. and Kirk, K. (2004) Of malaria, metabolism and membrane transport. *Trends Parasitol.* **20**, 590–596.
53. Staines, H.M., Alkhalil, A., Allen, R.J., De Jonge, H.R., Derbyshire, E., Egee, S., Ginsburg, H., Hill, D.A., Huber, S.M., Kirk, K., Lang, F., Lisk, G., Oteng, E., Pillai, A.D., Rayavara, K., Rouhani, S., Saliba, K.J., Shen, C., Solomon, T., Thomas, S.L., Verloo, P., and Desai, S.A. (2007) Electrophysiological studies of malaria parasite-infected erythrocytes: current status. *Int. J. Parasitol.* **37**, 475–482.
54. Desai, S.A., Krogstad, D.J., and McCleskey, E.W. (1993) A nutrient-permeable channel on the intraerythrocytic malaria parasite. *Nature* **362**, 643–646.
55. Izumo, A., Tanabe, K., Kato, M., Doi, S., Maekawa, K., and Takada, S. (1989) Transport processes of 2-deoxy-D-glucose in erythrocytes infected with Plasmodium yoelii, a rodent malaria parasite. *Parasitology* **98** *(Pt 3)*, 371–379.
56. Kirk, K., Horner, H.A., and Kirk, J. (1996) Glucose uptake in Plasmodium falciparum-infected erythrocytes is an equilibrative not an active process. *Mol. Biochem. Parasitol.* **82**, 195–205.
57. Goodyer, I.D., Hayes, D.J., and Eisenthal, R. (1997) Efflux of 6-deoxy-D-glucose from Plasmodium falciparum-infected erythrocytes via two saturable carriers. *Mol. Biochem. Parasitol.* **84**, 229–239.
58. Penny, J.I., Hall, S.T., Woodrow, C.J., Cowan, G.M., Gero, A.M., and Krishna, S. (1998) Expression of substrate-specific transporters encoded by Plasmodium falciparum in Xenopus laevis oocytes. *Mol. Biochem. Parasitol.* **93**, 81–89.
59. Woodrow, C.J., Burchmore, R.J., and Krishna, S. (2000) Hexose permeation pathways in Plasmodium falciparum-infected erythrocytes. *Proc. Natl. Acad. Sci. USA* **97**, 9931–9936.
60. Martin, R.E., Henry, R.I., Abbey, J.L., Clements, J.D., and Kirk, K. (2005) The 'permeome' of the malaria parasite: an overview of the membrane transport proteins of Plasmodium falciparum. *Genome Biol.* **6**, R26.
61. Seyfang, A. and Landfear, S.M. (1999) Substrate depletion upregulates uptake of *myo*-inositol, glucose and adenosine in *Leishmania. Mol. Biochem. Parasitol.* **104**, 121–130.
62. Nugent, P.G., Karsani, S.A., Wait, R., Tempero, J., and Smith, D.F. (2004) Proteomic analysis of Leishmania mexicana differentiation. *Mol. Biochem. Parasitol.* **136**, 51–62.
63. Landfear, S.M. (2008) Drugs and transporters in kinetoplastid protozoa. *Adv. Exp. Med. Biol.* **625**, 22–32.

Chapter 14

NMR Studies of Membrane Proteins

Gabriel A. Cook and Stanley J. Opella

Abstract

Nuclear magnetic resonance studies of membrane proteins yield valuable insights into their structure and topology. For example, the tilt angle and rotation of the helices in an ion channel can be determined by solid-state NMR spectroscopy in aligned lipid bilayers. Details about the structure of the protein in aligned phospholipids environments are immediately apparent from inspection of the SAMMY spectrum and the data can be further used for the determination of atomic resolution three-dimensional structures. SAR by NMR is a technique that is well suited for the field of membrane transporter proteins. The experiments on protein/phospholipid samples provide a unique insight into the interaction of drugs and the functional proteins.

The advances required to transform solid-state NMR from a spectroscopic technique to a generally applicable method for determining molecular structures included multiple-pulse sequences, double-resonance methods, and separated local field spectroscopy. It also required improvements in instrumentation, especially the use of high-field magnets and efficient probes capable of high-power radio-frequency irradiations at high frequencies. The pace of development is accelerating and the local field is being utilized in an increasing number of ways in spectroscopic investigations of molecular structure and dynamics. Applications to many helical membrane proteins are underway and promise to add to our understanding of membrane proteins in health and disease.

Key words: Nuclear magnetic resonance, membrane proteins, structure, topology, spectroscopy.

1. Introduction

Not only do membrane proteins represent a substantial fraction of the information in a genome but also they are responsible for many essential biological functions, some of which are unique (e.g., as membrane transporters). Consequently, mutations in genes that code for membrane proteins can cause human diseases.

Q. Yan (ed.), *Membrane Transporters in Drug Discovery and Development*, Methods in Molecular Biology 637,
DOI 10.1007/978-1-60761-700-6_14, © Springer Science+Business Media, LLC 2010

Thus, there are many reasons that the structural biology of membrane proteins is of considerable interest in biomedical research. However, because membrane proteins are not soluble in aqueous solution, only a few examples have been amendable to experimental structural analysis with X-ray crystallography and solution nuclear magnetic resonance (NMR) spectroscopy. The relatively few structures of helical membrane proteins that have been determined provide an initial view of the design principles involved in the use of helices as structural and functional elements, setting the stage for the interpretation of new structures of membrane transporters and other membrane proteins in terms of their functions. NMR spectroscopy is an extraordinarily powerful method for describing the structure and dynamics of proteins and other biopolymers and has the potential to be applied to membrane proteins. Furthermore, structure–activity relationship (SAR) studies by NMR have become a useful tool for the discovery and design of drugs that interact with proteins and also has the potential to be applied to membrane proteins.

2. NMR Spectroscopy

The overall rotational correlation time of a protein, the extent of alignment of the protein molecules in a sample, and strategy for assignment of the resonances to sites in the protein are the principal factors that affect NMR structural studies of proteins. For relatively small globular proteins, the sample conditions, instrumentation, experiments, and calculations that lead to structure determination are well established (1). The chief requirement for the structure determination of globular proteins is that samples can be prepared of isotopically labeled polypeptides that are folded in their native conformation and reorient relatively rapidly in solution. Such samples have been prepared for many hundreds of proteins, and it is likely that this can be done for thousands more of the polypeptide sequences found in genomes (2). In contrast, each of these factors needs to be revised or developed for applications of NMR to structure determination of membrane proteins (3).

The rotational correlation time problem is paramount. If the protein reorients rapidly enough in solution, as is the case for small membrane proteins in optimized micelles, then the standard methods of NMR can be applied and the structure determined. However, it is not possible to obtain NMR spectra of slowly orienting or immobile proteins such as membrane proteins in phospholipid bilayers, using conventional solution NMR methods and instruments. From the start, NMR studies of membrane proteins require crucial choices about samples, instrumentation, and experimental methods. It is feasible to apply both solution NMR and solid-state NMR to helical membrane proteins, depending primarily on the choice of lipid assemblies as well as other sample conditions.

3. Proteins and Lipids

Membrane proteins require the presence of lipids to maintain their native conformations and functions. **Figure 14.1** illustrates the types of lipid assemblies used in NMR structural studies of membrane proteins. The proteins are associated with lipids or mixtures of lipids that self-assemble into micelles, bicelles, or bilayers in aqueous solution. The properties of the lipid–protein complex determine the rotational correlation time of the protein molecules. Micelles can be prepared so that they contain a single polypeptide and reorient rapidly enough for solution NMR spectroscopy. Micelles are thought to be spherical aggregates of lipids with the hydrophobic chains on the interior. Samples for solution NMR spectroscopy require the use of relatively high temperatures and high concentration of lipids that form small micelles (4), most commonly sodium dodecyl sulfate (SDS), dodecylphosphocholine (DPC), or dihexanoyl phosphatidylcholine (DHPC). Bicelles are made from a mixture of long- and short-chain phospholipids that self-assemble into long-chain bilayers that are "capped" by the short-chain lipids (5). The molar ratios of

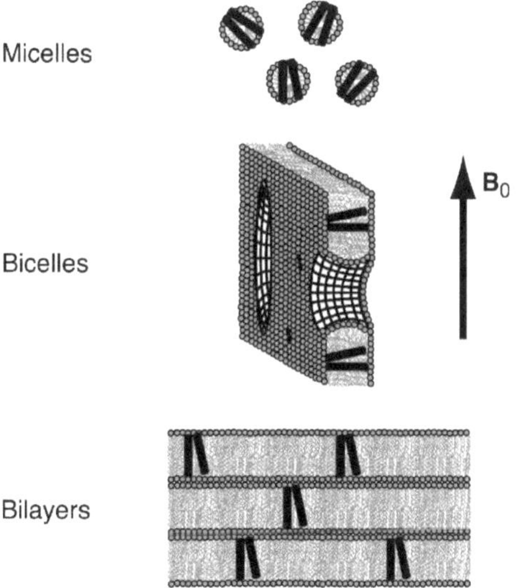

Fig. 14.1. Representation of proteins in micelles (*top*), bicelles (*middle*), and bilayers (*bottom*).

long-chain and short-chain lipids determine the size of the bicelles. Small isotropic bicelles can reorient rapidly without a preferred alignment while larger bicelles form an aligned liquid crystalline phase in the presence of a strong magnetic field with the bilayer normal perpendicular to the applied magnetic field. Bilayers are infinitely large in two dimensions, and two molecules thick in the third. In bilayers, nearly all of the residues in the polypeptide are immobile on timescales longer than milliseconds, and the relevant dipolar coupling and chemical shift interactions in individual backbone sites are not motionally averaged. As a result, bilayer samples of membrane proteins give very broad and poorly resolved spectra when instruments and methods appropriate for solution NMR are used. On the other hand, bilayer samples are suitable for the application of solid-state NMR methods where radio-frequency irradiations replace molecular reorientation as the principal mechanism of averaging the dipolar couplings responsible for the line broadening. The solid-state NMR approach that uses stationary, uniaxially aligned samples (6) is particularly well suited for membrane proteins in lipid bilayers because the proteins are immobilized and can be highly aligned (7). Spectroscopically, it is the combination of decoupling and heteronuclear dipolar interactions and sample alignment that results in narrow, single-line resonances and well-resolved, orientationally dependent spectra that provide the input for structure determination. There is a direct mapping of protein structure onto the solid-state NMR spectra of aligned samples, and the determination of complete three-dimensional structures from the spectra is feasible when multiple orientationally dependent frequencies are measured for nuclei at each residue.

Samples can be prepared with no, weak, or complete alignment of the protein molecules. Complete alignment is possible only with immobile proteins. Rapidly reorienting proteins can have either no or weak alignment. The vast majority of NMR studies have been performed on samples without molecular alignment. This is the case for globular proteins, which undergo effectively isotropic reorientation in solution. It is also the case for membrane proteins in micelles and isotropic bicelles. For these samples, structure determination is based on measurements that reflect internal molecular parameters, such as nuclear Overhauser enhancements (NOEs) that give short-range distance measurements and variations in isotropic chemical shifts associated with secondary structure (8). In the past few years, there has been a great deal of interest in weakly aligned samples of proteins. Because the proteins are reorienting rapidly, solution NMR methods are used for the measurement of residual dipolar couplings (RDCs). It is possible to weakly align membrane proteins in micelles and small bicelles through lanthanide ion (9, 10) and gel methods (11).

Unoriented, immobile samples, including membrane proteins in bilayers, can be studied using magic angle sample spinning (MAS) solid-state NMR spectroscopy (12), where distances and torsion angles are measured through spectroscopic parameters affected by both homonuclear and heteronuclear dipolar interactions. Immobile bilayer samples can be completely aligned to a degree rivaling that observed in single crystals of small peptides. Membrane proteins in bilayers can be aligned mechanically between glass plates, while membrane proteins in large bicelles can be aligned magnetically and then optionally "flipped" 90° so that the bilayer normal is parallel to the applied magnetic field through the addition of lanthanide ions (13).

The traditional approach to protein structure determination is based on the same overall principles, whether solution NMR or solid-state NMR methods are used and whether the sample is aligned. This involves the resolution of resonances through the use of isotopic labels and multidimensional NMR experiments, the measurement of spectral parameters associated with individual resonances (e.g., NOEs, J couplings, dipolar couplings, or chemical shift frequencies), the assignment of all resonances to specific sites in the protein, and then the calculation of structures. There are examples of the application of this approach to membrane proteins in micelles (8), bicelles (14), and bilayers (15). The availability of orientational information associated with individual resonances means that it is now possible to make effective use of limited amounts of assignment information (i.e., some residue-type assignments or a few sequential assignments). It may also be feasible to implement an "assignment-free" approach (16). The use of either limited or no assignment information prior to calculating structures greatly speeds the process of structure determination by NMR spectroscopy, especially in the case of membrane proteins, for which assignments are difficult to make in nearly all situations because of overlap of resonances and unfavorable relaxation parameters.

4. The Local Field

"The interaction between two nuclear spins depends on the magnitude and orientation of their magnetic moments and also the length and orientation of the vector describing their relative positions" (17). The local field, which results from the interaction of two or more proximate nuclei, provides a particularly direct source of structural information. Other nuclear spin interactions, especially the chemical shift, can also provide essential structural constraints, but their interpretation is generally more complicated and less firmly grounded. Taken together, the restrictions resulting from measurements of spectral parameters from several spin interactions, where at least one of them is a local field, are sufficient to determine the three-dimensional structures of proteins.

It is possible to study unoriented "powder" samples, including membrane proteins with the use of magic angle spinning. However, there is a second way to obtain high-resolution chemical shift spectra of solid samples, and that entails the use of single crystals, where depending on the space group, the molecules have only one or a few orientations relative to each other and the applied magnetic field. Fortunately, it is not necessary to have a single-crystal sample. One direction of molecular orientation is sufficient, as long as it is parallel to the magnetic field or the molecules undergo rotational diffusion about a single axis, to give all of the spectroscopic benefits of sample orientation. It is generally easier to prepare uniaxially oriented samples; each site on one molecule can be transformed into the identical site on another molecule through a combination of translation, inversion, and rotation operations about an axis parallel to the direction of the applied magnetic field. As is the case for single crystals, the spectra of oriented protein samples are characterized by narrow single-line resonances rather than powder patterns. The resonance frequencies reflect the orientations of the individual groups and, as a result, the solid-state NMR spectra of oriented samples are a rich source of structural information.

The SAMMY pulse sequence (18) is a high-resolution version of separated local field spectroscopy (19). It utilizes on-resonance "magic sandwich" pulses to accomplish CP transfer between two spins (i.e., ^{15}N and ^{1}H) while effectively decoupling the abundant spins (^{1}H) during the t1 evolution period of the pulse sequence. Very high-resolution two-dimensional heteronuclear dipolar spectra of solid samples can be obtained. The combination of narrow lines and favorable scaling factor has such a dramatic effect on the appearance of the spectra that it is now feasible to formulate solid-state NMR experiments where heteronuclear dipolar coupling frequencies provide a mechanism for resolution among similar chemical sites. In addition, dipolar coupling and chemical shift frequencies can be used in a number of complementary ways to enhance resolution in reduced and maximum-dimensional experiments and to provide qualitative and quantitative indications of molecular structure. Two-dimensional ^{1}H/^{15}N SAMMY spectra have narrow linewidths in both the heteronuclear dipolar coupling and the chemical shift frequency dimensions, enabling dipolar couplings to complement chemical shifts for resolution among sites as well as the measurement of readily interpretable orientationally dependent frequencies. SAMMY experiments are being used on a variety of systems, including membrane proteins.

5. Wheels and Waves

The secondary structure and topology of membrane proteins can be determined from the two-dimensional $^1H/^{15}N$ SAMMY spectra of uniformly ^{15}N-labeled samples in oriented bilayers. The characteristic "wheel-like" patterns of resonances observed in these spectra reflect helical wheel projections of residues in both transmembrane and in-plane helices and, hence, provide direct indices of secondary structure and topology of membrane proteins in phospholipid bilayers. We refer to these patterns as PISA (polarity index slant angle) wheels (20, 21).

The resonance frequencies in both the $^1H/^{15}N$ heteronuclear dipolar and the ^{15}N chemical shift dimensions in SAMMY spectra of aligned samples of membrane proteins depend on helix orientation, as well as on backbone dihedral angles, the magnitudes and orientations of the principal elements of the amide ^{15}N chemical shift tensor, and the N–H bond length. Consequently, it is possible to calculate spectra for any protein structure (22). The principles involved in the PISA wheel analysis of helices (23) are illustrated in **Fig. 14.2**. In **Fig. 14.2a** the projection down the axis of a helical wheel shows that the 3.6 residues per turn periodicity characteristic of an α-helix results in an arc of 100° between adjacent residues. The illustration of a peptide plane in **Fig. 14.2b** shows the orientations of the principal axes of the three operative spin interactions of the ^{15} N-labeled amide site. The 17° difference between the N–H bond axis and the σ33 principal element of the amide ^{15}N chemical shift tensor is the essential element of a PISA wheel. The striking "wheel-like" pattern of the resonances calculated from a two-dimensional SAMMY spectrum of the ideal helix is shown in **Fig. 14.2c**. A PISA wheel reflects the slant angle (tilt) of the helix and the assignment of the resonances reflects the polarity index (rotation) of the helix.

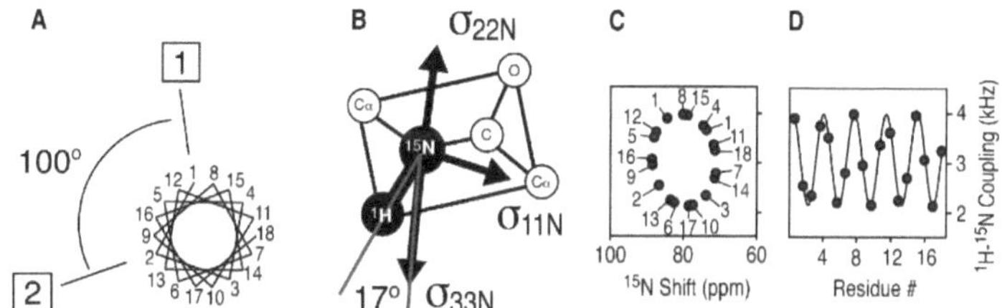

Fig. 14.2. Principles of PISA wheels. (**a**) Helical wheel showing the 100° arc between adjacent residues that is a consequence of the periodicity of 3.6 residues per turn in an α-helix; (**b**) orientations of the principal elements of the spin interaction tensors associated with ^{15}N in a peptide bond. (**c**) PISA wheel for an ideal α-helix; (**d**) dipolar wave for an ideal α-helix.

When the helix axis is parallel to the bilayer normal, all of the amide sites have an identical orientation relative to the direction of the applied magnetic field and, therefore, all of the resonances overlap with the same dipolar coupling and chemical shift frequencies. Tilting the helix away from the membrane normal results in variations in the orientations of the amide N–H bond vectors relative to the field. This is seen in the spectra as dispersions among both the heteronuclear dipolar couplings and the chemical shift frequencies. Essentially all transmembrane helices in membrane proteins are tilted with respect to the bilayer normal, and it is the combination of the tilt and the 17° difference between the tensor orientations in the molecular frame that make it possible to resolve many resonances from residues in otherwise uniform helices.

Dipolar waves can serve as maps of protein structure in NMR spectra of both weakly and completely aligned samples (23). The periodicity inherit in secondary structure elements is key to the use of both PISA wheels and dipolar waves as indices of secondary structure and topology in membrane proteins. **Figure 14.2d** illustrates the periodic-wavelike variations of the magnitudes of the static heteronuclear dipolar couplings as a function of residue number. Similar patterns are observed in RDCs measured for weakly aligned proteins. Simulations like those shown in **Fig. 14.3** of helices with tilt angles of 0, 10, 20, 30, and 40° can be used in conjunction with experimental spectra to determine the orientation of the helices without making any resonance assignments.

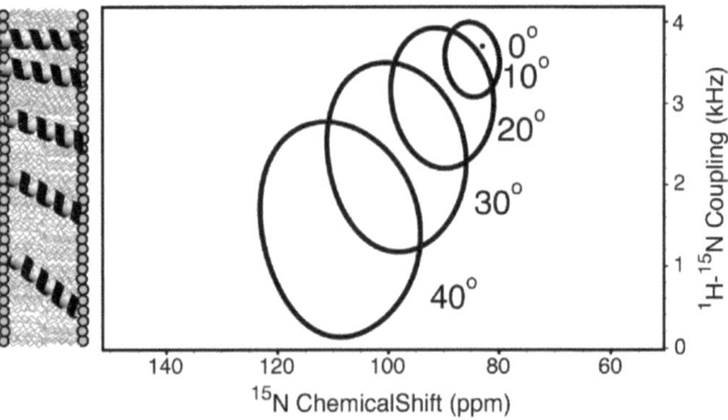

Fig. 14.3. Simulated PISA wheel patterns in "unflipped" bicelles undergoing rapid rotational diffusion for an α-helix with uniform dihedral angles $(\phi,\psi) = (-61°, -45°)$ tilted at 0, 10, 20, 30, and 40° with respect to the bilayer normal.

6. Application of NMR to an Ion Channel

The three-dimensional structures of a number of functional membrane proteins have been determined by solution (8, 24–32) and solid-state NMR (15, 33–37). In these cases, relatively large quantities of isotopically labeled protein required for NMR spectroscopy were prepared by expression of recombinant protein in *Escherichia coli*. The incorporation of purified proteins into lipid bilayers has been shown to yield functional channels as demonstrated by single-channel current recordings under voltage clamp conditions.

The experimental two-dimensional SAMMY spectrum in **Fig. 14.4** of selectively ^{15}N-leucine-labeled hepatitis C virus p7 protein in aligned bicelles has reasonable resolution, with each amide resonance characterized by ^{15}N chemical shift and ^{1}H-^{15}N dipolar coupling frequencies. The experimental SAMMY spectrum in **Fig. 14.4** is similar to that of the PISA wheel in **Fig. 14.2c** and, based on simulations, has at least one helix with a tilt angle of about 10°. Differences between an experimental spectrum and a calculated spectrum are a result of deviations between the experimentally determined backbone dihedral angles and those of an ideal α-helix and differences resulting from variations of the chemical shift tensors among the various amide sites.

The polarity of the resonances observed in the "wheel-like" pattern of a SAMMY spectrum provides a direct measure of the angle of the helix rotation about its long axis within the membrane. In principle, one well-resolved two-dimensional SAMMY

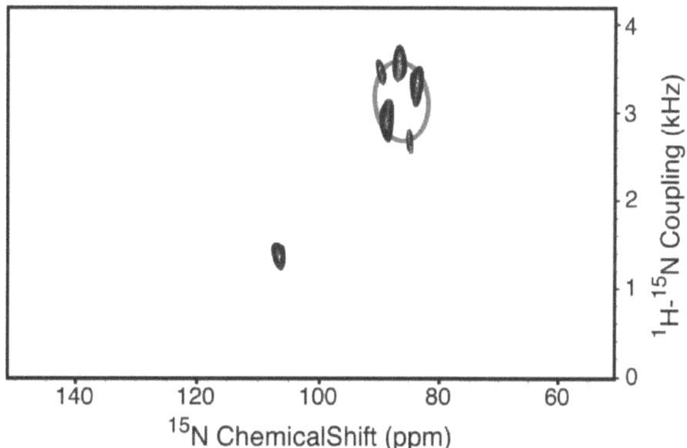

Fig. 14.4. Two-dimensional ^{1}H-^{15}N dipolar coupling and ^{15}N chemical shift solid-state SAMMY spectrum of selectively labeled ^{15}N-leucine HCV p7 in "unflipped" bicelles. A simulated wheel of 10° has been superimposed to show the approximate tilt angle of the transmembrane helix.

spectrum of an aligned sample of a uniformly ^{15}N-labeled protein provides sufficient information for complete structure determination. The orientationally dependent frequencies associated with each resonance depend on the magnitudes and orientations of the principal elements of the spin interaction tensors in the molecule and on the orientation of the molecular site with respect to the direction of the applied magnetic field. Because the orientation of the bilayer is fixed by the method of sample preparation and the properties of the nuclear spin interaction tensors are generally well characterized, each frequency reflects the orientation of a specific site in the protein with respect to the bilayer. The backbone structure of a protein is defined by the planes formed by the individual rigid peptide bonds and their directly bonded atoms. This is equivalent to the conventional description by vectors representing bonds between non-hydrogen atoms. The standard planar peptide geometry serves as the building block for the structure assembly process. The orientation of a peptide plane consists of the NMR frequencies measured from a SAMMY spectrum, the magnitudes and orientations of the principal elements of the amide ^{15}N and ^{1}H chemical shift tensors, and the N–H bond length. Once the orientations of all the peptide planes in the protein are determined from the experimental data, neighboring planes of fixed orientation are connected through their common α-carbon atom, with the only constraint of a fixed tetrahedral angle of 110°. Another program calculates the ϕ and χ dihedral angles for the two contiguous peptide plane combinations that satisfy tetrahedral angle geometry at the α-carbon.

The advantage of this direct mathematical analysis is that it is possible to determine the standard deviations in ϕ and μ based on uncertainties in the experimentally determined angles. Another important advantage of the method is that it results in the "piecewise" assembly of structures. Different parts of the protein structure can be determined independently. This is not possible in methods for which distance constraints between different regions of the protein must be established for three-dimensional structure determination. Because the orientation of individual peptide planes is determined relative to a unique external reference, any errors are not cumulative.

7. Drug Discovery and Development by NMR

The emergence of structure–activity relationship (SAR) studies by NMR (38) has shown great promise for the identification of lead compounds for drug development (39–45). NMR can be used to analyze the structural and chemical details of intermolecular interactions between proteins and small molecules. In solution NMR a simple correlation spectrum (e.g., HSQC) can display changes in chemical shifts, an indication of an effect on the local chemical environment. The use of solid-state NMR, although slightly more laborious in sample preparation, can indicate local as well as global changes in protein structure as a result of drug interaction. The magnetically aligned bilayer samples can provide information as to the orientation of the protein before and after the addition of drugs. Such samples avoid the time correlation problem associated with solution NMR, thus providing the opportunity to look at larger protein systems in environments similar to the ones used to characterize their channel activity. Molecules that bind specifically, albeit weakly or in mixtures, to the protein cause changes in the chemical shift and/or dipolar coupling frequencies of resonances associated with the residues in the binding site.

An appealing feature of SAR by NMR comes from its ability to identify compounds that bind to different parts of the protein. With only a basic outline of the protein architecture, it is possible to link two binding fragments to make a drug candidate that binds with high affinity and specificity. In solid-state NMR experiments, spectral resolution results from the orientation dependence of the frequencies in aligned samples; consequently, the resonances in the spectra are segregated by topology, and residues in loops can be readily differentiated from those in helices. The transmembrane helices of transporter proteins are the likely binding sites for channel blockers, and amphipathic cytoplasmic domains and basic inter-helical loops may provide separate binding sites.

Acknowledgments

This research was supported by grants from the National Institutes of Health; it utilized the Biomedical Technology Resource for NMR Molecular Imaging of Proteins, which is supported by grant P41EB002031; and G.A.C. received support from Training Grant 5T32DK00723332. We thank Gilead Sciences, Inc., for support.

References

1. Cavanagh, J., Fairbrother, W.J., Palmer III, A.G., and Skelton, N.J. (1995) *Protein NMR Spectroscopy*. Academic, San Diego, CA.
2. Wuthrich, K. (1998) The second decade into the third millennium. *Nat. Struct. Biol.* **5** *Suppl*, 492–495.
3. Opella, S.J. (1997) NMR and membrane proteins. *Nat. Struct. Biol.* **4** *Suppl*, 845–848.
4. McDonnell, P.A. and Opella, S.J. (1993) Effect of detergent concentration on multidimensional solution NMR spectra of mem-

brane proteins in micelles. *J. Magn. Reson.* **B 102**, 1205–1225.
5. Sanders, C.R., Hare, B.J., Howard, K., and Prestegard, J.H. (1994) Magnetically oriented phospholipid micelles as a tool for the study of membrane-associated molecules. Prog. NMR Spectrosc. **26**, 421–444.
6. Opella, S.J., Stewart, P.L., and Valentine, K.G. (1987) Protein structure by solid-state NMR spectroscopy. *Q. Rev. Biophys.* **19**, 7–49.
7. Marassi, F.M., Ramamoorthy, A., and Opella, S.J. (1997) Complete resolution of the solid-state NMR spectrum of a uniformly 15 N-labeled membrane protein in phospholipid bilayers. *Proc. Natl. Acad. Sci. USA* **94**, 8551–8556.
8. Almeida, F.C. and Opella, S.J. (1997) fd coat protein structure in membrane environments: structural dynamics of the loop between the hydrophobic trans-membrane helix and the amphipathic in-plane helix. *J. Mol. Biol.* **270**, 481–495.
9. Ma, C. and Opella, S.J. (2000) Lanthanide ions bind specifically to an added "EF-hand" and orient a membrane protein in micelles for solution NMR spectroscopy. *J. Magn. Reson.* **146**, 381–384.
10. Veglia, G. and Opella, S.J. (2000) Lanthanide ion binding to adventitious sites aligns membrane proteins in micelles for solution NMR spectroscopy. *J. Am. Chem. Soc.* **122**, 11733–11734.
11. Chou, J.J., Kaufman, J.D., Stahl, S.J., Wingfield, P.T., and Bax, A. (2002) Micelle-induced curvature in a water-insoluble HIV-1 Env peptide revealed by NMR dipolar coupling measurement in stretched polyacrylamide gel. *J. Am. Chem. Soc.* **124**, 2450–2451.
12. Griffin, R.G. (1998) Dipolar recoupling in MAS spectra of biological solids. *Nat. Struct. Biol.* **5** *Suppl*, 508–512.
13. Howard, K.P. and Opella, S.J. (1996) High-resolution solid-state NMR spectra of integral membrane proteins reconstituted into magnetically oriented phospholipid bilayers. *J. Magn. Reson.* **B 112**, 91–94.
14. Park, S.H., De Angelis, A.A., Nevzorov, A.A., Wu, C.H., and Opella, S.J. (2006) Three-dimensional structure of the transmembrane domain of Vpu from HIV-1 in aligned phospholipid bicelles. *Biophys. J.* **91**, 3032–3042.
15. Opella, S.J., Marassi, F.M., Gesell, J.J., Valente, A.P., Kim, Y., Oblatt-Montal, M., and Montal, M. (1999) Structures of the M2 channel-lining segments from nicotinic acetylcholine and NMDA receptors by NMR spectroscopy. *Nat. Struct. Biol.* **6**, 374–379.
16. Nevzorov, A.A. and Opella, S.J. (2003) Structural fitting of PISEMA spectra of aligned proteins. *J. Magn. Reson.* **160**, 33–39.
17. Abragam, A. (1961) *The Principles of Nuclear Magnetism*. Oxford University Press, Oxford, UK.
18. Nevzorov, A.A. and Opella, S.J. (2003) A "Magic Sandwich" pulse sequence with reduced offset dependence for high-resolution separated local field spectroscopy. *J. Magn. Reson.* **164**, 182–186.
19. Waugh, J.S. (1976) Uncoupling of local field spectra in nuclear magnetic resonance: determination of atomic positions in solids. *Proc. Natl. Acad. Sci. U S A.* **73**, 1394-1397.
20. Marassi, F. M., and Opella, S. J. (2000) A solid-state NMR index of helical membrane protein structure and topology. *J. Magn. Reson.* **144**, 150–155.
21. Wang, J., Denny, J., Tian, C., Kim, S., Mo, Y., Kovacs, F., Song, Z., Nishimura, K., Gan, Z., Fu, R., Quine, J.R., and Cross, T.A. (2000) Imaging membrane protein helical wheels. *J. Magn. Reson.* **144**, 162–167.
22. Bak, M., Schultz, R., Vosegaard, T., and Nielsen, N.C. (2002) Specification and visualization of anisotropic interaction tensors in polypeptides and numerical simulations in biological solid-state NMR. *J. Magn. Reson.* **154**, 28–45.
23. Mesleh, M.F., Veglia, G., DeSilva, T.M., Marassi, F.M., and Opella, S.J. (2002) Dipolar waves as NMR maps of protein structure. *J. Am. Chem. Soc.* **124**, 4206–4207.
24. Williams, K.A., Farrow, N.A., Deber, C.M., and Kay, L.E. (1996) Structure and dynamics of bacteriophage IKe major coat protein in MPG micelles by solution NMR. *Biochemistry* **35**, 5145–5157.
25. MacKenzie, K.R., Prestegard, J.H., and Engelman, D.M. (1997) A transmembrane helix dimer: structure and implications. *Science* **276**, 131–133.
26. Papavoine, C.H., Christiaans, B.E., Folmer, R.H., Konings, R.N., and Hilbers, C.W. (1998) Solution structure of the M13 major coat protein in detergent micelles: a basis for a model of phage assembly involving specific residues. *J. Mol. Biol.* **282**, 401–419.
27. Townsley, L.E., Tucker, W.A., Sham, S., and Hinton, J.F. (2001) Structures of gramicidins A, B, and C incorporated into sodium dodecyl sulfate micelles. *Biochemistry* **40**, 11676–11686.
28. Sorgen, P.L., Cahill, S.M., Krueger-Koplin, R.D., Krueger-Koplin, S.T., Schenck, C.C.,

and Girvin, M.E. (2002) Structure of the Rhodobacter sphaeroides light-harvesting 1 beta subunit in detergent micelles. *Biochemistry* **41**, 31–41.

29. Zamoon, J., Mascioni, A., Thomas, D.D., and Veglia, G. (2003) NMR solution structure and topological orientation of monomeric phospholamban in dodecylphosphocholine micelles. *Biophys. J.* **85**, 2589–2598.
30. Yushmanov, V.E., Mandal, P.K., Liu, Z., Tang, P., and Xu, Y. (2003) NMR structure and backbone dynamics of the extended second transmembrane domain of the human neuronal glycine receptor alpha1 subunit. *Biochemistry* **42**, 3989–3995.
31. Yushmanov, V.E., Xu, Y., and Tang, P. (2003) NMR structure and dynamics of the second transmembrane domain of the neuronal acetylcholine receptor beta 2 subunit. *Biochemistry* **42**, 13058–13065.
32. Howell, S.C., Mesleh, M.F., and Opella, S.J. (2005) NMR structure determination of a membrane protein with two transmembrane helices in micelles: MerF of the bacterial mercury detoxification system. *Biochemistry* **44**, 5196–5206.
33. Ketchem, R.R., Hu, W., and Cross, T.A. (1993) High-resolution conformation of gramicidin A in a lipid bilayer by solid-state NMR. *Science* **261**, 1457–1460.
34. Wang, J., Kim, S., Kovacs, F., and Cross, T.A. (2001) Structure of the transmembrane region of the M2 protein H(+) channel. *Protein Sci.* **10**, 2241–2250.
35. Valentine, K.G., Liu, S.F., Marassi, F.M., Veglia, G., Opella, S.J., Ding, F.X., Wang, S.H., Arshava, B., Becker, J.M., and Naider, F. (2001) Structure and topology of a peptide segment of the 6th transmembrane domain of the Saccharomyces cerevisae alpha-factor receptor in phospholipid bilayers. *Biopolymers* **59**, 243–256.
36. Marassi, F.M. and Opella, S.J. (2003) Simultaneous assignment and structure determination of a membrane protein from NMR orientational restraints. *Protein Sci.* **12**, 403–411.
37. Park, S.H., Mrse, A.A., Nevzorov, A.A., Mesleh, M.F., Oblatt-Montal, M., Montal, M., and Opella, S.J. (2003) Three-dimensional structure of the channel-forming trans-membrane domain of virus protein "u" (Vpu) from HIV-1. *J. Mol. Biol.* **333**, 409–424.
38. Hajduk, P.J., Meadows, R.P., and Fesik, S.W. (1997) Discovering high-affinity ligands for proteins. *Science* **278**, 497–499.
39. Hajduk, P.J., Meadows, R.P., and Fesik, S.W. (1999) NMR-based screening in drug discovery. *Q. Rev. Biophys.* **32**, 211–240.
40. Lepre, C.A., Moore, J.M., and Peng, J.W. (2004) Theory and applications of NMR-based screening in pharmaceutical research. *Chem. Rev.* **104**, 3641–3676.
41. Huth, J.R., Sun, C., Sauer, D.R., and Hajduk, P.J. (2005) Utilization of NMR-derived fragment leads in drug design. *Methods Enzymol.* **394**, 549–571.
42. Takeuchi, K. and Wagner, G. (2006) NMR studies of protein interactions. *Curr. Opin. Struct. Biol.* **16**, 109–117.
43. Yu, L., Sun, C., Song, D., Shen, J., Xu, N., Gunasekera, A., Hajduk, P.J., and Olejniczak, E.T. (2005) Nuclear magnetic resonance structural studies of a potassium channel-charybdotoxin complex. *Biochemistry* **44**, 15834–15841.
44. Watts, A. (2005) Solid-state NMR in drug design and discovery for membrane-embedded targets. *Nat. Rev. Drug Discov.* **4**, 555–568.
45. Luca, S., White, J.F., Sohal, A.K., Filippov, D.V., van Boom, J.H., Grisshammer, R., and Baldus, M. (2003) The conformation of neurotensin bound to its G protein-coupled receptor. *Proc. Natl. Acad. Sci. U S A* **100**, 10706–10711.

Chapter 15

Site-Directed Mutagenesis in the Study of Membrane Transporters

Audra A. McKinzie, Renae M. Ryan, and Robert J. Vandenberg

Abstract

One of the major goals in membrane transporter research is to understand how transporter proteins work at the molecular level. Ideally, this research would be carried out with a detailed knowledge of the three-dimensional structure of the protein. However, in the absence of atomic resolution structures for many membrane transporters other molecular tools need to be employed. In vitro site-directed mutagenesis is one method that has the capacity to provide both structural information and identification of the role of individual residues and/or regions of a protein that are involved in function.

Key words: Site-directed mutagenesis, membrane transporter, fusion-PCR, polymerase-based recombination, chimeric protein.

1. Introduction

One of the major goals in membrane transporter research is to understand how transporter proteins work at the molecular level. Ideally, this research would be carried out with a detailed knowledge of the three-dimensional structure of the protein. However, in the absence of atomic resolution structures for many membrane transporters other molecular tools need to be employed. In vitro site-directed mutagenesis is one method that has the capacity to provide both structural information and identification of the role of individual residues and/or regions of a protein that are involved in function.

Site-directed mutagenesis involves changing the genetic code for a protein sequence and generating a mutant protein with a subtly different structure. When this is combined with functional

Q. Yan (ed.), *Membrane Transporters in Drug Discovery and Development*, Methods in Molecular Biology 637,
DOI 10.1007/978-1-60761-700-6_15,

assays of the mutant, this method has the capacity to provide very useful information about the structural basis for transporter function. The utility of this approach derives from the fact that, in general, most point mutations do not dramatically alter function. The small number of mutations that do alter or destroy function of the transporter do so because specific contacts between the solute, ion, or substrate and the transporter are disrupted, or conformational changes required for the transport process are prevented (1). This raises a very important cautionary note in interpreting the significance of loss of function mutants. In addition to disruption of specific contacts between the solute and the transporter, loss of function can arise through misfolding of the protein or altered insertion of the protein into the membrane. Therefore, loss of function mutants are only useful if the cause of loss of function can be accurately ascertained. Even so, if a loss of function is due to lack of expression at the cell surface, no conclusions about the role of particular amino acid residues in the functional properties of the transporter can be made.

A number of site-directed mutagenesis strategies to study structure function relationships have been used and include mutating consecutive residues to alanine to scan through the amino acid sequence to identify residues involved in important functional properties; cysteine scanning mutagenesis in combination with chemical modifications of the introduced cysteine residues to probe accessibility of various regions to aqueous environments, and also their roles in determining functional properties of the transporter (2, 3); and tryptophan scanning to identify regions that may be associated with or interact with lipids within the membrane (4). In addition, conservative substitutions are very commonly used to address more directed questions concerning the structural basis for functional properties. In general, the methods used to construct single-point mutations are the same irrespective of the overall goal of the experiment.

The construction of chimeric transporters can be useful to identify functional domains of transporters (5, 6). Methods based on bacterial recombination (5) have been used to generate random hybrid molecules, with the resulting chimeric junction typically occurring in regions of high-sequence homology. While this method allows for the rapid generation of many different chimeras, the researcher has no control over the precise makeup of the final product or of the location of the specific junctions of domains of interest. Chimeras with more specific junctions can be constructed by traditional cloning methods following the introduction of restriction sites at both the acceptor and the donor sites flanking the domain of interest (6). The introduction of restriction sites is carried out by site-directed mutagenesis as described above and below. While this method offers greater control over the precise location of hybrid junctions, limitations can be imposed by the availability of unique restriction sites and the suitability of the DNA sequence for mutation to a particular recognition sequence at the desired location.

Polymerase-based recombination, also known as fusion-PCR, can be used to carry out directed joining of cloned DNA sequences and is ideally suited to the generation of simple or complex hybrid transporters (7, 8). Fusion-PCR eliminates the reliance on restriction sites and allows for precise engineering of chimeric junctions. Furthermore, the technique can be easily adapted to introduce insertions or deletions into molecules of interest.

In the following sections we shall describe the methods used in our laboratory to construct mutant transporters, both by site-directed mutagenesis and by fusion-PCR and also how the functional properties of the transporters are assessed (*see* **Note 1**).

2. Materials

2.1. Reagents

1. High fidelity DNA polymerase, 2 U/μL, includes reaction buffer.
2. 10 mM mixture dNTPs.
3. DpnI 20,000 U/mL.
4. Chemically competent *Escherichia coli* cells (*see* **Note 2**).
5. Bacterial growth medium (NZYM) broth. $NZYM^+$: Filter-sterilize 2 mL of 1 M D-glucose and 500 μL of 2 M $MgCl_2$ using a 0.2-μm filter and add to 100 mL of autoclaved NZYM.
6. Ampicillin: Make up to 100 mg/mL stock solution in water, filter-sterilize using a 0.2-μm filter, and store at –20°C. Then use at a concentration of 100 μg/mL in the broth or agar.
7. LB agar.
8. Agarose for gel electrophoresis. DNA fragments in the range of 0.5–5 kb are well separated on a 0.8% (wt/vol) gel.
9. SYBR Safe DNA Gel Stain add to melted agarose at a 1:25,000 dilution prior to pouring the gel to visualize DNA with a 470-nm Safe Imager transilluminator (*see* **Note 3**).
10. Tris–acetate (TAE) electrophoresis buffer, use at 1X. 50X: 242 g Tris base, 57.1 mL glacial acetic acid, 100 mL 0.5 M EDTA (pH 8.0).
11. Phenol:chloroform:isoamyl alcohol (25:24:1).
12. Chloroform (AR grade).

13. Molecular biology grade ethanol.
14. Molecular biology grade isopropanol.
15. 3 M sodium acetate: 40.8 g sodium acetate–H_2O, adjust to pH 5.3 with glacial acetic acid, and add H_2O to 100 mL.
16. DNase/RNase-free water.

2.2. Kits

2.2.1. Plasmid Miniprep Kit

There are numerous methods for the preparation of plasmid DNA from small volumes of liquid culture, including several commercially available kits that are fast, reliable, and affordable. Each method has advantages and disadvantages, and researchers should consider downstream requirements for purity of their DNA when selecting a procedure. Most of the commercially available miniprep kits are based on the traditional method of alkaline lysis followed by a matrix-based purification and have similar protocols and similar buffer components. Our laboratory currently uses the High Pure Plasmid Isolation Kit (Roche) which generates sufficient quantities of highly pure plasmid DNA for our purposes.

Buffer components:

1. Suspension buffer: 50 mM Tris–HCl (pH 7.5), 10 mM EDTA, 100 μg/mL RNaseA.
2. Lysis buffer: 0.2 M NaOH, 1% SDS.
3. Binding buffer: 4 M guanidine hydrochloride, 0.5 M potassium acetate (pH4.2).
4. Wash buffer I: 5 M guanidine hydrochloride, 20 mM Tris–HCl (pH 6.6), and 37% ethanol.
5. Wash buffer II: 20 mM NaCl, 2 mM Tris–HCl (pH 7.5), and 80% ethanol.
6. Elution buffer: 10 mM Tris–HCl (pH 8.5).

2.2.2. RNA Transcription Kit

Our laboratory characterizes transporter function by expression in *Xenopus laevis* oocytes. In our experience the mMessage mMachine® High Yield Capped RNA Transcription Kit (Ambion) produces high-quality mRNA suitable for this application.

Kit components:

1. Enzyme mix (may be SP6, T7, or T3 RNA polymerase) (*see* **Note 4**).
2. 10X reaction buffer (specific for each RNA polymerase).
3. 2X NTP/CAP mix.
4. Turbo DNase (2 U/μL).
5. Ammonium acetate stop mix: 5 M ammonium acetate, 100 mM EDTA.

6. Nuclease-free water.
7. Gel loading buffer: 95% formamide, 0.025% xylene cyanol, 0.025% bromophenol blue, 18 mM EDTA, 0.02% SDS.

3. Methods

3.1. Overview of Site-Directed Mutagenesis Method

This method uses a double-stranded DNA vector containing the insert to be mutated and two synthetic complementary oligonucleotides with the desired mutation. Some of the advantages of this method are that it does not require the use of specialized vectors, any subcloning steps, or single-strand rescue and it can be completed in 3 days. After denaturation of the double-stranded DNA, the oligonucleotides are allowed to anneal to the insert DNA (step 1, **Fig. 15.1**) and DNA polymerase is used to synthesize the complementary strands generating the mutated plasmid with staggered nicks (step 2, **Fig. 15.1**). The parental (wild-type) plasmid is digested with DpnI endonuclease, which digests only methylated DNA leaving the in vitro synthesized mutant plasmid intact (step 3, **Fig. 15.1**). The plasmid is then transformed into a strain of competent bacterial cells appropriate for the vector. In our experience this results in greater than 80% of colonies containing plasmids with the desired mutation, which can be screened either by restriction digestion or by DNA sequencing.

3.1.1. Primer Design

1. Both of the mutagenic primers must contain the desired mutation and anneal to the same sequence on opposite strands. Primers should be 20–35 bases in length, and the melting temperature (T_m) of the primers should be $\geq 78°C$. The following formula can be used to estimate the T_m:

   $T_m = 81.5 + 0.41(\%GC) - 675/N - \%$ mismatch where N is the primer length in bases.
2. The mutation should be in the middle of the primer with 10–15 correct bases on either side. Ideally, the GC content of the primer should be 50%, but 40–80% is an acceptable range. The primers should contain G or C bases on both the 5′ and 3′ termini to promote annealing and ensure efficient extension by DNA polymerase.
3. Restriction enzymes can be used to confirm that a mutant has been generated. Ideally a unique restriction site can be engineered into the mutation site and the presence of the restriction site is confirmation that the mutation has been successful. If a restriction site cannot be engineered into the mutation site, it can be engineered into the primer surrounding the mutation without changing the amino acid sequence.

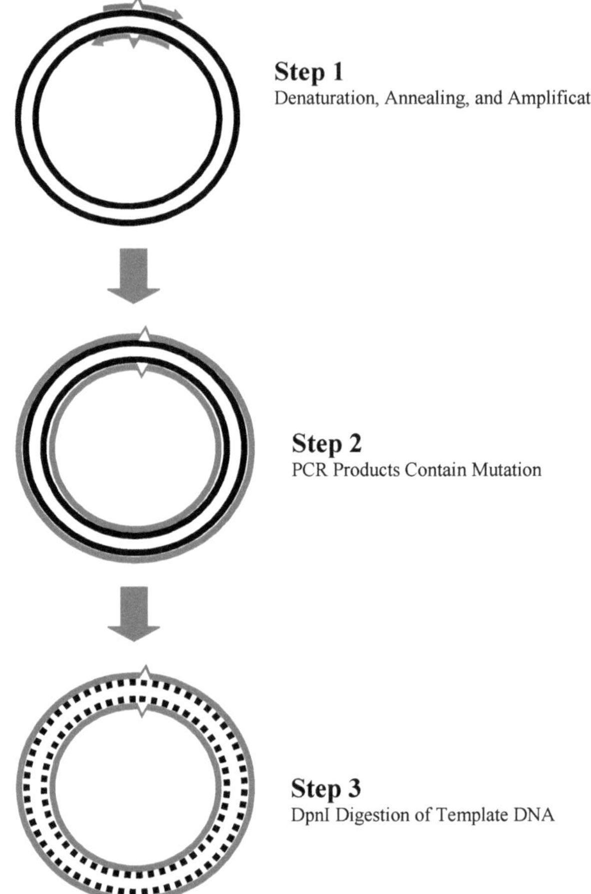

Fig. 15.1. Flow diagram for generating site-directed mutants. Plasmid DNA containing the gene of interest is amplified by PCR using complementary primers containing the desired mutation. The template DNA is eliminated by digestion with DpnI prior to bacterial transformation.

The assumption is that if the primer is incorporated into plasmid to generate the restriction site, then the desired mutation has also been generated. Sequencing should always be performed to confirm any mutation and the lack of any extraneous mutations that may arise in the PCR steps.

4. Make a 100 μM stock of primers with DNase/RNase-free H_2O or Tris–HCl (pH 8.0). Prepare 20 μM working stocks by performing a 1:5 dilution of this stock.

3.1.2. Detailed Site-Directed Mutagenesis Method

The cDNA to be mutated can be subcloned into any plasmid and ideally one that may also be used for expression in the cell type that best suits the functional analysis of the transporter. In our case we use pOTV (Oocyte Transcription Vector) because it is designed for efficient production of cRNA and generates high expression levels in *X. laevis* oocytes.

1. Thaw all components on ice and prepare the sample reaction on ice as below:
   10 μL of 5X velocity reaction buffer
   X μL (5–50 ng) of dsDNA template
   1 μL 20 μM sense primer
   1 μL 20 μM antisense primer
   1 μL of 10 mM dNTP mix
   0.5–1.0 μL velocity DNA polymerase (2 U/μl)
   DNAse/RNase-free water to a final volume of 50 μL
2. If mineral oil is required in your PCR machine, then it should be overlayed on top of the reaction solution.
3. Suggested PCR cycling parameters are listed in the table below.

No. of cycles	Temp (°C)	Time
1	98	30 s
16–22 (*see* **Note 5**)	98	30 s
	T_m (primers) – 5	30 s
	72	15–30 s/kb of plasmid length
1	72	5 min (optional)

4. Make sure the reaction tube is cooled to at least 37°C before continuing.
5. (Optional) Remove 5 μL reaction mix for agarose gel analysis.
6. Add 1 μL DpnI restriction enzyme (10 U/μL) directly to each amplification reaction. Thoroughly mix the contents of the tube by gently pipetting up and down. Incubate the tubes at 37°C for 2 h to digest parental (non-mutated) DNA.
7. (Optional) Remove 5 μL for agarose gel analysis (*see* **Note 6**).
8. Proceed to **Section 3.3** for further protocols.

3.2. Overview of Construction of Chimeric Transporters by Fusion-PCR

This method uses PCR-based genetic recombination to generate linear DNA fragments composed of two or more distinct cDNAs with specific junction sites in order to create plasmids that will express chimeric proteins. Previously cloned genes, cDNA, or genomic DNA may serve as the source material. Chimeric oligonucleotides are used to amplify individual fragments of each gene of interest while introducing overlapping identity sequence between the fragments to be joined. When these individual fragments are combined into a fusion-PCR reaction, the overlapping sequences anneal to each other and are able to serve as priming sites that allow 3′ extension by DNA polymerase resulting in a hybrid template. The hybrid template is then further amplified with oligonucleotides at distal ends of the molecules to generate an insert suitable for cloning by traditional methods, such as TA, blunt end, or through the introduction of specific restriction sites. The accuracy of the resulting chimeric junction(s) is then confirmed by DNA sequencing.

The schematic in **Fig. 15.2** illustrates a strategy for creating a cDNA that is chimeric for a central region of a transporter, with fragments A and C originating from one gene and fragment B from another. In individual PCR reactions, fragment A is amplified with primers I and II, fragment B with primers III and IV, and fragment C with primers V and VI (step1, **Fig. 15.2**). In this example, primers II, III, IV, and V are chimeric for the desired junctions in that the 3′ half of each primer is complementary to the target amplicon and contains a 5′ tail identical to the terminus of the sequence that will comprise the junction. Primers I and VI are specific to the 5′ and 3′ ends of the desired insert and contain restriction sites to facilitate subsequent cloning of the end product. The resulting fragments A, B, and C are combined into a

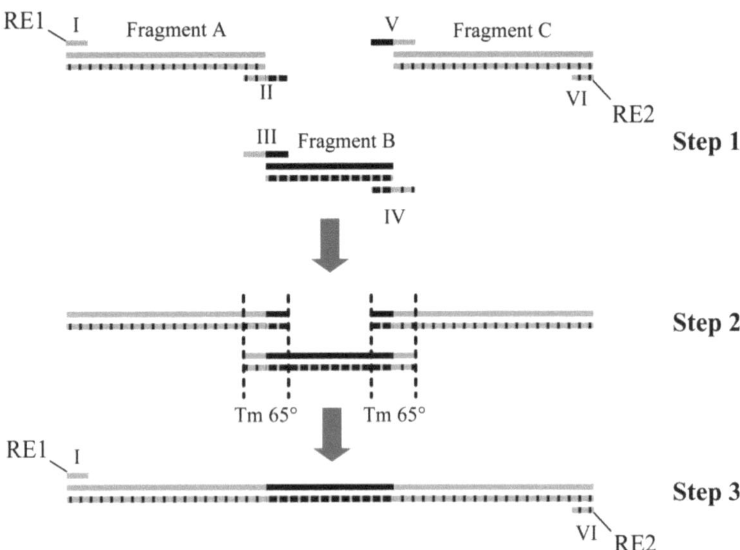

Fig. 15.2. A schematic summary of the procedure for constructing a transporter chimeric for a central domain by fusion-PCR. Fragments A and C are from a single gene product and fragment B is from another gene product. I–VI refer to primers; note that primers II, III, IV, and V are chimeric. Unique restriction enzyme sites (RE1 and RE2) are incorporated into primers I and VI at the 5′ and 3′ ends of the desired insert. See text for more details.

separate PCR reaction and subjected to standard cycling conditions in the absence primers (step 2, **Fig. 15.2**). This allows the overlapping ends of the fragments to anneal to the overlapping ends of adjacent fragments to generate a hybrid template. The hybrid template is then 'fused' and further amplified by primers I and VI (step 3, **Fig 15.2**).

3.2.1. Primer Design

Once the precise location of the desired chimeric junctions has been determined by analysis of the protein sequences of each transporter, primers are designed that will amplify the individual fragments to be fused. When designing primers, be sure to pay attention to the location of codons to avoid introduction of frameshifts in the final product. Select sequences that adhere to the usual rules for primer design, i.e., 18–24 bases with a T_m between 55 and 65°C, and containing a G or C at the 3′ terminus. Note that % G/C content is restricted by the target sequence and can therefore not be controlled. Once forward and reverse primers have been selected which will amplify the individual fragments, incorporate sequence into the 5′ portion of each oligonucleotide that corresponds to the sequence to be fused. This sequence should also adhere to the usual rules for primer design so that the overlapping sequence of the resulting fragments is suitable for annealing under normal PCR conditions.

For example, the 3′ portion of the antisense primer II (**Fig. 15.2**) is specific for sequence encoding the carboxy-terminus of fragment A and has an annealing temperature of 65°C, yet it contains a 5′ tail that is specific for the sequence encoding the amino-terminal of fragment B. Likewise, the 3′ portion of the sense primer III is specific for sequence encoding the amino-terminal of fragment B and has an annealing temperature of 65°C, yet it contains a 5′ tail that is specific for the sequence that encodes the carboxy-terminal of fragment A. The resulting overlap between fragments A and B formed by the incorporation of primers II and III, respectively, also has an annealing temperature of 65°C. To facilitate cloning of the resulting chimeric transporter into the oocyte transcription vector pOTV for further experimentation (see below), we routinely incorporate restriction sites into the 5′ tails of primers I and VI. However, these primers are also designed with an annealing temperature of 65°C.

3.2.2. Detailed Methods for Fusion-PCR

1. Thaw all components on ice and prepare the sample reaction as below for each fragment that will comprise the final chimera:

   10 μL of 5X velocity reaction buffer

   X μL (10 pg–50 ng) of dsDNA template

   1 μL 20 μM sense primer (primer I, III, or V)

1 μL 20 μM antisense primer (primer II, IV, or VI)

1 μL of 10 mM dNTP mix

0.5–1.0 μL velocity DNA polymerase (2 U/μL)

DNase/RNase-free water to a final volume of 50 μL

2. If mineral oil is required in your PCR machine, then it should be overlayed on top of the reaction solution.
3. Suggested PCR cycling parameters are listed in the table below.

No. of cycles	Temp (°C)	Time
1	98	30 s
20–25	98	30 s
	60	30 s
	72	15–30 s/kb of target DNA
1	72	5 min (optional)

4. Analyze 5 μL of each reaction on a 0.8% agarose gel to confirm generation of fragments of expected size and to estimate yield (*see* **Note 7**).
5. Assemble the following fusion-PCR reaction using approximately equal molar ratios of each of the individual fragments:

   10 μL of 5X velocity reaction buffer

   X μL fragment A

   X μL fragment B

   X μL fragment C

   1 μL of 10 mM dNTP mix

   0.5–1.0 μL velocity DNA polymerase (2 U/μL)

   DNase/RNase-free water to a final volume of 50 μL
6. Perform 10 cycles of PCR as above, adjusting the extension time as necessary to account for the length of the fusion product.
7. Assemble the following fusion-amplification reaction:

   10 μL of 5X velocity reaction buffer

   1–5 μL fusion-PCR reaction (step 5)

   1 μL 20 μM distal sense primer (primer I)

   1 μL 20 μM distal antisense primer (primer VI)

   1 μL of 10 mM dNTP mix

0.5–1.0 μL velocity DNA polymerase (2 U/μL)

DNase/RNase-free water to a final volume of 50 μL

8. Perform 20–25 cycles of PCR as above, adjusting the extension time as necessary to account for the length of the fusion product.
9. Analyze 5 μL of each reaction on a 0.8% agarose gel to confirm generation of fusion product of expected size and to estimate yield.

The resulting chimeric product should be cloned into a vector suited to downstream applications. However, a detailed discussion of the various cloning techniques is beyond the scope of this chapter.

3.3. Additional Protocols for Isolation of Mutant Transporters

Once a mutant transporter has been generated by either site-directed mutagenesis or fusion-PCR followed by subcloning into a suitable vector, the resulting plasmid is transformed into a suitable strain of bacteria for selection and amplification.

3.3.1. Transformation of Competent Cells

1. The α-select chemically competent cells should be stored at –80°C. Allow them to gently thaw on ice immediately before use. For each sample reaction add 25–50 μL of competent cells to pre-chilled 13 mL Falcon® 2059 polypropylene tubes.
2. Add 1 μL of the sample reaction to 23–50 μL of competent cells, swirl to mix (do not pipette up and down), and incubate on ice for 30 min.
3. Heat shock each tube for 45 s at 42°C by placing the tubes in a heated water bath (*see* **Note 8**) followed by a 2 min incubation on ice.
4. Add 500 mL of $NZYM^+$ broth and shake at 37°C for 1 h at 225–250 rpm.
5. Plate 100 and 400 μL of each transformation on pre-warmed LB agar plates containing the appropriate antibiotic for the plasmid vector used (in our case we use ampicillin at 100 μg/mL). Incubate the plates at 37°C for >16 h.
6. Pick 6–8 colonies for further analysis to screen for the desired mutations. Scrape a single colony with a sterile tip and place into 5 mL of $NZYM^+$ broth (*see* **Note 9**) containing the appropriate antibiotic in 13 mL Falcon® 2059 polypropylene tubes, 50 mL Falcon® tubes, or any other sterile vessel suitable for liquid cultures. Shake at 37°C at 225–250 rpm for >16 h.

3.3.2. Purification of Plasmid DNA

A number of plasmid purification kits are available and individual researchers may have preferences of one over another. In our laboratory we use the High Pure Plasmid Isolation Kit (Roche).

1. If desired, make a glycerol stock of the bacterial culture that contains the mutant plasmids. Mix 700 μL of bacterial culture and 300 μL of 50% sterile glycerol and store at −70°C.
2. Pellet 3–5 mL of bacterial culture by centrifugation at room temperature at >6,000 × *g* for 1–5 min. Carefully remove and discard all of the supernatant.
3. Thoroughly resuspend the pellet in 250 μL of suspension buffer by gently pipetting the solution up and down.
4. Transfer to 1.5 mL Eppendorf tubes or equivalent.
5. Add 250 μL of lysis buffer directly into the resuspended cells and mix gently by inversion until solution is clear. Let stand 5 min at room temperature.
6. Add 350 μL of binding buffer and invert 6–10 times. Incubate on ice for 10 min.
7. Spin for 10 min at room temperature and 14,000 × *g* and transfer supernatant to spin columns that have been pre-equilibrated with 200 μL of binding buffer.
8. Place column in 2 mL collection tube and spin 30 s at 14,000 × *g*. Discard flow through. Replace column in 2 mL collection tube.
9. Wash column with 500 μL wash buffer I. Spin 30 s at 14,000 × *g*. Discard flow through. Replace column in 2 mL collection tube.
10. Wash column with 700 μL wash buffer II. Spin 30 s at 14,000 × *g*. Discard flow through. Replace column in 2 mL collection tube.
11. Spin 1–2 min at 14,000 × *g* to remove residual wash buffer. Transfer column to clean 1.5 mL Eppendorf tube.
12. Add 100 μL elution buffer to center of column. Let stand at room temperature for 5 min.
13. Spin 1 min at 14,000 × *g* to elute.
14. Run 5 μL of the plasmid sample on a 0.8% (w/v) agarose gel to check the yield and quality of the DNA. Two bands should be present, corresponding to the supercoiled and relaxed plasmid DNA.
15. For more precise measurement of yield, DNA should be quantified by spectrophotometric methods.

3.3.3. Restriction Digest to Check for Mutation

If restriction sites have been engineered into the DNA with the mutation, check if the purified DNA samples contain the mutation by digesting each sample with the appropriate restriction enzyme. Also, digest wild-type plasmid template DNA as a control. Restriction enzyme digestion may also be employed to check for the presence of inserts following the traditional cloning methods used to clone chimeric transporters.

1. Mix 5 μL of miniprep plasmid DNA with 1 μL of restriction enzyme, 2 μL of 10X restriction enzyme buffer (supplied with the restriction enzyme), and 12 μL of H_2O. Incubate at 37°C for 60 min.
2. Check the restriction digest by electrophoresis of 10 μL of the sample on a 0.8% (w/v) agarose gel made up in TAE buffer and containing 1:25,000 dilution of SYBR Safe DNA Stain.
3. The DNA should also be sequenced for confirmation of the desired mutation and to confirm that no other mutations have been introduced.

3.4. In Vitro Transcription of RNA for Expression in X. laevis Oocytes

Although a number of kits are available for production of RNA for injection into *X. laevis* oocytes we have found that the mMessage mMachine® kit from Ambion works very well.

3.4.1. The pOTV Vector

Linearized plasmid DNA and PCR products that contain an RNA polymerase promoter site can be used as templates for in vitro transcription with mMessage mMachine® kit. We use pOTV (Oocyte Transcription Vector) (9) which is derived from the pBluescript II SK +/− vector (Stratagene) and contains a number of useful features. First, it contains the *X. laevis* β-globin 5′ and 3′ untranslated sequences flanking several unique restriction sites for insertion of the cDNA (10). After the 3′ untranslated β-globin sequence, there are a number of additional unique restriction sites to be used for linearization of the plasmid (see below). Immediately upstream of the 5′ untranslated β-globin sequence is a T7 promoter which is used for RNA synthesis using T7 RNA polymerase. Other vectors containing an insert under the control of an RNA polymerase promoter can be used and will give varying levels of expression depending on the nature of the 5′ and 3′ untranslated sequences.

3.4.2. Linearization of DNA for Transcription Reaction

Prior to transcription, the plasmid DNA must be linearized by restriction enzyme digestion because circular plasmids will generate very long, heterogeneous RNA. The restriction enzyme used should not cut the 5′ untranslated sequence, the cDNA, or the 3′ untranslated sequence and ideally cut the plasmid only once after the 3′ untranslated sequence.

1. Linearize 1.0 μg of plasmid DNA with 2 μL of restriction endonuclease, 6 μL of 10X reaction buffer, and ddH_2O to a final volume of 60 μL.
2. Incubate for 1.5–2 h in a 37°C water bath.

3. Purify the linear DNA using a phenol:chloroform extraction procedure. Bring each sample to 200 μL with sterile H_2O and add 200 μL of phenol:chloroform:isoamyl alcohol (25:24:1). Vortex for 20 s and centrifuge at room temperature at 14,000×*g* for 5 min.
4. Transfer the upper aqueous phase to a new 1.5 mL Eppendorf tube. Add an equal volume of chloroform:isoamyl alcohol (24:1), vortex for 20 s, and centrifuge as above for 1 min.
5. Transfer the upper aqueous phase to a new tube. Add 1/10 the volume of 3 M sodium acetate and 2.5X the total volume of 100% molecular biology grade ethanol (*see* **Note 10**). Precipitate the DNA by placing at –20°C for 20 min. Centrifuge at 4°C and 14,000×*g* for 20 min.
6. Carefully remove the supernatant and wash the pellet with 200 μL of 70% ethanol (*see* **Note 10**). Centrifuge as above for 5 min. Carefully remove all liquid with a pipette tip and air-dry the pellet.
7. Dissolve the pellet in 7 μL of nuclease-free H_2O. Run 1 μL on a 0.8% agarose gel to confirm linearization and recovery.

3.4.3. In Vitro Transcription of RNA

Most eukaryotic mRNA molecules have a 5,7-methyl guanosine residue or cap structure which functions in the protein synthesis initiation process and protects the mRNA from degradation by intracellular nucleases. This can be achieved in vitro by substituting the cap analog ($m^7G(5')ppp(5')G$) for a portion of the GTP in the reaction. Any in vitro transcripts which are to be microinjected into oocytes, used for transfection experiments or for in vitro splicing reactions, should be capped.

The following protocol is adapted from the mMessage mMachine® kit instruction manual (Ambion).

1. Prepare the transcription reaction in a 1.5 mL Eppendorf tube on ice as follows:
   6 μL linearized DNA
   10 μL 2X NTP/CAP mix
   2 μL 10X reaction buffer
   2 μL 10X enzyme mix (add last)
   The final volume of the reaction is 20 μL.
2. Incubate the reaction at 37°C for 1.5–2 h in a 37°C oven (do not use a water bath as condensation may develop and reduce the yield).
3. Remove the DNA template by addition of 1 μL Turbo DNase and mix well. Incubate at 37°C for 15 min.
4. Add 115 μL nuclease-free water plus 15 μL ammonium acetate stop solution. Mix thoroughly.
5. Purify the linear RNA using a phenol:chloroform extraction procedure. Add 150 μL of phenol:chloroform:isoamyl

alcohol (25:24:1). Vortex for 20 s and centrifuge at room temperature at 14,000×*g* for 5 min.

6. Transfer the upper aqueous phase to a new 1.5 mL Eppendorf tube. Add an equal volume of chloroform:isoamyl alcohol (24:1), vortex for 20 s, and centrifuge as above for 1 min.
7. Transfer the upper aqueous phase to a new tube. Add 150 μL isopropanol.
8. Precipitate the RNA by placing at –20°C for 20 min. Centrifuge at 4°C and 14,000×*g* for 20 min.
9. Carefully remove the supernatant and wash the pellet with 200 μL of 70% ethanol (*see* **Note 10**). Centrifuge as above for 5 min. Carefully remove all liquid with a pipette tip and air-dry the pellet.
10. Dissolve the pellet in 20 μL of nuclease-free water. Run 1 μL on a 0.8% agarose gel to confirm linearization and estimate yield.
11. Store RNA in aliquots at –80°C.

3.5. Determination of Functional Properties

In all experiments investigating the effects of particular mutations of a transporter, it is desirable to compare the results to those obtained with the wild-type transporter measured under the same experimental conditions. In our laboratory we routinely use the *X. laevis* oocyte expression system to study the functional properties of transporters. This expression system is particularly useful for studying transporters, because oocytes are sufficiently large cells to conduct both electrophysiology and radiolabelled flux measurements on single cells. This is particularly useful when studying the relationships between electrical properties of the transporters and their ability to translocate solutes across the membrane, which allows a detailed exploration of the functional states of transporter proteins. Further details concerning the functional analysis of transporters can be found in **Chapter 16** (*X. laevis* oocytes) and **Chapter 19** (electrophysiological studies).

4. Notes

1. Many companies sell site-directed mutagenesis kits – e.g., Stratagene, Promega, Clonetech, Finnzymes. However, substantial cost savings and improved results can often be obtained by purchasing individual components separately or from preferred suppliers. The latest generation of modified DNA polymerases often incorporates hot-start technology, high fidelity, and improved processivity

(processivity is the term used to describe how long a polymerase will maintain fidelity as it replicates the template) compared to *Taq* polymerase, making them especially suitable for site-directed mutagenesis of plasmids or for the generation of large fragments of hybrid DNA.

2. Be sure to select a bacterial strain suitable to the vector being used.
3. DNA stained with SYBR Safe may be visualized using a UV light source. However, if the DNA needs to be isolated and purified from an agarose gel for downstream applications, the use of UV light should be avoided as it damages the DNA, rendering it unsuitable for downstream enzymatic manipulation such as restriction enzyme digestion or ligation.
4. Be sure to select the enzyme appropriate for the promoter present in the DNA template.
5. Fourteen cycles should be used for point mutations, 16 cycles for single amino acid changes, and 18 cycles for multiple amino acid deletions or insertions.
6. Comparison of the pre-DpnI reaction to the post-DpnI reaction ensures that any bands present are indeed the result of PCR amplification and not residual vector. If no band is present, increase the number of cycles, increase the template concentration, or reduce the annealing temperature. If a smear is present, increase the annealing temperature or redesign oligos to increase specificity.
7. If no band is present, increase the number of cycles, increase the template concentration, or reduce the annealing temperature. If multiple bands are present, increase the annealing temperature or reduce the amount of template DNA. Alternatively, specific bands may be gel isolated and purified for use in the fusion-PCR reaction.
8. The heat pulse has been designed for Falcon 2059 tubes and the use of other tubes may give different results.
9. LB broth can also be used.
10. The 100% ethanol and 70% ethanol solutions should be stored at –20°C.

References

1. Kaback, H.R., Sahin-Toth, M., and Weinglass, A.B. (2001) The kamikaze approach to membrane transport. *Nat. Rev. Mol. Cell Biol.* **2**, 610–620.
2. Seal, R.P. and Amara, S.G. (1998) A reentrant loop domain in the glutamate carrier EAAT1 participates in substrate binding and translocation. *Neuron* **21**, 1487–1498.
3. Grunewald, M., Bendahan, A., and Kanner, B.I. (1998) Biotinylation of single cysteine mutants of the glutamate transporter GLT-1

from rat brain reveals its unusual topology. *Neuron* **21**, 623–632.
4. Monks, S.A., Needleman, D.J., and Miller, C. (1999) Helical structure and packing orientation of the S2 segment in the Skaker K^+ channel. *J. Gen. Physiol.* **113**, 415–423.
5. Buck, K. and Amara, S.G. (1994) Chimeric dopamine-norepinephrine transporters delineate structural domains influencing selectivity for catecholamines and 1-methyl-4-phenylpyridinium. *Proc. Natl. Acad. Sci. USA* **91**, 12584–12588.
6. Mitrovic, A.D., Amara, S.G., Johnston, G.A.R., and Vandenberg, R.J. (1998) Identification of functional domains of glutamate transporters. *J. Biol. Chem.* **273**, 14698–14706.
7. Yon, J. and Fried, M. (1989) Precise gene fusion by PCR. *Nucleic Acids Res.* **17**, 4895.
8. Shevchuk, N.A., Bryskin, A.V., Nusinovich, Y.A., Cabello, F.C., Sutherland, M., and Ladisch, S. (2004) Construction of long DNA molecules using long PCR-based fusion of several fragments simultaneously. *Nucleic Acids Res.* **32**, e19.
9. Arriza, J.L., Fairman, W.A., Wadiche, J.I., Murdoch, G.H., Kavanaugh, M.P., and Amara, S.G. (1994) Functional comparisons of three glutamate transporter subtypes cloned from human motor cortex. *J. Neurosci.* **14**, 5559–5569.
10. Kreig, P.A. and Melton, D.A. (1984) Functional messenger RNAs are produced by SP6 in vitro transcription of cloned cDNAs. *Nucleic Acids Res.* **12**, 7057–7070.

Chapter 16

Xenopus laevis Oocytes

Stefan Bröer

Abstract

Xenopus oocytes are a versatile expression system particularly suited for membrane transporters and channels. Oocytes have little background activity and therefore offer a very high signal-to-noise ratio for transporter and channel characterization. This chapter provides an overview of the basic methods used for the analysis of membrane transporters in this system, including preparation of oocytes, assays of transport activity, protocols for immunostaining and fluorescence microscopy, and other assays to study surface expression.

Key words: Two-electrode voltage clamp, flux assay, surface biotinylation, transport, microtransplantation, efflux, ion-sensitive electrodes.

1. Introduction

During oogenesis, Xenopus oocytes accumulate large amounts of storage proteins that – after fertilization – provide the developing embryo with building blocks and energy metabolites. The size of the fully developed oocyte (diameter 1.2 mm) is largely governed by the stored amounts of egg yolk protein. The mature oocyte thus is equipped to initiate protein synthesis, cell growth, and replication after fertilization. The size of the oocyte, the large reserve of storage proteins, and its ability to synthesize protein on demand make the oocyte an almost ideal single-cell expression system. Some basic physical properties of this expression system are listed in **Table 16.1**.

The use of Xenopus oocytes as an in vitro expression system was initiated in 1971 by the observation of John B. Gurdon that mRNA injected into Xenopus oocytes was translated into

Q. Yan (ed.), *Membrane Transporters in Drug Discovery and Development*, Methods in Molecular Biology 637,
DOI 10.1007/978-1-60761-700-6_16,

Table 16.1
Physicochemical properties of oocytes

Property	Value	References
Water accessible volume	368 ± 21 nl	(1)
Water permeability	$1–4 \times 10^{-4}$ cm/s	(2)
Surface area	18–30 mm^2	(3)
Membrane potential	–30 to –60 mV	(3)
Buffering capacity	20 mM/pH unit at pH 7.0	(4)
Intracellular pH	7.4 ± 0.1	(4)
Na^+ concentration	4–10 mM	(3)
K^+ concentration	76–120 mM	(3)
Cl^- concentration	24–50 mM	(3)
Ca^{2+} concentration	<0.3 μM	(3)
ATP	3–5 mM	(5)

protein (6). It was, however, not before 1982 that Xenopus oocytes were used to express membrane proteins as pioneered by studies on the acetylcholine receptor in Miledis and Numas group (7, 8). Because of the large amount of stored protein, oocytes do not depend on extracellular resources for nutrition. As a result only limited numbers of endogenous membrane transporters are expressed. Background transport activity is particularly low for some charged substances such as glutamate or phosphate. Upon injection of mRNA, large amounts of transporter proteins are synthesized. Figures of 10^{10}–10^{11} proteins/oocyte have been reported for the glucose transporter SGLT1 and other membrane proteins (9). Remarkably, the synthesis of 10^{11} transporters in 4 days translates into the incorporation of roughly 300,000 transporter molecules per second. More recently direct microtransplantation of membrane preparations into oocytes has been reported, which resulted in functional expression of membrane-embedded proteins, albeit at levels lower than those obtained by mRNA expression (10, 11).

In addition to the strength of this expression system in the mechanistic and structural characterization of transport proteins, oocytes are also well equipped with signal transduction pathways that can be exploited to study regulation of transport proteins.

Protein kinase A and adenylate cyclase are endogenously expressed in oocytes. Progesterone which is used to induce maturation in oocytes drops cAMP levels in minutes by a mechanism that is thought to involve a decrease of PKA activity (12). By contrast, PKA can be induced by 8-bromo-cAMP, 3-isobutylmethylxanthine (IBMX), or forskolin. Alternatively, β-adrenergic receptors can be expressed in oocytes that couple to G_s-proteins. The α-subunit of the G_s-proteins subsequently activates the adenylate cyclase resulting in the production of cAMP. PKA-coupled pathways have been monitored in oocytes by expression of the CFTR protein, the chloride conductance of which is activated by PKA (13).

The α-subunit of oocyte endogenous G_q proteins activates phospholipase C. Phospholipase C in turn splits phopshatidylinostol-4,5-bisphosphate to generate diacylglycerol and inositol-1,4,5-trisphosphate (IP_3). The two molecules activate different pathways. Diacylglycerol activates protein kinase C. Inositol 1,4,5-trisphosphate by contrast binds to the intracellular IP_3 receptor channel ($InsP_3$-receptor 1) that resides on the endoplasmatic reticulum, resulting in the release of Ca^{2+} from the ER. The elevation of intracellular Ca^{2+} results in the opening of Ca^{2+}-sensitive chloride channels. Similar to monitoring cAMP levels by expression of CFTR, Ca^{2+} levels have been monitored by measuring chloride conductance (14). The chloride channel can, for example, be activated by lysophosphatidic acid, which binds to endogenous receptors (15). The activity of the Ca^{2+}-activated chloride channel is also modulated by overexpression of foreign proteins in the oocyte membrane.

Pathways that are activated by tyrosine receptor kinases are also endogenous to Xenopus oocytes. The oocyte contains a MAP kinase pathway that is activated during oocyte maturation, e.g., when treated with progesterone (16).

A second pathway related to tyrosine kinases, which has been described in oocytes, is the PI3-kinase-PDK-Akt pathway (16). Signaling induced by the Akt-related protein kinase Sgk1 is also recognized by the oocyte. Sgk1, when coexpressed with the epithelia Na^+-channel, promotes trafficking of the channel to the plasma membrane (17).

Thus it appears possible to simulate signal transduction pathways in oocytes that are occurring in quite different cell types in vivo, such as neurons or epithelial cells of the kidney. However, it has to be kept in mind that regulation in oocytes may not coincide with regulation in every cell type. For example, downregulation of transporters by protein kinase C has been reported frequently in oocytes, but may not occur in all cell types in which the transporter is expressed in vivo (18).

2. Materials

2.1. Oocyte Preparation

1. 3-aminobenzoic acid ethyl ester (MS222), 1.5 g/l prepare fresh.

2. Resorbable suture: Dexon, 1 metric, 5/0 USP, Needle HR13, 45 cm.
3. Non-resorbable suture: Dafilon 0.7 metric, 6/0, Needle DSM 11, 45 cm.
4. Gentamicin sulfate, stock solution 50 mg/ml, store at – 20°C.
5. Collagenase D from Clostridium histolyticum EC 3.4.24.3; 0.3 U/mg. Prepare fresh (*see* **Note 1**).
6. Oocyte Ringer 2 without $CaCl_2$ ($OR2^-$) in mM: NaCl, 82.5; KCl, 2.5; $MgCl_2$, 1; Na_2HPO_4, 1; *N*-[2-hydroxyethyl]piperazine-N'-[2-ethanesulfonic acid] (HEPES), 5. Titrated with NaOH to pH 7.8. A 10X stock can be prepared, stable. Use filter sterilization not autoclaving because phosphate will precipitate.
7. Oocyte Ringer 2 with $CaCl_2$ ($OR2^+$) $OR2^-$ supplemented with 1.0 mM $CaCl_2$, stable.
8. ND96 composition in mM: NaCl, 96; KCl, 2; $MgCl_2$, 1; $CaCl_2$, 1.8; HEPES, 5. Titrated with NaOH to pH 7.4, stable.

2.2. Flux Measurements

1. Nunc-Immunotubes (Nalge Nunc, Naperville, IL, USA).
2. Plastic Pasteur pipettes (Bacto Laboratories, Liverpool, NSW, Australia)

2.3. Electrophysiological Recordings

1. Glass capillaries (GC150F-10) (Harvard Apparatus Ltd., Edenbridge, UK).
2. Tributylchlorosilane (Sigma-Fluka, Buchs, Switzerland).
3. Tetrachlorocarbon (CCl_4) (Sigma-Fluka, Buchs, Switzerland).
4. 5% tributylchlorosilane in CCl_4. The solution is stable for 3 months in the dark but evaporates quickly.
5. Proton Ionophore Cocktail A (Sigma-Fluka, Buchs, Switzerland).
6. 0.1 M Na-citrate pH 6.0 (any company). Titrate Na-citrate with HCl to provide chloride ions for the Ag/AgCl electrode.

2.4. Membrane Preparation

1. Pefabloc (AEBSF 4-(2-aminoethyl)-benzenesulfonyl fluoride hydrochloride) (Roche, Mannheim, Germany), store at 4°C, unstable in aqueous solutions.
2. Homogenization buffer 1, composition in mM: Tris, 50; NaCl, 100; EDTA, 1; Pefabloc, 1; titrate with HCl to pH 7.5. Pefabloc is unstable and needs to be added shortly before usage.

2.5. Membrane Fractionation

1. Homogenization buffer 2, composition in mM: sucrose, 320; Tris, 50; EDTA 1; Pefabloc, 1; titrate with HCl to pH 7.5, prepare fresh.
2. Gradient buffers: (i) 2.0 M sucrose, (ii) 1.3 M sucrose, (iii) 1.0 M sucrose, and (iv) 0.6 M sucrose each in TE-buffer, prepare fresh.
3. TE-buffer, composition in mM: Tris, 50; EDTA, 1; $MgCl_2$, 5; titrate with HCl to pH 7.5, stable.

2.6. Surface Biotinylation

1. Phosphate-buffered saline, composition in mM: NaCl, 137; KCl, 2.7; Na_2HPO_4, 5; titrate with HCl to pH 8.0. Stable, pH 8.0 important for reaction.
2. EZ-link Sulfo-NHS-lc-Biotin (0.5 mg/ml dissolved in phosphate-buffered saline; use 1 ml/10 oocytes, prepare fresh each time) (Pierce, Rockford, USA).
3. Immunopure, immobilized streptavidin gel (Pierce, Rockford, USA), store at 4°C.
4. Lysis buffer, composition in mM: NaCl, 150; Tris, 20; 1% Triton X-100, titrate with HCl to pH 7.5, stable but light sensitive.

2.7. Microtransplantation

1. Glycine buffer, composition in mM: glycine, 200; NaCl 150; ethyleneglycolbistetraacetic acid (EGTA), ethylenediaminetetraacetic acid (EDTA), 50; sucrose 300 (final pH should be 9.0, titrate with NaOH, keep in 5 ml aliquots at 4°C).
2. Protease inhibitor: Dissolve protease inhibitor cocktail (Sigma P2714) in 100 ml water. Keep 5 ml aliquots at –20°C.
3. Homogenizer Pro 200 with 5 mm generator (Pro Scientific, Inc., Oxford, Connecticut, USA)
4. Storage buffer, composition in mM: glycine 5, pH 7.4

2.8. Immunohistochemistry

1. Phosphate-buffered saline (PBS), composition in mM: NaCl, 137; KCl, 2.7; Na_2PO_4, 4.3; KH_2PO_4, 1.4 (final pH should be 7.1, stable).
2. Dent's fixant, composition in %: Methanol, 80; DMSO, 20; stable.
3. 3.7% Paraformaldehyde in PBS, prepare fresh, titrate to pH 7.4. It may be necessary to warm up the solution until paraformaldehyde is fully dissolved.
4. 90% PBS, 10% normal donkey serum, store frozen at –20°C.
5. Technovit 7100 (Heraeus Kulzer, Wehrheim, Germany), stable.

6. Embedding solution 1: 1:1 mixture of Technovit 7100 and ethanol.
7. Embedding solution 2: Technovit 7100 plus hardener 1 (1 g hardener 1 (dibenzoylperoxide) per 100 ml Technovit 7100, solution stable at 4°C for 1 month).
8. Embedding solution 3: 15 ml Technovit 7100, 0.15 g hardener 1 plus 1 ml hardener 2, prepare fresh each time.
9. Embedding capsules (BEEM capsules, Proscitech, Turingowa, Queensland, No. RB001).
10. Embedding mould (Tissue-Tek, Cryomold, Miles, Inc., Elkhart, IN, USA, No. 4565).

3. Methods

3.1. Preparation of Oocytes

1. Submerse frog in 1.5 g/l MS222 for 15′–30′. Anesthesia is complete when the frog can be turned upside down without eliciting any reaction.
2. Place frog on a sheet of wet paper or cloth on ice with belly up.
3. Make a small incision (1 cm) slightly off the middle line of the belly. Using a sterile scalpel, first cut through the skin, then through the muscular layer.
4. Remove parts of the ovary with tweezers and scissors and place ovary in $OR2^-$ buffer.
5. Close incision with one stitch of reabsorbable suture in the muscular layer.
6. Close incision with two stitches of non-absorbable suture in the skin.
7. Let the frog recover under a moist tissue, when fully awake submerse in water.
8. Prepare a collagenase solution (1 mg/ml) in a 100 ml Erlenmeyer flask (*see* **Note 1**). Using tweezers and scissors cut ovary into smaller pieces (about 20–30 oocytes each) and let them drop into the Erlenmeyer flask containing collagenase in $OR2^-$ buffer.
9. Incubate the ovary at 28°C for 2–4 h under slight agitation (*see* **Note 2**).
10. After digestion is completed, oocytes are thoroughly washed with copious volumes of $OR2^-$ buffer (at least 1 l). Small oocytes (stages I–III) can be discarded during the washing procedure.
11. Finally, oocytes are washed three times in $OR2^+$ buffer.

12. On the next day, stages V and VI oocytes are manually selected for injection (*see* **Note 3**).
13. Oocytes are stored at 16–18°C in sterile cell culture dishes (35 mm) in OR2$^+$ buffer. For long-term storage gentamicin (10 mg/l) is added. Change buffer daily to increase life expectancy of oocytes. Remove dead or damaged oocytes.

3.2. Substrate Uptake

1. Transfer 7–10 oocytes into a Nunc Immunotube (*see* **Note 4**).
2. Flush oocytes two times with 4 ml transport buffer, e.g., ND96.
3. Remove supernatant completely. This is important because otherwise the radioactivity may be diluted by an unknown factor. If complete aspiration appears too risky use an automatic pipette with a yellow tip to remove the residual liquid.
4. Add 100 μl of transport buffer containing the labeled substrate (*see* **Note 5**).
5. Incubate oocytes for an appropriate time (usually between 5 and 30 min depending on the expression level of the transporter).
6. Add 4 ml ice-cold transport buffer to terminate transport and remove >95% of the supernatant by aspiration.
7. Flush oocytes three times with 4 ml ice-cold transport buffer and aspirate the supernatant each time.
8. After the third removal add 1 ml ice-cold transport buffer.
9. Transfer individual oocytes with a plastic Pasteur pipette to scintillation vials.
10. Dissolve oocytes by addition of 200 μl 10% SDS.
11. Vortex briefly to homogenize oocytes, add > 1.5 ml scintillation fluid.

3.3. Substrate Efflux, Preloading Method

1. Transfer 7–10 oocytes into a Nunc Immunotube.
2. Flush oocytes two times with 4 ml transport buffer (room temperature).
3. Remove supernatant completely (*see* **Section 3.2**, step 3).
4. Preloading: Add 100 μl of transport buffer containing the labeled substrate. Incubate oocytes until radioactivity is equilibrated (usually between 30 and 120 min, *see* **Note 6**). Use a low-substrate concentration for preloading to maximize the specific activity (e.g., 10 μM).
5. Flush oocytes three times with 4 ml ice-cold transport buffer. Aspirate >95% of the supernatant the first two times. The third time supernatant must be removed completely.

6. Add 1 ml of room temperature transport buffer to the oocytes to initiate efflux. Extracellular substrates should preferably be added at saturating concentrations.
7. Take samples at different times from supernatant for counting.

3.4. Injection Method

1. Inject oocytes with 10–40 nl radiolabeled substrate. The radiolabeled substrate can be used as delivered by the manufacturer. Incubate 10 min to allow diffusion of substrate in the oocyte and closure of the injection spot. If transporter allows net movement of substrates (i.e., no antiporter) injection should be performed in ice-cold transport buffer to avoid loss of substrate.
2. Transfer 7–10 oocytes to a Nunc Immunotube.
3. Flush oocytes two times with 4 ml transport buffer (ice-cold when transporter mediates net movement).
4. Remove supernatant completely.
5. Add 1 ml of room temperature transport buffer to the oocytes to initiate efflux. Extracellular substrates should preferably be added at saturating concentrations.

3.5. Voltage Clamp Recordings

The two-electrode voltage clamp (TEVC) technique allows the control of the membrane potential (clamping) to measure currents flowing through ion channels, electrogenic transporters, or pumps. For the technicalities of the electrophysiological setup, the reader is referred to reviews in this area (19). Two glass microelectrodes are impaled into the oocyte, a membrane potential recording electrode and a current-delivering electrode. The membrane potential electrode connects to a feedback amplifier where the signal is compared to the command voltage given by a generator. A variable current is applied through the current-delivering electrode, across the membrane, and to the bath-grounding electrode that aligns the membrane potential with the command potential. All electrogenic ion or substrate fluxes across the membrane are now measured as a deflection from the baseline current. By convention, upward deflections correspond to the influx of anions (or efflux of cations) and downward deflections correspond to the influx of cations (efflux of anions). The setup is described for a Geneclamp 500 amplifier.

1. Pull capillaries to form microelectrodes (fire-polish the back end to avoid damage to the $AgCl_2$ coating of the $Ag/AgCl_2$ electrode).
2. Back-fill electrodes with 3 M KCl and insert into $Ag/AgCl_2$ electrode holder (*see* **Note 7**). The microelectrode with the higher input resistance is the voltage-recording electrode,

the microelectrode with the lower input resistance is the current passing electrode in the voltage clamp modus (*see* **Note 8**).

3. Switch amplifier to the setup mode (current clamp).
4. Submerse electrodes in the bath. Cancel voltage offset to ±1 mV.
5. Measure resistance of microelectrodes, which should be between 0.5 and 3 MΩ.
6. Impale oocyte with both microelectrodes (*see* **Note 9**).
7. Set command voltage to the resting potential of the oocyte (about –40 mV). Switch amplifier to the voltage clamp mode and adjust voltage to the resting potential.
8. Apply test pulse (e.g., –10 mV square wave from the resting potential) and increase gain until output looks like the command pulse. Oscillations can be counteracted by introducing a phase shift (stability setting).
9. Put command voltage to the desired value (e.g., –60 mV) and apply substances.

3.6. Ion-Sensitive Electrodes

Ion-sensitive electrodes are particularly useful to investigate H^+-coupled transporters (4). Movement of Na^+ ions can be followed as well; however, changes of the intracellular Na^+ concentration in oocytes are rather slow even when expressing very fast transporters. Ion-selective electrodes are made out of the same borosilicate glass as membrane potential recording electrodes. In order to generate ion selectivity the tip is filled with a specific ionophore. Due to the hydrophobic nature of the ionophore the glass capillary has to be coated with silane to generate a hydrophobic surface to which the ionophore adheres. As ion-selective electrodes record both the membrane potential and the electrochemical gradient of the respective ion species a second standard membrane potential recording electrode has to be used to subtract the membrane potential from the combined signal. Double-barreled electrodes can be used to combine both electrodes in one, but those demand a special pulling device. Due to the large size of the oocyte two separate electrodes can be used instead. Ion-selective electrodes have a very high resistance (10^{10}–10^{12} Ω), thus headstages or preamplifier with even higher input resistance have to be used. The preparation of the ion-sensitive electrodes is described in the following:

1. Pull capillaries as for membrane potential recording.
2. Back-fill the first 2–3 mm of each capillary with silane solution.
3. Place the capillaries tip to the center on a hot plate at 450°C.
4. Incubate for 5 min to evaporate CCl_4.

5. Back-fill ionophore solution (for pH use: Proton Ionophore Cocktail A) into the silanized tip. Use less ionophore than silanizing solution.
6. Layer 0.1 M Na-citrate pH 6.0 on top and fill capillary up half way.
7. Insert capillary into Ag/AgCl electrode holder.

Capillaries should be calibrated with solutions of defined pH. The response of the electrode should be instantaneous and be >50 mV/pH unit.

3.7. Simple Oocyte Membrane Preparation

1. 25 oocytes are homogenized in 1 ml homogenization buffer 1 by trituration in a blue tip of an automatic pipettor.
2. Spin the homogenate in a microcentrifuge at 2000×*g* for 10 min at 4°C (*see* **Note 10**).
3. Transfer the supernatant and spin it at 140,000×*g* for 30 min at 4°C.
4. Discard the supernatant.
5. Dissolve pellets in 30–50 μl homogenization buffer 1 supplemented with 4% SDS
6. Add sample buffer and subject to PAGE.

3.8. Fractionation of Oocyte Membranes

1. 150 oocytes are homogenized in 2 ml homogenization buffer 2 by 15 strokes of a tight-fitting pestle in a chilled Dounce homogenizer (20).
2. The homogenate is centrifuged twice at 1000×*g* for 10 min at 4°C (*see* **Note 10**).
3. Transfer the supernatant on top of a discontinuous sucrose gradient: 2.0 M (2 ml), 1.3 M (3.2 ml), 1.0 M (3.2 ml), 0.6 M (2.0 ml) all solution prepared in TE-buffer containing 5 mM $MgCl_2$.
4. The gradient is centrifuged in a Beckman SW41 swing out rotor at 40,000 rpm for 4 h at 4°C. Fractions (1 ml) are collected starting from the bottom. Each fraction is diluted 4-fold with 0.15 M sucrose in TE-buffer.
5. Pellet membranes from each fraction by centrifugation at 48,000 rpm in a Beckmann SW 50.1 rotor for 3 h at 4°C.
6. Resuspend pellet in 80 μl TE-buffer or in 15 μl SDS–PAGE sample buffer (*see* **Note 11**).

3.9. Biotinylation of Oocyte Surface Proteins

1. Wash 5–10 oocytes three times with 4 ml ice-cold PBS pH 8.0. The pH is important for the succinimide ester formation.
2. Incubate 5–10 oocytes in 0.5 ml Sulfo-NHS-lc-Biotin solution for 10–30 min at room temperature.

3. Wash oocytes 4X with 4 ml ice-cold PBS pH 8.0.
4. Transfer oocytes into a 1.5 ml reaction tube.
5. Lyse oocytes by incubation in 1 ml of lysis buffer for 30–60 min on ice, invert tube from time to time, do not vortex! (*See* **Note 12.**)
6. Spin down at top speed in table-top centrifuge for 15 min at 4°C.
7. Transfer supernatant into fresh 1.5 ml reaction tubes.
8. Add 50 μl streptavidin-coated agarose particles (*see* **Note 13**).
9. Incubate at 4°C for 1 h with slight agitation.
10. Spin down 10 min at maximum speed in a table-top centrifuge. Carefully remove most of the supernatant, do not remove any pellet (pellet difficult to see).
11. Wash pellets four times with 1 ml lysis buffer, each time repeating step 10.
12. Resuspend pellet in 20 μl SDS–PAGE sample buffer, boil for 5 min, and use for gel electrophoresis.

3.10. Microtransplantation of Membrane Preparations

1. Prepare 4 ml of glycine buffer by adding 40 μl protease inhibitor stock immediately before use (10).
2. Homogenize 0.2–0.5 g tissue in 4 ml ice-cold glycine buffer for 2 min (setting 5) on ice.
3. Transfer the homogenate to 1.5 ml reaction tubes and centrifuge at 9500×*g* for 15 min at 4°C.
4. Centrifuge the supernatant at 100,000×*g* for 2 h at 4°C in a Beckmann Optimax table-top ultracentrifuge (Rotor TLA 100.2). Discard supernatant.
5. Wash pellet twice with distilled water.
6. Resuspend pellet in 400 μl storage buffer. Prepare 10–50 μl aliquots and use directly or store at –80°. Determine protein concentration.
7. Inject each oocyte with 100 nl membrane preparation (1–2 mg protein/ml) 1 day after preparation.
8. Measure transport activity 12 h post-injection.

3.11. Immunohistochemical Analysis of Transporter Expression

This protocol describes the procedure of how to prepare oocytes for immunofluorescence analysis.

1. Transfer oocytes to HPLC glass vials. Solutions can be easily exchanged with Pasteur pipettes. All volumes are given as per vial.
2. Fix oocytes in 1 ml Dent's fixant for 2 h at room temperature (or overnight at –20°C).

3. Wash oocytes with 0.5 ml each of solution A (90% methanol in H_2O), solution B (70% methanol in H_2O), solution C (50% methanol in PBS), and solution D (30% methanol in PBS). Apply each solution for 10 min.
4. Wash oocytes three times for 10 min with 1 ml PBS.
5. Incubate oocytes at 4°C overnight (or 3 h at room temperature) with primary antibody in 90% PBS, 10% normal donkey serum (or goat serum). Dilute antibody as appropriate.
6. Wash oocytes with PBS using 2 ml aliquots in each step: three times for 5 min, three times for 15 min, three times for 30 min, and finally two times for 1 h.
7. Incubate oocytes with secondary antibody in 90% PBS, 10% normal donkey serum (or goat serum) for 1 h in the dark at room temperature. Use dilution as recommended by the manufacturer, the volume can be kept as small as convenient (e.g., 200 μl).
8. Wash oocytes six times with 2 ml PBS.
9. Wash oocytes overnight with 2 ml PBS.
10. Postfixate oocytes with 3.7% paraformaldehyde in PBS for 30 min.
11. Wash oocytes twice with 2 ml PBS for 15 min.
12. Dehydrate oocytes by incubating in 0.5 ml ethanol solutions of increasing concentrations each for 15 min: solution A2 30% ethanol in PBS, solution B2 (50% ethanol in PBS), solution C2 (70% ethanol in H_2O), solution D2 (90% ethanol in H_2O), and finally in 100% ethanol.
13. Embed oocytes in acrylic resin by infiltration for 2 h at room temperature with 0.5 ml embedding solution 1 (Technovit 7100/ethanol mixture (1:1)).
14. Exchange infiltration solution to 0.5 ml embedding solution 2 (Technovit 7100/hardener 1). Incubate for 2 h at room temperature.
15. Renew embedding solution 2 and incubate overnight at 4°C.
16. Prefill disposable embedding capsules with 100 μl embedding solution 3 (15 ml embedding solution 2 + 1 ml hardener 2).
17. Transfer oocytes on top of the embedding solution.
18. Add more embedding solution until capsules are filled to completion. Close capsules. Polymerization requires about 2 h.
19. The shape of the capsules is suitable for mounting the specimen on a microtome.
20. Cut 5 μm thick slices using a standard microtome.

3.12. Immunostaining of Oocytes for Confocal Microscopy

1. Transfer oocytes to HPLC glass vials. Solutions can be easily exchanged with Pasteur pipettes. All volumes are given as per vial.
2. Fix oocytes in 1 ml 4% paraformaldehyde (in PBS) for 30 min at room temperature.
3. Wash oocytes three times for 10 min each in 1 ml PBS at room temperature, shake occasionally.
4. Permeabilize oocytes by incubation in 1 ml 100% methanol (20 min at room temperature).
5. Wash oocytes three times for 10 min each in 1 ml PBS at room temperature, shake occasionally.
6. Block non-specific binding sites by incubation in 3% BSA, 1% normal goat serum, 0.1% Triton X-100 in PBS for 1 h at room temperature.
7. Stain with primary antibody (dilution as optimized) in 1 ml PBS containing 1% BSA, 0.01% Triton X-100 over night at 4°C, or for 2 h at room temperature.
8. Wash oocytes three times for 10 min each in 1 ml PBS at room temperature, shake occasionally.
9. Stain with secondary antibody (dilution as recommended) in 1 ml PBS containing 1% BSA, 0.01% Triton X-100 for 2 h in the dark at room temperature.
10. Wash oocytes three times for 10 min each in 1 ml PBS at room temperature, shake occasionally.
11. Analyze oocytes by confocal microscopy (*see* **Note 14**).

4. Notes

1. For long-term survival of oocytes (>5 days) it is critical to use collagenase with low tryptic activity. Companies usually provide information about tryptic activity of batches that are in stock. Collagenase D is formulated for low tryptic activity. Digestion sometimes takes a couple of hours with collagenase D. In that case spiking the digestion solution with a few grains of collagenase A is helpful.
2. Digestion is complete when (i) most oocytes float around separately, (ii) most oocytes are devoid of adhering blood vessels, and (iii) the follicular cell layer is removed. Oocytes are still surrounded by the vitelline layer (glycoprotein matrix) at this stage.
3. Good oocytes are large in size (diameter > 1 mm) and have a brown color on the animal pole. Pigmentation is fine-grained, and poles are separated by a sharp border. Injection may be performed on the 2 days following

preparation. Life expectancy of oocytes injected on the second day is slightly shorter than for those injected on the first day. We use custom-made perspex egg cups for injection. The design is identical to a 100 mm petri dish but with a thicker bottom plate (5 mm). Holes (0.5 mm deep and 1.1 mm wide) are drilled into the bottom plate (15 × 12, distance 3 mm).

4. Flux measurements are best carried out in 5 ml Nunc-Immunotubes or γ-counter tubes. Oocytes are transferred from the storage culture dishes to the tubes using disposable plastic Pasteur pipettes. The transport buffer ND96 can be aspirated using a glass Pasteur pipette with a tilted end attached to a vacuum pump.
5. As a first approximation, the transport buffer should contain 5–10 kBq of labeled substrate/100 μl. This usually gives a well-detectable result. Unlabeled substrate is added as necessary. For uncharacterized transporters 10 μM is a good concentration to start with.
6. Preloaded substrates are usually not metabolized quickly by the oocyte. Amino acids such as isoleucine or glutamine, e.g., are stable for at least 2 h.
7. Sometimes leaking of 3 M KCl can cause oocytes to swell. If the holding currents are not too large 300 mM KCl can be used as an electrode filling solution. $Ag/AgCl_2$ electrodes can easily be prepared from silver wire by bathing 2/3 of its length in 12% NaOCl for 20 min.
8. The current passing headstage/microelectrode 2 should be able to pass currents of up to 100 μA. The voltage-recording electrode 1, however, does not need to pass large currents. The headstage with a lower output resistance should thus be chosen for the current passing electrode. For example, a headstage with Ro=1 MΩ is best suited to serve as ME2 and a headstage with Ro=10 MΩ as ME1.
9. Penetration can easily be monitored in the current clamp mode (where both electrodes monitor the membrane potential) by the rapid deflection of the electrical potential which is usually in the range of about –35 to –45 mV, but may be different in protein expressing oocytes. In ringer solution the membrane potential of the oocyte is mainly dominated by a K^+ diffusion potential generated by potassium channels and a contribution of the endogenous Na,K-ATPase to the membrane potential. Thus, elevation of KCl in the buffer effectively depolarizes Xenopus oocytes.
10. Spinning down homogenized oocytes generates three layers of material: (i) the egg yolk pellet, (ii) the

homogenization buffer containing cytosolic components, and (iii) a top layer of white lipids. It is very difficult to avoid transfer of lipids to the next step completely. Avoid repetitive pipetting. Use a blue tip punch through the floating lipids and remove the clear supernatant as complete as possible. Wipe outside of tip to minimize transfer of lipids. The more oocytes are used, the thicker the lipid layer becomes.

11. According to (20), the rough ER is found in fractions 2–3, the plasma membrane is detected in fraction 5, and the trans-golgi network is in fractions 9–10.
12. Vortexing results in homogenization of the oocyte generating a layer of lipids after centrifugation that is difficult to remove (*see* **Note 10**).
13. Mix content of bottle well to resuspend agarose particles. Cut the end of a yellow tip to transfer the suspension.
14. The laser light of the confocal microscope only penetrates the outer surface of the oocytes. As a result membrane proteins residing in the ER or Golgi are not visible (21). This can be important when analyzing trafficking mutations. It is advisable to combine this method with other methods of analyzing expression.

References

1. Stegen, C., Matskevich, I., Wagner, C.A., Paulmichl, M., Lang, F., and Broer, S. (2000) Swelling-induced taurine release without chloride channel activity in Xenopus laevis oocytes expressing anion channels and transporters. *Biochim. Biophys. Acta.* **1467**, 91–100.
2. Zeuthen, T., Meinild, A.K., Loo, D.D., Wright, E.M., and Klaerke, D.A. (2001) Isotonic transport by the Na+-glucose cotransporter SGLT1 from humans and rabbit. *J. Physiol.* **531**, 631–644.
3. Weber, W. (1999) Ion currents of Xenopus laevis oocytes: state of the art. *Biochim. Biophys. Acta.* **1421**, 213–233.
4. Broer, S., Schneider, H.P., Broer, A., Rahman, B., Hamprecht, B., and Deitmer, J.W. (1998) Characterization of the monocarboxylate transporter 1 expressed in Xenopus laevis oocytes by changes in cytosolic pH. *Biochem. J.* **333**, 167–174.
5. Broer, A., Hamprecht, B., and Broer, S. (1998) Discrimination of two amino acid transport activities in 4F2 heavy chain-expressing Xenopus laevis oocytes. *Biochem. J.* **333**, 549–554.
6. Gurdon, J.B., Lane, C.D., Woodland, H.R., and Marbaix, G. (1971) Use of frog eggs and oocytes for the study of messenger RNA and its translation in living cells. *Nature* **233**, 177–182.
7. Mishina, M., Kurosaki, T., Tobimatsu, T., Morimoto, Y., Noda, M., Yamamoto, T., Terao, M., Lindstrom, J., Takahashi, T., Kuno, M., et al. (1984) Expression of functional acetylcholine receptor from cloned cDNAs. *Nature* **307**, 604–608.
8. Barnard, E.A., Miledi, R., and Sumikawa, K. (1982) Translation of exogenous messenger RNA coding for nicotinic acetylcholine receptors produces functional receptors in Xenopus oocytes. *Proc. R Soc. Lond. B Biol. Sci.* **215**, 241–246.
9. Zampighi, G.A., Kreman, M., Boorer, K. J., Loo, D. D., Bezanilla, F., Chandy, G., Hall, J. E., and Wright, E. M. (1995) A method for determining the unitary functional capacity of cloned channels and transporters expressed in Xenopus laevis oocytes. *J. Membr. Biol.* **148**, 65–78.
10. Miledi, R., Eusebi, F., Martinez-Torres, A., Palma, E., and Trettel, F. (2002) Expres-

sion of functional neurotransmitter receptors in Xenopus oocytes after injection of human brain membranes. *Proc. Natl. Acad. Sci. USA* **99**, 13238–13242.

11. Miledi, R., Palma, E., and Eusebi, F. (2006) Microtransplantation of neurotransmitter receptors from cells to Xenopus oocyte membranes: new procedure for ion channel studies. *Methods Mol. Biol.* **322**, 347–355.
12. Maller, J.L., Butcher, F.R., and Krebs, E.G. (1979) Early effect of progesterone on levels of cyclic adenosine 3′:5′-monophosphate in Xenopus oocytes. *J. Biol. Chem.* **254**, 579–582.
13. Uezono, Y., Bradley, J., Min, C., McCarty, N.A., Quick, M., Riordan, J.R., Chavkin, C., Zinn, K., Lester, H.A., and Davidson, N. (1993) Receptors that couple to 2 classes of G proteins increase cAMP and activate CFTR expressed in Xenopus oocytes. *Receptors Channels* **1**, 233–241.
14. Landau, E.M. and Blitzer, R.D. (1994) Chloride current assay for phospholipase C in Xenopus oocytes. *Methods Enzymol.* **238**, 140–154.
15. Guo, Z., Liliom, K., Fischer, D.J., Bathurst, I.C., Tomei, L.D., Kiefer, M.C., and Tigyi, G. (1996) Molecular cloning of a high-affinity receptor for the growth factor-like lipid mediator lysophosphatidic acid from Xenopus oocytes. *Proc. Natl. Acad. Sci. USA* **93**, 14367–14372.
16. Ferrell, J.E., Jr. (1999) Xenopus oocyte maturation: new lessons from a good egg. *Bioessays* **21**, 833–842.
17. Wagner, C.A., Ott, M., Klingel, K., Beck, S., Melzig, J., Friedrich, B., Wild, K. N., Broer, S., Moschen, I., Albers, A., Waldegger, S., Tummler, B., Egan, M.E., Geibel, J.P., Kandolf, R., and Lang, F. (2001) Effects of the serine/threonine kinase SGK1 on the epithelial Na(+) channel (ENaC) and CFTR: implications for cystic fibrosis. *Cell Physiol. Biochem.* **11**, 209–218.
18. Trotti, D., Peng, J.B., Dunlop, J., and Hediger, M.A. (2001) Inhibition of the glutamate transporter EAAC1 expressed in Xenopus oocytes by phorbol esters. *Brain Res.* **914**, 196–203.
19. Levis, R.A. and Rae, J.L. (1992) Constructing a patch clamp setup. *Methods Enzymol.* **207**, 14–66.
20. Corey, J.L., Davidson, N., Lester, H.A., Brecha, N., and Quick, M.W. (1994) Protein kinase C modulates the activity of a cloned gamma-aminobutyric acid transporter expressed in Xenopus oocytes via regulated subcellular redistribution of the transporter. *J. Biol. Chem.* **269**, 14759–14767.
21. Chubb, S., Kingsland, A.L., Broer, A., and Broer, S. (2006) Mutation of the 4F2 heavy-chain carboxy terminus causes y+ LAT2 light-chain dysfunction. *Mol. Membr. Biol.* **23**, 255–267.

Chapter 17

Measurement of Intracellular pH

Frederick B. Loiselle and Joseph R. Casey

Abstract

The activity of most cellular processes is sensitive to pH. Cells therefore tightly control cytosol pH within narrow bounds. Measurement of cytosolic pH is of interest in studying many processes, including pH regulatory transport proteins. Key approaches that have been used to determine intracellular pH include pH-sensitive microelectrodes, nuclear magnetic resonance, and pH-sensitive fluorescent proteins. Here we review these approaches while providing details on the use of pH-sensitive fluorescent dyes to measure cytosolic pH.

Key words: Intracellular pH, fluorescent dye, fluorescent protein.

1. Introduction

Intracellular pH regulation is critical for most cellular processes including cell volume regulation, vesicle trafficking, cellular metabolism, cell membrane polarity, muscular contraction, and cytoskeletal interactions (1–6). Changes of intracellular pH (pH_i) affect the concentration of intracellular messengers like Ca^{2+} and cAMP and thus influence cellular signaling (7, 8). In addition, some growth-stimulating signals including epidermal growth factor, platelet-derived growth factor, insulin, vasopressin, and serum activate sodium–proton exchange activity to induce cellular alkalinization, which is important in cell activation, growth, and proliferation (9–13).

Techniques to measure intracellular pH include H^+ permeable microelectrodes, nuclear magnetic resonance (NMR) analysis of metabolites whose resonance frequency is influenced by pH,

Q. Yan (ed.), *Membrane Transporters in Drug Discovery and Development*, Methods in Molecular Biology 637,
DOI 10.1007/978-1-60761-700-6_17,

and emission/excitation of weak acid fluorescent dyes (14, 15). pH microelectrodes are essentially microscopic versions of the pH electrodes in routine laboratory use to measure solution pH. Microelectrode-based measurement of intracellular pH involves inserting an electrode small enough so as not to damage the cell, usually 1 μm or less in diameter, through the plasma membrane and into the cytoplasm (15). The electrode is filled with a solution of low pH, producing a H^+gradient across the electrode wall that generates a potential proportional to the concentration of H^+ outside and the permeability to H^+. The most popular types of electrodes are the recessed tip microelectrode and the double-barrel microelectrode (16, 17). The recessed tip electrode allows for small bore tips, while the double-barreled electrode allows for the independent measurement of membrane potential and pH. This is a major advantage as single-barreled electrodes measure the sum of the pH response and the membrane potential.

NMR spectroscopy is potentially the most powerful method for intracellular pH measurement and provides the added advantage of being amenable to tomography, which allows for two- or three-dimensional imaging. In NMR, the behavior of nuclei with non-zero spin in a strong magnetic field is measured. pH-dependent protonation/deprotonation alters the electronic environment of nuclei, which shifts the position of the NMR peak. For measurement of intracellular pH by NMR, spectroscopy of phosphate groups is most useful, because of the high abundance of phosphate-containing compounds in cells and the pK_a of phosphate groups is in the physiological range (15). In practice, ^{31}P-NMR is most effective to study pH_i in whole tissues, for example, to measure changes of pH_i in isolated hearts during ischemia (18). In these experiments the relative concentrations of protonated and deprotonated forms of a phosphate group can be determined. Since the concentration ratio of the protonated and deprotonated forms is a function of pH, this allows easy calculation of pH_i. However, using NMR to measure intracellular pH requires complex equipment, methodology, and data analysis. The technique also suffers from low sensitivity, which requires high concentrations of cells or long data acquisition times (15).

Fluorescence spectroscopy is a highly sensitive method of pH_i measurement. The basis for this approach is that cells can be loaded with a fluorescent dye (fluorophore) whose fluorescence varies with pH. Fluorescent molecules will absorb photons of light of the appropriate wavelength (excitation wavelength), causing the molecule to rise to an excited state. Some time later the energy will be emitted as heat combined with a photon released at a longer wavelength (emission wavelength) than the photon it absorbed.Fluorometers are spectroscopic devices that allow one to illuminate a sample with light at a particular excitation wavelength and monitor and quantify light at the emission wavelength.

For measurement of pH_i, fluorescence spectroscopy has several technical and practical advantages over other methods. NMR may be the least invasive technique. However, its low sensitivity and technical complexity make it a less desirable choice. Microelectrode techniques offer the advantage of potentially recording both pH changes and total ion flux at the same time, which can be useful when looking at electrogenic H^+ or H^+-equivalent transporters. However, the difficulties encountered in the preparation of microelectrodes, experimental cell size limits, and the invasiveness of the technique make fluorescence spectroscopy the best option for most applications (14). Fluorescent pH-responsive indicators are also more ion selective, thereby reducing background signal observed with microelectrodes due to ions other than H^+. Finally, fluorescence allows either a whole population of cells or several individual cells to be studied at the same time by combining cell imaging techniques with fluorescence spectroscopy.

Several indicator fluorophores are available that span the physiological organellar and cytosolic pH range (**Table 17.1**). In general, fluorescent indicator dyes are loaded into the cell as an acetoxymethyl ester (AM), which is readily permeable to cell membranes (19). Esterification with AM groups converts negatively charged, membrane-impermeant acids into neutral, membrane-permeant ester analogues. Once inside the cell the AM groups are cleaved by ubiquitous intracellular esterases, which release a charged species that cannot exit the cell (**Fig. 17.1**) (19). pH indicators can be delivered to specific cellular com-

Table 17.1
Intracellular pH indicators and their characteristics

Parent fluorophore	Useful pH range	Typical measurement
SNAFL indicators	7.2–8.2	Excitation ratio 490/540 nm or emission ratio 540/640 nm
SNARF indicators	7.0–8.0	Emission ratio 580/640 nm
HPTS (pyranine)	7.0–8.0	Excitation ratio 450/405 nm
BCECF	6.5–7.5	Excitation ratio 490/440 nm
Oregon green dyes	4.2–5.7	Excitation ratio 510/450 nm or excitation ratio 490/440 nm
Rhodols (including NERF dyes)	4.0–6.0	Excitation ratio 514/488 nm or excitation ratio 500/450 nm
LysoSensor probes	3.5–8.0*	Excitation ratio 340/380 nm
pH-sensitive GFP mutants	5.1–8.1	Single measurement: excitation 440 nm or 480 nm; emission 480 nm or 535 nm

*Several probes of differing pK_a are available.

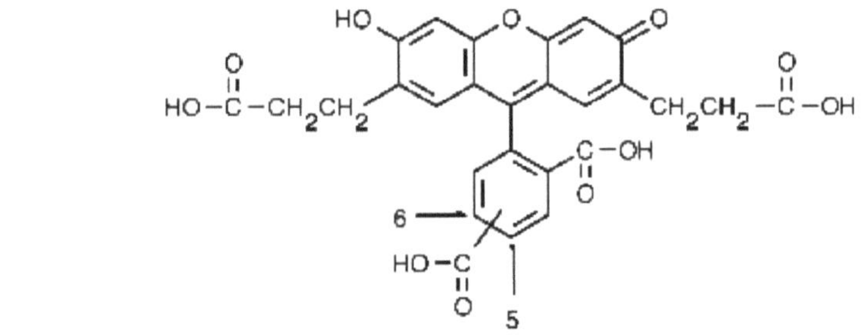

Fig. 17.1. Molecular structure of the pH-sensitive fluorescent dye, BCECF acid. Figure courtesy of Molecular Probes Inc.

partments either by conjugation to targeting molecules or by partitioning into more acidic compartments by protonation, for example, lysosensor probes (20, 21). Recently pH-sensitive green fluorescent proteins (GFPs) have been developed that allow for precise compartmental targeting via fusion proteins (20, 22, 23).

A gradual loss of fluorescence signal may occur during the course of an experiment to measure pH_i. The pH indicator dye may slowly leak out of the cell, be released from lysed cells, or may be bleached by exposure to light during the experiment. Ratiometric indicator dyes allow for correction of the fluorescence signal for these losses. Some pH indicator dyes display a peculiar spectral property when one examines emitted fluorescent light over a range of pH values. The amount of emitted light is the same at all pH values at one particular excitation wavelength, called the isosbestic point. The significance of this pH-independent wavelength is that it can be used to normalize fluorescence data. That is, the dye can be excited at both the pH-sensitive and the pH-insensitive wavelengths (**Fig. 17.2**) (24, 25). Dividing the fluorescence signal at the pH-sensitive wavelength by the fluorescence signal at the insensitive wavelength gives a fluorescence ratio that can reduce or eliminate variability due to dye loss. BCECF, SNAFL, and SNARF pH indicator dyes all exhibit a spectral isosbestic point that is insensitive to pH change.

The pH indicator, BCECF, is ideally suited to study pH_i because its pK_a of 6.98 allows measurements in the physiological pH range 6.0–8.0 (26). With four to five negative charges at pH 7–8, BCECF is well retained and not very susceptible to leakage. When excited at 505 nm the wavelength of peak emission intensity does not change as a function of pH for BCECF. Fluorescence excitation scans at a range of pH values show an isosbestic point around 440 nm, which allows this wavelength to be used for ratiometric normalization, as discussed above. The maximum pH-sensitive excitation wavelength of 505 nm is well resolved from the pH-insensitive excitation wavelength (440 nm).

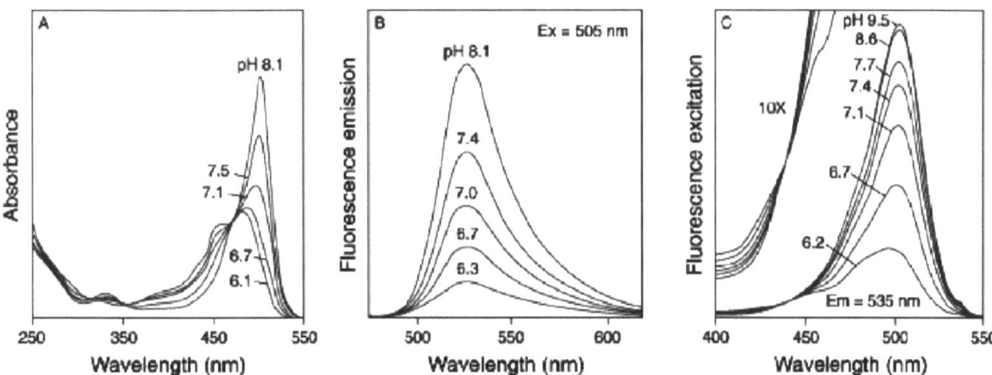

Fig.17.2. pH-dependent spectra of BCECF. (**a**) Absorption spectra at a range of pH values. (**b**) Emission spectra at a range of pH values. (**c**) Excitation spectra at a range of pH values. The excitation spectrum around the isosbestic point has been enlarged 10-fold (**c**). Ex is excitation wavelength. Em is emission wavelength. Figure courtesy of Molecular Probes Inc.

Typically, intracellular pH measurements are made by determining the ratio of emission intensity, at 535 nm, when BCECF is excited at 440 (pH-insensitive) and 505 (pH-sensitive) nm.

Changes of pH_i following an acid or alkaline load are a function of both the amount of acid or base added to the cell and the cellular buffering capacity. Buffer capacity is a measure of the ability of a cell to withstand addition of removal of H^+, without change of pH, in much the same way that buffered solutions resist pH changes. Buffer capacity is the sum of all the proton buffers within a cell and is usually divided into CO_2 and non-CO_2-dependent buffering such that

$$\beta_{\text{total}} = \beta_i + \beta_{CO_2}$$

Intrinsic buffer capacity, β_i, is provided by metabolites in acid base equilibria such as PO_4^{3-} containing molecules and ionizable amino acid side chains (27). β_{CO_2} depends on P_{CO_2} and HCO_3^- concentration. Under most in vivo and in vitro experimental conditions, P_{CO_2} is constant. Thus, β_{CO_2} depends only on HCO_3^- concentration. For example, in mammalian muscle (pH_i 7.1 and $P_{CO_2} = 37$ mm Hg) $\beta_i = 40$ mM and $\beta_{CO_2} = 29$ mM or about 40% of β_{total} (Roos and Boron, (27), *#1037*). Most cells have buffer capacities that are highest below pH 7, due to the pK_as of carbonic acid, amino acid side chains, and phosphate groups. Above pH 7, cellular buffer capacities drop off dramatically. Thus, the buffer capacity varies not only with cell type but also with pH. In order to relate changes of pH_i to the amount of acid or base added to a cell, β_{total} must be experimentally determined over the experimental pH range, as described in **Section 3.4**.

Measurement of changes of pH_i can report on the transmembrane flux of H^+ equivalents (H^+, HCO_3^-, etc.) due to facilitated transport. Ionic flux is related to the measured pH_i change and the buffer capacity of the cell. The H^+-equivalent flux (J_{H^+}) is related to the buffer capacity by

$$J_{H+} = (d_p H/dt) \bullet \beta_{total}$$

$J_H{}^+$ is the flux of protons in millimolar H^+ per unit time, dpH/dt is the rate of pH change, and β_{total} is the total buffer capacity.

The following is an overview of the methodology for measurement of pH_i employed by our laboratory (*see* **Section 3**.3). Our system is appropriate for the measurement of pH_i in primary and immortalized cell lines that adhere to glass coverslips. Cells are trypsinized, then plated onto polylysine-coated glass coverslips, and allowed to adhere. In the case of transiently transfected cells, transfection is performed after cells are plated. Cells are allowed to grow to a cell density of between 25 and 90% confluency and are then loaded with BCECF-AM dye to facilitate ratiometric pH_i measurement. Cells are transferred to a modified fluorescence cuvette in a fluorometer and perfused with physiological buffer. The perfusion solution can be changed to achieve acid or alkaline loading or to add pharmacological agents (*see* **Sections 3.3.6** and **3.3.7**). Some H^+-equivalent transporters can be studied by alternatively adding and removing substrate from the perfusion solution. For example, plasma membrane Cl^-/HCO_3^- anion exchange is studied by switching from Cl^--containing perfusion buffer to Cl^--free buffer, thereby driving HCO_3^- movement coupled to Cl^- movement (*see* **Section 3.3.5**). BCECF in the cells is excited at 440 and 505 nm and the resulting fluorescence emission at 535 nm recorded by a photomultiplier.

At the end of each experiment the fluorescence ratio data is calibrated to reflect pH_i, using the nigericin/high-potassium method (*see* **Section 3.3.8**) [Thomas, 1979 #768]. Together the H^+/K^+ ionophore, nigericin, and strongly pH-buffered high-concentration K^+ solutions of known pH values break down the pH gradient across the plasma membrane and equilibrate pH_i with pH_o (extracellular pH). Measurement of the fluorescence ratio data (i.e., excitation at 505 nm/excitation at 440 nm) at three different pH values for the nigericin solutions allows a standard curve to be plotted, with a linear relationship between fluorescence ratio (x-axis) and pH (y-axis). The standard curve, with the form pH = slope × (fluorescence ratio) + y-intercept, can be applied to fluorescence data to determine pH_i values that were recorded. In membrane transport experiments, the rate of H^+-equivalent transport is assessed from the initial rate of pH_i change. To determine substrate flux, the cellular buffer capacity (**Section 3.4**) under the experimental conditions must be known.

2. Materials

2.1. Tissue Culture

2.1.1. Coverslip Preparation

1. #1.5–2 glass coverslips 22 × 22 mm (Fisher Scientific).
2. Tungsten carbide glass scoring rod (or glass cutting stylus).
3. 5 M NaOH.
4. 100% ethanol.
5. PBS pH 7.4: 140 mM NaCl, 3 mM KCl, 6.5 mM Na_2HPO_4, 1.5 mM KH_2PO_4, pH 7.5.
6. 1 × Polylysine in PBS, pH 7.4 (store aliquots of polylysine as 100 × stock (10 mg/ml in water) at –20°C).
7. Sterile PBS, pH 7.4.

2.1.2. Tissue Culture

1. Cell culture medium (e.g., DMEM, store in the dark at 4°C).
2. Cell culture grade trypsin solution (store as 10 × stock at 4°C).
3. Serum-free medium (store in the dark at 4°C).
4. Sterile PBS, pH 7.4.
5. Light microscope with ×10 and ×20 objectives.

2.2. BCECF Loading

1. BCECF-AM stock solution: 1 mM BCECF-AM in DMSO (aliquots stored in darkness at –20°C are stable for months).

2.3. Fluorometer Apparatus

2.3.1. Cuvette Design

See **Fig. 17.3** for construction details and component dimensions.

1. Disposable, plastic fluorometer cuvette 10 × 10 mm internal dimensions. The height of the cuvette can be adjusted by cutting the cuvette so that the coverslip, attached to the cuvette lid, will be in the middle of the excitation beam path when placed in the fluorometer sample chamber.
2. Tygon microtubing 0.04–0.06 in. internal diameter.
3. Plexiglas cuvette lid.
4. Petroleum jelly.

2.3.2. Perfusion Pump

1. Peristaltic perfusion pump with at least two lines (inflow and outflow) capable of 3–4 ml/min flow.
2. Waste container to collect perfusion solutions.

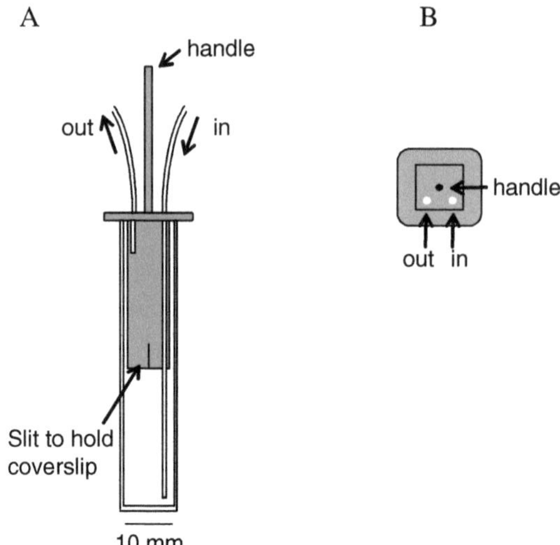

Fig.17.3. Schematic diagrams of perfusion cuvette. (**a**) side view of perfusion cuvette. The inflow line empties into the cuvette near the bottom and is arranged to be distant from the coverslip to reduce turbulence in the light paths. The outflow line draws near the top of the cuvette such that the coverslip is always submerged in perfusion buffer. The coverslip lid slit should be sized to hold the coverslip snugly. Petroleum jelly may be applied to the slit to hold the coverslip in place. (**b**) Top view of cuvette lid.

2.3.3. Gas System

1. 5% CO_2/95% air gas cylinder.
2. 100% O_2 gas cylinder (for Hepes buffer).
3. 2-stage gas cylinder regulators (appropriate for gas type).
4. 1–5 l/min gas flow monitor (optional).
5. Tubing (3/8 in. internal diameter) and "T" valves (to divide gas flow among several perfusion solution bottles).
6. Bubble stones (small aquarium type).

2.3.4. Perfusion Solutions

1. Ringer's buffer: 5 mM glucose, 5 mM K gluconate, 1 mM Ca gluconate, 1 mM $MgSO_4$, 10 mM Hepes (light sensitive), 140 mM transport anions (e.g., NaCl or Na gluconate), 2.5 mM NaH_2PO_4 (add after dilution to prevent $Ca_3(PO_4)_2$ precipitation), 25 mM $NaHCO_3$ (add just before adjusting pH), pH 7.4. Bubble with 95% air/5% CO_2. Solution can be prepared and stored as a 10 × stock without transport ions and $NaHCO_3$. pH should be adjusted with HCl or NaOH (or KOH for Na^+-free solutions) just before use.
2. Acid loading solutions: (a) NH_4Cl prepulse technique: Ringer's buffer containing 40 mM NH_4Cl. Bubble with 95% air/5% CO_2. (b) Hepes/HCO_3^- prepulse technique. (c) Hepes buffer is Ringer's buffer without $NaHCO_3$. Bubble

with 100% O_2 to remove any CO_2. (d) Bicarbonate buffer is Ringer's buffer without Hepes. Bubble with 95% air/5% CO_2.

2.3.5. Fluorometer Hardware and Software

1. Photon Technologies International RCR DeltaScan fluorometer, or equivalent. Fluorometer needs to be able to make dual excitation wavelength measurements either by slewing between wavelengths or with two excitation monochromators and a chopper.
2. FeliX version 1.21 Photon Technology International, or equivalent.

2.4. pH Calibration

1. 10,000 × nigericin stock: 10 mM nigericin in ethanol. (Store separate aliquots at –20°C. During experiment store on ice.)
2. pH calibration solutions: Ringer's buffer with 140 mM KCl substituted for transport ions and made to 30 mM Hepes (store in the dark at room temperature).

3. Methods

3.1. Tissue Culture

3.1.1. Coverslip Preparation

1. Either by hand or with a graphics program draw a 7 × 11 mm grid to use as a template for coverslip cutting. Tape the template to a table. Place the coverslip on the template and score the surface with tungsten carbide rod into the grid and gently break apart into rectangles. Minor imperfections at the edges are acceptable.
2. Place 10–15 coverslips in a 100 mm petri dish and prepare the coverslips for polylysine coating by sequential washing with 5 M NaOH, water, 100% ethanol, and twice with PBS. Perform washes in a petri dish for 10 min each step.
3. Coat each coverslip with several drops of 1 × polylysine solution, leaving a domed droplet on each coverslip.
4. Sterilize overnight in a tissue culture hood, using the overhead UV lamps. Place petri dish lids inside up beside the dishes. In the morning coverslips should look salt encrusted.
5. In a tissue culture hood, under sterile conditions, wash the coverslips with sterile PBS for 10 min or until salt grains cannot be seen under a light microscope (cells will not be able to grow on salt grains). Rinse with PBS twice before using the coverslips (*see* **Note 1**).

3.1.2. Preparation of Cells

The assay will work for most cell types that can be grown on the polylysine-coated glass surface, including primary and immortalized cell lines, whether transfected or not.

1. Deplate and split apart the cells by trypsinolysis (or the splitting method appropriate for that cell type).
2. Place coverslips in dishes (60 mm dish holds up to 10 coverslips) and plate cells on top. Confluency at the time of assay should be between 25 and 90%. At this density, approximately 10^4 cells will be in beam path of the fluorometer. Stable cell lines can be plated down the night before and transiently transfected cells should be plated and transfected so as to give maximal expression on the day of experimentation (*see* **Note 2**.)

3.2. BCECF Loading

This procedure is performed immediately prior to the experiment to measure pH_i. Serum contains esterases that may cleave the acetoxymethyl ester (AM) groups on BCECF-AM. Therefore, all steps of the loading procedure should be carried out in serum-free medium.

1. Add 2 ml serum-free culture medium to a 35 mm petri dish. Add 4 μl of 1 mM BCECF-AM and swirl to mix.
2. Remove a cell-covered coverslip from the petri dish by gently holding the coverslip by one corner with forceps. Pass the coverslip through three sequential 60 mm petri dishes, containing 8 ml of serum-free media. Then place the coverslip in the dye-containing dish.
3. Incubate 10–15 min at 37°C and use in pH_i measurement experiment immediately. (Also *see* **Notes 3–6** on BCECF loading.)

3.3. Fluorometer Assay

3.3.1. Perfusion Preparation

1. Prepare perfusion buffers. Buffers should be stored in foil-covered or opaque flasks, since light-generated breakdown products of Hepes can be cytotoxic (*see* **Note 7**).
2. Perfusion buffers should be bubbled with 5% CO_2 (or 95% O_2 for Hepes buffer acid load method) for at least 15 min prior to experiments to allow an equilibrium to be established (*see* **Note 8**).
3. Monitor perfusion buffer pH constantly during assays and adjust as needed with 1 M NaOH (or KOH).
4. Apply a thin layer of petroleum jelly to the underside of the cuvette lid. This will help to hold the coverslip in place.
5. Perfuse the cuvette for at least 5 min with Ringer's buffer before starting a new experiment to ensure all traces of

nigericin have been removed. The perfusion flow rate is usually 3.5 ml/min and should be checked each day. (*See* **Note 9** on leaks and balancing inflow versus outflow rates.)

3.3.2. Fluorometer Apparatus Startup

1. Ensure all system components are turned off. Then turn on the lamp power supply and ignite the lamp. Power surge during lamp ignition may damage computer and other electrical components if they are on prior to lamp ignition. The remaining system components may now be turned on. (*See* **Note 10** on lamp life.)
2. Set up the fluorometer system software so that fluorescence from the pH-sensitive excitation wavelength is divided by fluorescence from the pH-insensitive wavelength. Start fluorometer system software and under acquire select excitation ratio. Record both the excitation and the emission monochromator position settings. Load the appropriate excitation ratio file and acquire new preparation.
3. If your excitation monochromators have slits, ensure that they are set to the same slit width (about 1.5 nm to start).
4. Recalibrate pH meter at the start of each day.

3.3.3. Fluorometer Assay Data Collection

1. Remove coverslip from BCECF-AM solution and insert into cuvette lid holder (**Fig. 17.3**). Replace the cuvette lid, ensuring that the cells are facing the light source and rotated 45° toward the photomultiplier (**Fig. 17.4**). Start perfusion with the appropriate buffer.

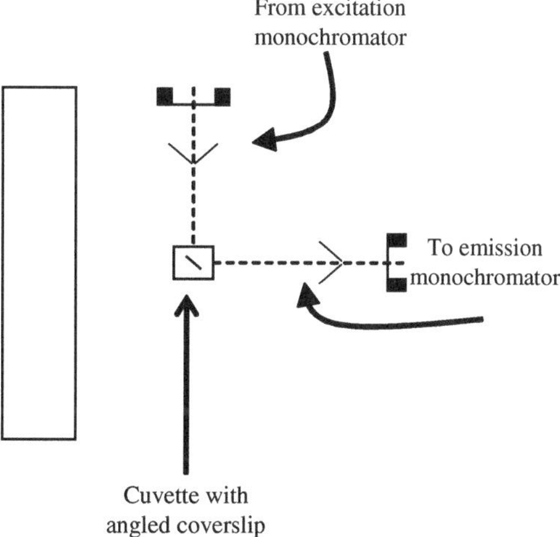

Fig. 17.4. Schematic diagram of fluorometer sample chamber. The excitation and emission monochromator beam paths are set at right angles to reduce fluorescent noise. The coverslip is rotated 45° to both paths and the angle may be adjusted to increase or decrease the intensity of light received by the photomultiplier.

2. Start fluorescence data collection using fluorometer software. Check absolute fluorescence for pH-sensitive and pH-insensitive wavelengths. Values should be between 4.0 to 10.0×10^5 counts/s and 0.5 to 2.5×10^5 counts/s, respectively. Adjust monochromator slits as necessary. (*See* **Note 11** on maximum slit widths.)
3. To change perfusion solutions, stop the peristaltic pump and move the intake line to the new flask. Take care not to bump the line so as to avoid introducing bubbles into the line. (*See* **Note 12** on bubbles.) (Also *see* **Notes 13–16** on other assay considerations.)

3.3.4. Types of Transport Assay

Cells regulate their pH_i using regulatory transport proteins including the monocarboxylate/H^+ co-transporters (lactate/H^+ co-transporter), Na^+/H^+ exchangers (NHE), Na^+-dependent and Na^+-independent Cl^-/HCO_3^- exchangers, and Na^+/HCO_3^- co-transporters. Depending on the direction and mechanism of transport being studied, different methods of altering intracellular pH must be employed. For example, Cl^-/HCO_3^- exchange is normally studied by alternately removing or adding Cl^- to establish a Cl^- gradient across the plasma membrane (**Fig. 17.5**a). The other families of transporters can be studied by loading cells with acid (**Sections 3.3.6** and **3.3.7**) and monitoring recovery of cell pH. Several techniques are available to acidify the cytosol, including direct microinjection of acid, or passing current through microelectrodes. The most common methods are to incubate cells with weak acids or bases. Two acids used in our laboratory are NH_4Cl and H_2CO_3 (**Fig. 17.5 b,c**). The following methodologies are all employed in our laboratory during studies of pH_i regulatory transporters.

3.3.5. Cl^-/HCO_3^- Exchange Assay

The identification of a Cl^-/HCO_3^- exchanger is based on coupled Cl^-/HCO_3^- exchange. Transport of Cl^- in one direction is coupled to movement of HCO_3^- in the opposite direction across the membrane and concomitant change of pH_i(28).

1. Perfusion is initiated with Cl^- containing Ringer's buffer until pH_i is stable. Switching to Cl^--free (Na gluconate replaces NaCl) Ringer's buffer stimulates HCO_3^- uptake and alkalosis.
2. After a plateau of fluorescence is reached, the perfusion solution is switched back to Cl^--containing Ringer's, which stimulates HCO_3^- efflux and acidosis.
3. Transport rates are measured by linear regression of the initial rates of change of pH_i.

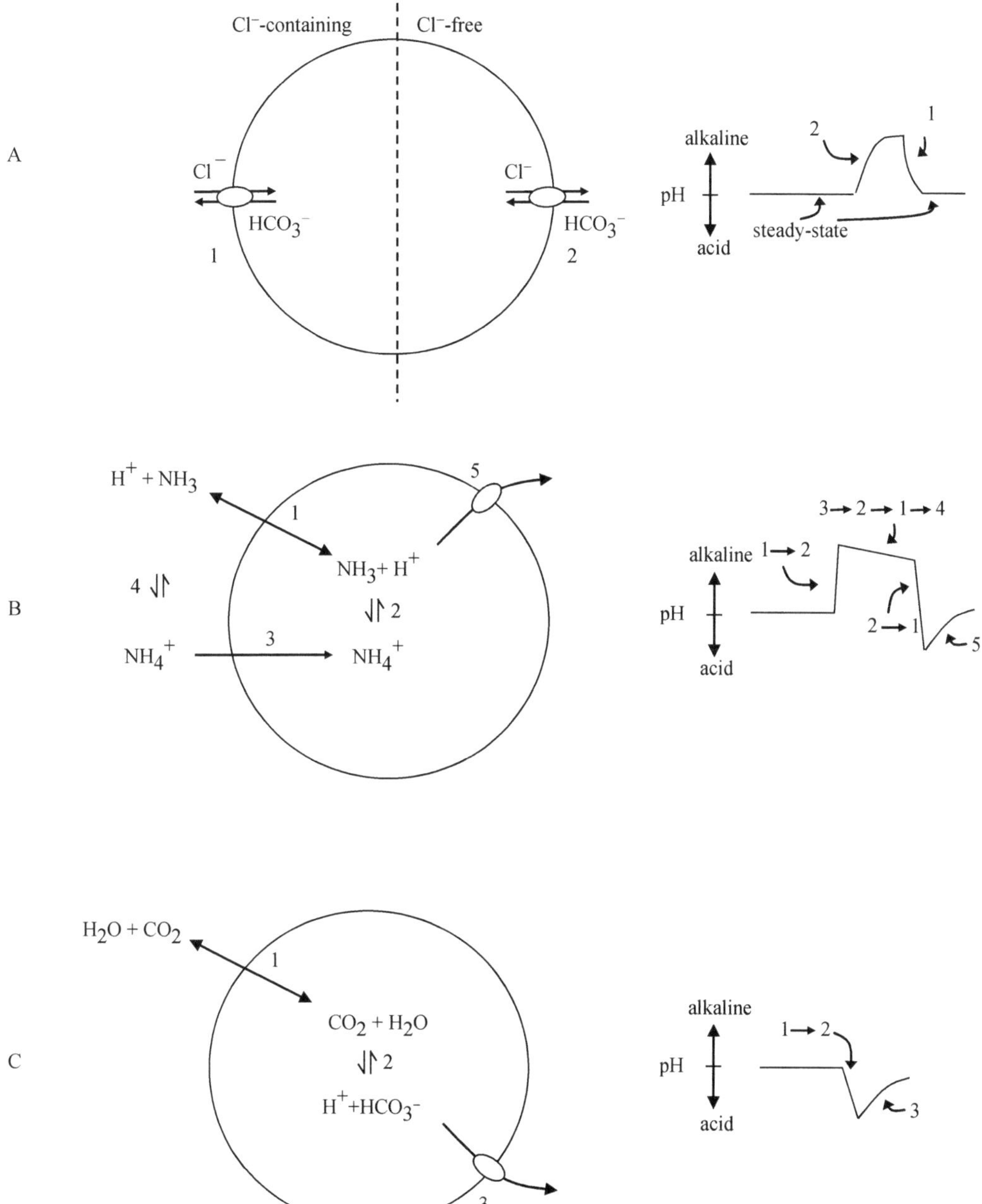

Fig.17.5. Manipulation of cell pH. (**a**) Cl^-/HCO_3^- exchange assay. Cells are equilibrated with Ringer's buffer containing Cl^- (*1*), then switched to Ringer's buffer without Cl^- (*2*). This creates an outwardly directed Cl^- gradient. If a Cl^-/HCO_3^- exchanger is expressed in the plasma membrane, then the cell will undergoes alkalinization due to the uptake of HCO_3^-. Similarly, when the cell is switched back to Ringer's buffer containing Cl^- the process is reversed and an acid load results. (**b**) Manipulation of cell pH with NH_4Cl. Cells are placed in an NH_4Cl solution. Extracellular NH_3 diffuses passively into the cell (*1*) and equilibrates with NH_4^+ (*2*). Extracellular NH_4^+, which is much less permeable, then begins to diffuse into the cell driving NH_3 plus H^+ out of the cell, thereby effectively setting up a H^+ shuttle (*3*). When the perfusion solution is switched to Ringer's buffer without NH_4Cl an acid load results and transport can be assessed as pH recovery (*5*). (**c**) Cell acidification with Hepes/bicarbonate technique. Initially the cells are made nominally free of CO_2 by perfusion with Hepes buffer bubbled with 95% 0_2 to remove dissolved CO_2. The perfusion solution is then switched to CO_2-containing buffer resulting in an acid load (*1* and *2*). Transport is then assessed as pH recovery (*3*).

3.3.6. NH_4Cl Prepulse Technique of Acid Loading

Cytosolic acidification is useful so that the cell's pH recovery mechanisms may be studied (29). Acidification may be performed by the NH_4Cl prepulse method outlined here or the Hepes/HCO_3^- technique (*see* **Section 3.3.7**)

1. Perfusion is initiated with physiological Ringer's buffer and switched to Ringer's buffer containing 40 mM NH_4Cl. Extracellular NH_4^+ is in equilibrium with NH_3, which enters the cell and combines with a H^+ to form NH_4^+, causing the cell to alkalinize (**Fig. 17.5 b**). Eventually the poorly permeable NH_4^+ will begin to diffuse into the cell, likely driven by the large electrical gradient across the plasma membrane. For each NH_4^+ that enters the cell an NH_3then leaves the cell leaving a H^+ behind causing a slow acidification. In the presence of external NH_4Cl, any NH_4^+ that enters the cell causes a proton to disproportionately accumulate in the cell so that when the extracellular NH_4Cl is removed the cell will acidify below its starting pH, before NH_4Cl exposure.
2. After 5 min of treatment with NH_4Cl-containing Ringer's buffer, the perfusion solution is switched back to Ringer's buffer. At this point the intracellular NH_4^+ dissociates into NH_3 plus H^+ and the NH_3 diffuses out of the cell, resulting in an acid load in proportion to the extent of additional NH_4^+ loaded during the incubation. The amount of acid loaded will depend on the permeability of the plasma membrane to NH_4^+, the initial concentration of NH_4^+, and the duration of the prepulse.
3. Following recovery drugs may be added and a second acid load performed to assay the effect of the drug on the ion transport rate.

3.3.7. Hepes/HCO_3^- Prepulse Technique of Acid Loading

1. Perfusion is initiated with Hepes buffer bubbled with 95% O_2, which drives any $CO-2/HCO_3^-$ from the cells (**Fig. 17.5 c**) (30).
2. After collecting at least 100 s of fluorescence equilibrium plateau data, switch to bicarbonate buffer. CO_2 will diffuse into the cell, combine with H_2O and establish an equilibrium with HCO_3^- plus H^+, causing the cell to acidify. Proton extrusion by a transporter, for example NHE, can then be measured as the rate of pH recovery.
3. Following recovery, drugs may be added and a second acid load performed to assay their effect on transport rate.

3.3.8. pH Calibration

1. Prepare, 10 ml of each pH calibration solution per experiment at pH values approximately 6.5, 7.0, and 7.5. The exact pH of each solution should be recorded. Immediately

before pH calibration, add nigericin to 10 ml of each pH calibration solution to a final concentration of 1 μM.

2. Turn off peristaltic pump. Move the inflow line to the calibration solution and turn on the pump. When the calibration solution has all been pumped stop the pump again and wait for the new fluorescence level to stabilize. Note some cell lines exhibit a transient alkalosis above the plateau during calibration. Record at least 100 s of stable fluorescence data before switching to the next calibration solution. The general order of calibration is pH 7.5, 6.5, and finally 7.0.
3. To calibrate the experiments, plot calibration solution pH versus fluorescence ratio value for all three points and perform linear regression on the line. The slope (multiplier) and *y*-intercept (offset) are then used to perform a linear scaling of the fluorescence ratio.

3.4. Determination of Buffer Capacity

Intracellular buffer capacity (β_{total}) is not constant. β_{total} is influenced by pH_i such that a plot of β_{total} versus pH_i appears exponential with little buffer capacity at pH_i greater than 7.2 and progressively higher buffer capacity at pH less than 7.0 (31). In part, this is seen because the pK_a of most ionizable groups is well below physiological pH values. Since each cell type has a different concentration of buffering components the buffer capacity of each must be calculated at pH_i values encountered within a given experimental protocol. For example, transiently transfected cells differ in their buffer capacity from untransfected cells. Buffer capacity is determined by adding a known amount of weak acid or base to the experimental system and measuring the change of pH_i observed. Below is a method to determine buffer capacity over the pH range 6.5–8.0. Note at acidic pH_i values it is difficult to measure β_{total} accurately. Thus, during experiments performed at acidic pH_i care should be taken to ensure that the extent of acid loading in the control and experimental samples is the same. In this way, the β_{total} will be the same for both conditions and H^+-equivalent fluxes can be compared directly.

1. Data collection: Prepare Ringer's buffers containing 0, 1, 5, 10, and 20 mM NH_4Cl. Mount a cell-covered coverslip in the fluorescence cuvette. Perfused cuvette with Ringer's buffer without NH_4Cl until fluorescence is stable. Sequentially perfuse with Ringer's buffer containing 20, 10, 5, 1, and 0 mM NH_4Cl. Perfuse with each solution for enough time to give a fluorescence reading that is stable for 200 s. The perfusion interval should be consistent for each solution. At the end of the experiment, pH_i should be calibrated by the nigericin high-K^+ method (*see* **Section 3.3.4**). Data should be collected for 4–8 cell-covered coverslips.

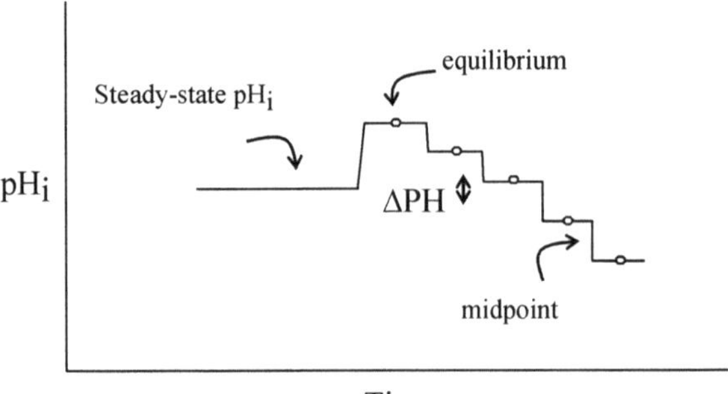

Fig.17.6. Determination of cellular buffer capacity. Example data for buffer capacity determination. Cells are perfused with Ringer's buffer until a steady-state pH_i is achieved. Buffer is then changed to Ringer's buffer plus 20 mM NH_4Cl and 200 s of steady fluorescence recorded. The equilibrium pH_i is calculated as the average pH_i over the steady interval. The buffer is then changed to the next concentration of NH_4Cl, a new equilibrium pH_i determined and the midpoint pH_i is identified as the pH_i between two consecutive equilibrium pH_i values. The process is repeated for the remaining NH_4Cl concentration steps and the data analyzed, as in **Section 3.4**.2. *Circles* represent example positions of equilibrium pH_i. × represents the position of midpoint pH_i values.

2. Data analysis: Prepare the table shown below. For each $[NH_4Cl]_o$ determine the equilibrium pH_i value from the average pH_i over the stable fluorescence range (**Fig. 17.6**). For each equilibrium pH_i value, calculate $[NH_4Cl]_i$ using the equation $[NH_4Cl]_i = ([NH_4Cl]_o \times 10^{(9.02 - pHi)}) / (1 + 10^{(9.02 - pHo)})$. To determine $\Delta[NH_4Cl]$ calculate the difference between sequential NH_4 Cl concentration steps (e.g., 20 – 10 mM = 10 mM, for the first step). Midpoint pH_i corresponds to the pH_i value halfway between equilibrium pH_i values (**Fig. 17.6**). Intrinsic buffer capacity (β_i) is calculated from the equation $\beta_i = \Delta[NH_4Cl]_i / \Delta pH_i$. Once the data from the 4–8 cell-covered coverslips has been analyzed calculate the mean and standard error of midpoint pH_i and buffer capacity for each $\Delta[NH_4Cl]_o$ step. Plot β_i (*y*-axis) against pH_i (*x*-axis). $\beta_{total} = \beta_i + \beta_{CO_2}$ where $\beta_{CO_2} = 2.3[HCO_3^-]_i$. $[HCO_3^-]_i$ is a function of, pH_i, temperature, and the pK_a for CO_2 (6.37). For example, at room temperature, physiological P_{CO_2} (5%), pH_i 7.4, $[HCO_3]_i$ is 25 mM so that β_{CO_2} equals 57.5 mM.

$[NH_4Cl]_o$ (mM)	Equilibrium pH_i	$[NH_4Cl]_i$ (mM)	$\Delta[NH_4Cl]_I$ (mM)	ΔpH_i	Midpoint pH_i	β_i (mM/pH)
20						
10						
5						
1						
0						

4. Notes

1. Occasionally variability in polylysine coating is seen due to uneven drying of the solution on coverslips. Some care should be taken to choose coverslips that have an even coating of salt crystals.
2. Cells more than 90% confluent may be stressed due to the accumulation of metabolic waste products and can exhibit unusual transport characteristics. This effect can be reduced by changing media a few hours before performing transport experiments. Coverslips with too few cells will have a low fluorescence signal, resulting in a decreased signal:noise ratio.
3. Preparations of BCECF contain three different molecular species that are all hydrolyzed to BCECF acid by intracellular esterases. A mixture of forms II and III shows better solubility and cell loading characteristics than pure form I. Therefore, commercial preparations are usually of the mixture. Since the forms differ in molecular weight, the effective molecular weight of each lot is reported on the packaging and should be noted for preparing stock solutions.
4. BCECF-AM loading efficiency varies from batch to batch. Therefore, the time of loading and/or concentration of BCECF-AM can be varied to achieve effective loading.
5. BCECF-AM is light sensitive and should be stored in a lightproof container when thawed.
6. Hydrolyzed BCECF-AM. Rarely BCECF-AM can hydrolyze to BCECF acid during storage. If so, cell loading will be inefficient because BCECF is poorly membrane permeant. Since BCECF-AM is only weakly fluorescent, cleavage can be assessed by fluorescence before and after exposure to serum or NaOH. If BCECF-AM is hydrolyzed to BCECF its fluorescence will not rise upon hydrolysis.
7. Rinsing perfusion lines with water or buffer is critical because nigericin contamination can ruin an experiment by causing H^+ leakage. Between experiments lines should be rinsed with ethanol, water, and then the appropriate experimental buffer. All perfusion tubing should be washed once a week in 10% sodium dodecyl sulfate (SDS). Rinse well with water to remove the SDS.
8. Check gas levels before the start of experiments to ensure sufficient supply. All gas components should be checked for leaks by spraying with dilute detergent solution and looking for bubbles, as even small leaks can rapidly consume gas supply. Gas lines for bubble stones should be labeled,

and not used interchangeably between different buffers. In addition they should be changed at least once a month.

9. Peristaltic pump failure. The pump should be checked for leaks and the flow rate calibrated at least once a month to ensure constant flow. In the event of pump failure, check the fuse within the unit (have spares available). It is essential that the outflow line pumps slightly faster than the inflow line to ensure the cuvette does not overflow and buffer leak into the scanning chamber. This can be accomplished by setting the flow rate of both lines to 3.5 ml/min and then relaxing the resistance tensioner slightly on the outflow line.
10. Fluorescence lamp. Fluorometers use a high-intensity light source, which will decline in light output with use. The decline in lamp function can be measured in two ways: (i) As the lamp ages the voltage across the lamp will rise. If your system provides data on voltage across the lamp, keep a logbook for the fluorometer and record the initial value when the lamp is installed. The lamp manufacturer or fluorometer supplier may be able to tell you the voltage level that represents an aged lamp. (ii) The light intensity will drop with time. One way to assess light intensity of the lamp in a reproducible, objective way is to measure the light output in a water Raman assay. Place pure water in a quartz fluorescence cuvette. Use a single excitation wavelength of 350 nm. Perform an emission wavelength scan from 360 to 450 nm. Set excitation slits to a standard setting (e.g., 5 nm) each time a water Raman spectra are collected. Compare the peak light output with a new lamp (in our system this is about 1×10^5 counts/s) to light output from an aged lamp. If output drops by 30% or more, a new lamp may be needed.
11. Fluorescence signal. The optimal fluorescence level is determined by two factors. First, the amount of fluorescent light measured is determined by a photomultiplier tube (PMT) that counts photons. The PMT will saturate its capacity to count beyond a certain maximum level (check specifications for your system) and can be damaged by a high light level. Second, the amount of light measured must be above a certain level, or the signal to noise ratio will be unacceptable. This is determined empirically as noise level will be very high at low fluorescence levels. Typically, experiments are performed at fluorescence levels that are in the range of 10–100% of maximum counting level of the PMT. Occasionally the fluorescence signals fall outside of the optimal range. This may be due to problems with dye loading, cell number, monochromator slit width, or rotation of the coverslip in relation to the excitation or emission

monochromator beam path. Several remedies are possible and should be performed in the following order: (i) Adjust excitation and emission monochromator slit width. Most monochromators have a pair of slits for each monochromator. These slits determine the range of wavelengths that the monochromator will allow to pass. That is, slits of 2.0 nm will allow wavelengths of light ± 1 nm of the set wavelength to pass through the monochromator. Wider slits allow more light to pass and a higher fluorescence signal. However, this comes at the cost of potentially collecting light from wavelengths whose fluorescence is pH independent; thus a slit width of more than 2.5 nm is not recommended. Slits should be adjusted so that pairs are identical. (ii) Adjust rotation of coverslip (toward emission monochromator if signal is too high and away if too low). (iii) Start a new coverslip with a different cell density or with higher or lower BCECF-AM loading.

12. Bubbles in the fluorescence cuvette. Bubbles can be introduced into the cuvette if the intake line sucks up air, for example, when a buffer container runs dry or forgetting to turn off the pump while changing buffers. Bubbles in the cuvette cause an extremely noisy fluorescence signal from both the pH-sensitive and the pH-insensitive wavelengths. The experiment can usually be salvaged by pausing data collection to allow the bubble to pass through the cuvette; fluorescence signals will likely return to normal.

13. Electrical problems. On rare occasions changes in electrical demands can cause a small sharp decrease in fluorescence signal at pH-sensitive and pH-insensitive wavelengths. This can be remedied by immediately stopping and restarting data collection. A new trace will begin and the data can be spliced when transferred to a graphing program. Most often electrical problems arise when turning the peristaltic pump on an off. Try to have fluorometer system components and computer wired on a different circuit from peripheral electrical components (peristaltic pump).

14. Coverslip falls out of groove in cuvette lid. Start a new assay.

15. Gradual change of pH-insensitive wavelength may indicate photobleaching of the fluorescent dye, dye leakage, or cell death or loss. Photobleaching can be reduced by closing excitation monochromator slits to reduce light intensities. Dye leakage may indicate a problem with AM hydrolysis, and cell death may indicate a problem with perfusion buffers. Cell loss from the coverslip may occur if solution flow is too turbulent in the cuvette or if cells do not adhere sufficiently strongly to coverslip. Some alternative to polylysine coating may be required.

16. The wavelength calibration of each fluorometer may vary so that it is worthwhile to determine the spectral characteristics of pH-sensitive dyes on your system. Using a fixed excitation wavelength (e.g., 505 nm) perform a scan of the emission spectrum (e.g., 510–540 nm). Likewise once optimal emission wavelength is determined, use this wavelength to determine optimal excitation wavelength. In experiments use the wavelengths that provide the highest fluorescence levels. Note that excitation and emission wavelengths must be separated by at least 5 nm to prevent excessive excitation light from spilling over into the emission wavelength, which can damage the PMT.

Acknowledgments

We would like to thank Dr. Bernardo Alvarez and Deborah Sterling for discussion and critical review of the chapter and the Academic Press and Molecular Probes, Inc., for permission to reprint figures from previous publications.

References

1. Busa, W.B. and Nuccitelli, R. (1984) Metabolic regulation via intracellular pH. *Am. J. Physiol.* **246**, R409–R438.
2. Ritter, M., Woll, E., Haussinger, D., and Lang, F. (1992) Effects of bradykinin on cell volume and intracellular pH in NIH 3T3 fibroblasts expressing the ras oncogene. *FEBS Lett.* **307**, 367–370.
3. Edmonds, B.T., Murray, J., and Condeelis, J. (1995) pH regulation of the F-actin binding properties of Dictyostelium elongation factor 1 alpha. *J. Biol. Chem.* **270**, 15222–15230.
4. Nosek, T.M., Fender, K.Y., and Godt, R.E. (1987) It is diprotonated inorganic phosphate that depresses force in skinned skeletal muscle fibers. *Science* **236**, 191–193.
5. Hao, W., Luo, Z., Zheng, L., Prasad, K., and Lafer, E.M. (1999) AP180 and AP-2 interact directly in a complex that cooperatively assembles clathrin. *J. Biol. Chem.* **274**, 22785–22794.
6. Joseph, D., Tirmizi, O., Zhang, X.L., Crandall, E.D., and Lubman, R.L. (2002) Polarity of alveolar epithelial cell acid-base permeability. *Am. J. Physiol. Lung Cell Mol. Physiol.* **282**, L675–L683.
7. Brokaw, C.J. (1987) Regulation of sperm flagellar motility by calcium and cAMP-dependent phosphorylation. *J. Cell Biochem.* **35**, 175–184.
8. Speake, T. and Elliott, A.C. (1998) Modulation of calcium signals by intracellular pH in isolated rat pancreatic acinar cells. *J. Physiol.* **506** *(Pt 2)*, 415–430.
9. Li, X., Galli, T., Leu, S., Wade, J.B., Weinman, E.J., Leung, G., Cheong, A., Louvard, D., and Donowitz, M. (2001) Na^+-H^+ exchanger 3 (NHE3) is present in lipid rafts in the rabbit ileal brush border: a role for rafts in trafficking and rapid stimulation of NHE3. *J. Physiol.* **537**, 537–552.
10. Yan, W., Nehrke, K., Choi, J., and Barber, D.L. (2001) The Nck-interacting kinase (NIK) phosphorylates the Na^+-H^+ exchanger NHE1 and regulates NHE1 activation by platelet-derived growth factor. *J. Biol. Chem.* **276**, 31349–31356.
11. Kaloyianni, M., Bourikas, D., and Koliakos, G. (2001) The effect of insulin on Na^+-H^+ antiport activity of obese and normal subjects erythrocytes. *Cell Physiol. Biochem.* **11**, 253–258.
12. Aharonovitz, O. and Granot, Y. (1996) Stimulation of mitogen-activated protein kinase and Na^+/H^+ exchanger in human platelets. Differential effect of phorbol ester and vasopressin. *J. Biol. Chem.* **271**, 16494–16499.
13. Yip, J.W., Ko, W.H., Viberti, G., Huganir, R.L., Donowitz, M., and Tse, C.M. (1997) Regulation of the epithelial brush bor-

der Na^+/H^+ exchanger isoform 3 stably expressed in fibroblasts by fibroblast growth factor and phorbol esters is not through changes in phosphorylation of the exchanger. *J. Biol. Chem.* **272**, 18473–18480.
14. Wray, S. (1988) Smooth muscle intracellular pH: measurement, regulation, and function. *Am. J. Physiol.* **254**, C213–225.
15. Kotyk, A.S. and Slavík, J. (1989) *Intracellular pH and Its Measurement.* CRC Press, Boca Raton, FL, USA.
16. Thomas, J.A., Buchsbaum, R.N., Zimniak, A., and Racker, E. (1979) Intracellular pH measurments in Ehrlich ascites tumor cells utilizing spectroscopic probes generated in situ. *Biochemistry* **18**, 2210–2218.
17. de Hemptinne, A. (1979) A double-barrel pH micro-electrode for intracellular use [proceedings]. *J. Physiol.* **295**, 5P–6P.
18. Kupriyanov, V.V., Xiang, B., Sun, J., Jilkina, O., and Deslauriers, R. (2002) Effects of regional hypoxia and acidosis on Rb^+ uptake and energetics in isolated pig hearts: ^{87}Rb MRI and ^{31}P MR spectroscopic study. *Biochim. Biophys. Acta.* **1586**, 57–70.
19. Tsien, R.Y. (1981) A non-disruptive technique for loading calcium buffers and indicators into cells. *Nature* **290**, 527–528.
20. Llopis, J., McCaffery, J.M., Miyawaki, A., Farquhar, M.G., and Tsien, R.Y. (1998) Measurement of cytosolic, mitochondrial, and Golgi pH in single living cells with green fluorescent proteins. *Proc. Natl. Acad. Sci. USA* **95**, 6803–6808.
21. Luby-Phelps, K. (1989) Preparation of fluorescently labeled dextrans and ficolls. *Methods Cell Biol.* **29**, 59–73.
22. Miesenbock, G., De Angelis, D.A., and Rothman, J.E. (1998) Visualizing secretion and synaptic transmission with pH-sensitive green fluorescent proteins. *Nature* **394**, 192–195.
23. Jankowski, A., Kim, J.H., Collins, R.F., Daneman, R., Walton, P., and Grinstein, S. (2001) In situ measurements of the pH of mammalian peroxisomes using the fluorescent protein pHluorin. *J. Biol. Chem.* **276**, 48748–48753.
24. Bright, G.R., Fisher, G.W., Rogowska, J., and Taylor, D.L. (1989) Fluorescence ratio imaging microscopy. *Methods Cell Biol.* **30**, 157–192.
25. Silver, R.B. (1998) Ratio imaging: practical considerations for measuring intracellular calcium and pH in living tissue. *Methods Cell Biol.* **56**, 237–251.
26. Paradiso, A.M., Tsien, R.Y., and Machen, T.E. (1984) Na^+-H^+ exchange in gastric glands as measured with a cytoplasmic-trapped, fluorescent pH indicator. *Proc. Natl. Acad. Sci. USA.* **81**, 7436–7440.
27. Roos, A. and Boron, W.F. (1981) Intracellular pH. *Physiol. Rev.* **61**, 296–434.
28. Tang, X.B., Fujinaga, J., Kopito, R., and Casey, J.R. (1998) Topology of the region surrounding Glu681 of human AE1 protein, the erythrocyte anion exchanger. *J. Biol. Chem.* **273**, 22545–22553.
29. Burnham, C.E., Amlal, H., Wang, Z., Shull, G.E., and Soleimani, M. (1997) Cloning and functional expression of a human kidney $Na^+:HCO_3^-$ cotransporter. *J. Biol. Chem.* **272**, 19111–19114.
30. Camilion de Hurtado, M.C., Alvarez, B.V., Perez, N.G., Ennis, I.L., and Cingolani, H.E. (1998) Angiotensin II activates Na^+-independent Cl^-–HCO_3^- exchange in ventricular myocardium. *Circ. Res.* **82**, 473–481.
31. Sterling, D. and Casey, J. R. (1999) Transport activity of AE3 chloride/bicarbonate anion-exchange proteins and their regulation by intracellular pH. *Biochem. J.* **344** *(Pt 1)*, 221–229.

Chapter 18

Measurement of Plasma Membrane Calcium–Calmodulin-Dependent ATPase (PMCA) Activity

Tamer M. A. Mohamed, Florence M. Baudoin-Stanley, Riham Abou-Leisa, Elizabeth Cartwright, Ludwig Neyses, and Delvac Oceandy

Abstract

The plasma membrane calcium–calmodulin-dependent ATPase (PMCA) is a calcium-extruding enzymatic pump that ejects calcium from the cytoplasm to the extracellular compartment. Although in excitable cells such as skeletal and cardiac muscle cells PMCA has been shown to play only a minor role in regulating global intracellular calcium concentration, increasing evidence points to an important role for PMCA in signal transduction, in particular in the nitric oxide signaling pathway. Moreover, recent evidence has shown the functional importance of PMCA in mediating cardiac contractility and vascular tone. Here we describe a method in determining PMCA activity in the microsomal membrane preparation from cultured cells that overexpress specific isoform of PMCA by using modified coupled enzyme assay.

Key words: Calcium pump, microsome, ATPase, enzyme assay.

1. Introduction

The plasma membrane calcium–calmodulin-dependent ATPase (PMCA) is a ubiquitous calcium transporter that pumps calcium from the cytoplasm to the extracellular matrix. In most cell types, the role of PMCA is crucial in maintaining low intracellular calcium concentration. Four different PMCA genes (named PMCA1–4) and more than 20 splice variants have been described.

Q. Yan (ed.), *Membrane Transporters in Drug Discovery and Development*, Methods in Molecular Biology 637, DOI 10.1007/978-1-60761-700-6_18,

The structure of PMCA, isoform diversity, splice variants, and tissue distribution have previously been reviewed in great detail (1).

Unlike in non-excitable cells, where PMCA plays a major role in calcium extrusion, in excitable cells such as cardiac and skeletal muscle PMCA has only a minor role in the regulation of cytosolic calcium. In these cells, calcium is mainly removed through the sarcoplasmic reticulum calcium ATPase (SERCA) and the sodium calcium exchanger (NCX) (2). However, increasing evidence suggests that PMCA plays a more important role in signal transduction. Two main characteristics of a signaling molecule displayed by PMCA are the ability to interact with signaling proteins (3) and its localization in caveolae (4). Recent evidence from our laboratory suggested that PMCA has a role in mediating nNOS activity (5, 6) and demonstrated the physiological relevance of this novel signaling complex in regulating cardiac contractility (7, 8).

As PMCA plays an essential role in cell function development of an assay to measure its activity is very important. The use of human erythrocyte ghosts was common in the preparation of purified PMCA for this purpose (9). However, this will yield mainly PMCA4 as this is the main isoform expressed in human red blood cells. Moreover, a large amount of human blood is needed to obtain a substantial amount of good-quality purified PMCA. In this chapter we describe a method for measuring PMCA activity in the microsomal membrane preparation from cultured cells that overexpress PMCA. Measurement of PMCA activity is based on a coupled enzyme ATPase activity. The coupled enzyme assay is based on regeneration of PMCA-dependent release of ADP by pyruvate kinase, which converts phosphoenolpyruvate to pyruvate. Then, pyruvate is converted to lactate by lactate dehydrogenase using NADH (*see* **Fig. 18.1**). Detection of NADH decline

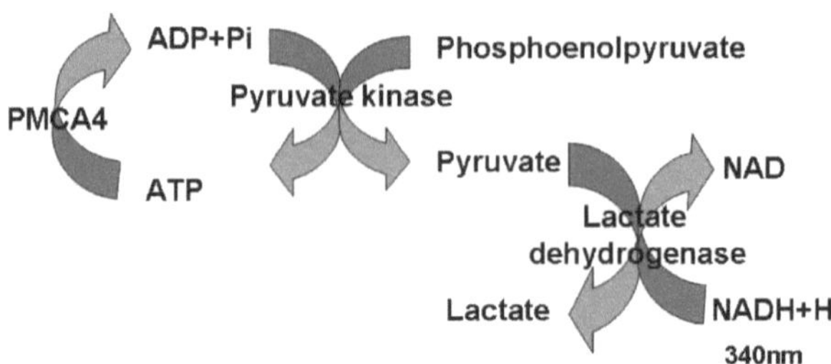

Fig. 18.1. Schematic diagram of PMCA activity assay. The phosphorylation of the aspartate residue in the catalytic domain of PMCA by ATP results in the generation of ADP + Pi. Then, the ADP is used by pyruvate kinase to generate pyruvate from phosphoenolpyruvate. Pyruvate produced by this reaction will be converted to lactate by lactate dehydrogenase. This reaction is coupled to the oxidation of NADH to NAD. The degradation of NADH can be monitored at 340 nm wavelength.

rate at 340 nm will determine the activity of the pump. An adaptation of this method in 96-well format is also discussed, which could be useful for the development of a high-throughput assay system.

2. Materials

2.1. Cell Culture and Infection with Adenovirus

1. Dulbecco's modified Eagle's medium (DMEM).
2. Fetal bovine serum (FBS).
3. Penicillin/streptomycin.
4. Non-essential amino acids.
5. Human embryonic kidney cells (HEK293) (ATCC).
6. Trypsin–EDTA (1X) liquid (0.05% trypsin, 0.53 mM EDTA tetra sodium salt).
7. AdEasy Adenoviral Vector System.
8. BJ5183 electroporation competent *Escherichia coli* cells.
9. LB broth medium.
10. Lipofectamine^{TM} 2000.
11. PMCA4 recombinant adenovirus (*see* **Section 3** for the cloning and generation of adenovirus) (*see* **Note 1**).

2.2. Microsomal Membrane Preparation

1. Harvest solution: 1X PBS, 0.26% 2 mg/ml aprotinin, 0.11% 2 mg/ml leupeptin, and 0.1% 0.1 M phenylmethylsulfonyl fluoride (PMSF).
2. Hypotonic solution: 10 mM Tris–HCl, pH7.5, 1 mM $MgCl_2$, 0.5 mM EGTA, 2 mM DTT, 0.2% 2 mg/ml aprotinin, and 0.05% 2 mg/ml leupeptin.
3. Homogenate solution: 10 mM Tris–HCl, pH 7.5, 2 mM DTT, 0.38 M sucrose, 0.3 M KCl, 0.2% 2 mg/ml aprotinin, and 0.05% 2 mg/ml leupeptin.
4. Final solution: 10 mM Tris–HCl, pH 7.5, 1 mM DTT, 0.19 M sucrose, 0.15 M KCl, 0.2% 2 mg/ml aprotinin, 0.05% 2 mg/ml leupeptin, and 0.02 mM $CaCl_2$.

2.3. Coupled Enzyme Assay

1. Coupled enzyme assay reaction mixture: 50 mM HEPES–Tris, pH 7.4, 160 mM KCl, 2 mM $MgCl_2$, 5 mM NaN_3, 1 μg/ml alamethicin, 1 mM ATP–Tris, 1 mM phosphoenolpyruvate, 1 U/ml pyruvate kinase, 0.6 mmol/L NADH, 1 U/ml lactate dehydrogenase.

2. Calcium and calmodulin mixture: 5 μg of calmodulin and 4 μM-free calcium (*see* **Note 2**).
3. Spectrophotometer able to read at 340 nm wavelength.
4. Microplate reader with 340 nm wavelength filter.

3. Methods

3.1. Cell Culture for HEK293 Cells

Maintained human embryonic kidney (HEK293) cells in Dulbecco's modified Eagle's medium (DMEM) supplemented with 10% fetal bovine serum (FBS), 1% penicillin/streptomycin, and 1% non-essential amino acids. Cells are passed when they reach 75–90% confluence. To passage the cells, remove the medium from the tissue culture flask and wash the cells twice with phosphate-buffered saline (PBS). Then, add 5 ml of trypsin–EDTA (1X) liquid to each flask and incubate for 5 min at 37°C. To neutralize trypsin, add 10 ml of the culture medium to each flask. Cells from one flask are split either into two or three flasks and topped up to 25 ml medium in each flask.

3.2. Generation of Recombinant Adenovirus for PMCA Overexpression

1. AdEasy Adenoviral Vector System is used to clone the PMCA4b cDNA to the adenoviral vector. The PMCA4b cDNA is cloned to the shuttle vector pShuttle-CMV to obtain pShuttle-CMV-PMCA4b. As a control, an adenovirus overexpressing LacZ is used. To generate recombinant adenoviral vectors carrying CMV-PMCA4b or CMV-LacZ constructs, linearize pShuttle-hPMCA4b and pShuttle-CMV-LacZ using *Pme*I restriction enzyme. Precipitate linearized vectors (using three volumes of absolute ethanol), then dephosphorylate, and gel purify.
2. Mix 1 μg of linearized/dephosphorylated shuttle vector and 100 ng of pAdEasy-1 with 40 μl BJ5183 electro-competent *E. coli* on ice.
3. Electroporate cells with a single pulse using Gene-pulser set at 200 Ω, 2.5 kV, and 25 μF.
4. Add 1 ml of LB broth medium to each transformation and incubate at 37°C for 1 h.
5. Plate cells onto LB agar plates containing 50 μg/ml kanamycin and incubate overnight at 37°C.
6. Select small colonies from the plates (large colonies usually represent re-linearization of the shuttle vector). Grow selected colonies in small scale cultures (~5 ml of LB broth with kanamycin).
7. Check plasmids for homologous recombination by digestion with *Pac*I on a 0.75% agarose gel.

3.3. Generation of Recombinant Adenovirus Particles

1. Plate HEK293 cells at a density of 0.7×10^6 cells per 25 cm^2 flask 24 h prior to transfection.
2. Transfect 5 μg of linearized adenoviral plasmids into HEK293 cells using 10 μl Lipofectamine™ 2000.
3. Replace the medium 24 h after transfection and then every 48 h. Cells producing adenovirus will appear as patches of rounding dying cells. These cells will lyse and release adenoviral particles which subsequently infect neighboring cells, then plaques start to form. When the majority of cells are detached, a primary stock of adenovirus can be prepared.
4. Collect floating and attached cells in the growth medium by pipetting it up and down.
5. Pellet cells by centrifugation (1,000×*g* for 5 min).
6. Resuspend each flask of cells in 0.5 ml PBS and then perform four rounds of rapid freezing and thawing by transferring between a dry ice/methanol bath and a 37°C water bath (for 5 min each).
7. Remove cell debris by centrifugation at 13,000×*g* for 10 min. The supernatant contains the recombinant adenovirus.
8. Use 10 μl of the primary stock to infect a 175 cm^2 flask of 70% confluent HEK293 cells, change culture medium every 2–3 days until plaques formed.
9. Once the majority of cells have detached, harvest the cells and isolate adenovirus by freeze–thawing. Use the secondary stock to infect 20 flasks of HEK293 cells and repeat the procedure above to provide a tertiary stock.
10. Use freezing and thawing method to purify the adenovirus particles from the tertiary stock (*see* **Note 3**).
11. In order to determine the number of adenovirus particles in the tertiary preparations, plate HEK293 cells at 5×10^3 cells/well (in 96-well plate) in a total volume of 100 μl medium. After 24 h remove the medium and replace with 100 μl of serially diluted adenovirus stocks (1×10^{-2} to 7.63×10^{-12}).
12. Analyze each dilution in triplicate. After 24 h, add 100 μl of fresh DMEM (with 10% FBS) to each well. Change the medium on day 4 and monitor plaque formation each day.
13. Eight days after infection, use the final dilution which shows plaque formation as the endpoint of the assay. From knowing the number of adenovirus particles the infection process can be standardized (*see* **Table 18.1**).

Table 18.1
Determination of adenovirus PFU from serial dilutions

Dilution	PFU/mL	Dilution	PFU/mL
10^{-2}	1×10^{3}	7.81×10^{-9}	1.28×10^{9}
10^{-3}	1×10^{4}	3.91×10^{-9}	2.56×10^{9}
10^{-4}	1×10^{5}	1.95×10^{-9}	5.12×10^{9}
10^{-5}	1×10^{6}	9.77×10^{-10}	1.02×10^{10}
10^{-6}	1×10^{7}	4.88×10^{-10}	2.05×10^{10}
5×10^{-7}	2×10^{7}	2.44×10^{-10}	4.1×10^{10}
2.5×10^{-7}	4×10^{7}	1.22×10^{-10}	8.19×10^{10}
1.25×10^{-7}	8×10^{7}	6.1×10^{-11}	1.64×10^{11}
6.25×10^{-8}	1.6×10^{8}	3.05×10^{-11}	3.28×10^{11}
3.12×10^{-8}	3.2×10^{8}	1.53×10^{-11}	6.55×10^{11}
1.56×10^{-8}	6.4×10^{8}	7.63×10^{-12}	1.31×10^{12}

3.4. Microsomal Membrane Preparation

1. Plate HEK293 cells in 100 mm tissue culture plates, at a density of 4.5×10^{6} cells/plate, 24 h prior to virus infection.
2. Infect cells in 10 ml of DMEM plus 10% FBS by 25 MOI (*see* **Note 1**) of the indicated recombinant adenovirus, PMCA4b or LacZ (control) virus, for 48 h.
3. After 48 h, remove the medium and wash cells three times with PBS.
4. After washing, harvest cells in 5 ml of harvest medium and then centrifuge cells at $3000\times g$ for 10 min at 4°C.
5. Incubate the cell pellet in 3 ml hypotonic solution for 10 min in ice.
6. Then homogenize cells using Dounce homogenizer for 40 strokes before the addition of 3 ml homogenate solution.
7. Homogenize cells slowly for another 20 strokes to seal the vesicles.
8. Spin cell homogenate at 3500× g for 20 min to remove cell debris.
9. Then, add 60 μl 0.25 mM EDTA and 1.08 ml of 2.5 M KCl to the supernatant.
10. Centrifuge supernatant at $100{,}000\times g$ for 40 min at 4°C to pellet the microsomes. Finally resuspend the pellet in 0.4 ml final solution and measure protein concentration using standard bicinchoninic acid (BCA) protein assay kit (*see* **Note 4**).

3.5. Coupled Enzyme Assay Measuring Ca^{+2}-Dependent ATPase Activity

1. All measurements are carried out at 37°C using Ultrospec 3000 Spectrophotometer setup at 340 nm wavelength.
2. Add 500 μl of coupled enzyme assay reaction mixture to a quartz cuvette and pre-warmed to 37°C. Then, record the baseline activity for 20 min after addition of 20 μg microsomal preparation to the reaction mixture.
3. To start the ATPase activation, add 4 μM calcium followed by the addition of 5 μg calmodulin to the reaction. Then, record the decrease of NADH at 340 nm absorbance over 5 min following each addition.
4. To stop the Ca^{+2}-dependent ATPase activity, add 2 mM EGTA and follow the reaction for another 5 min.
5. To stop the calcium- and magnesium-dependent activity, add 1 mM EDTA to the reaction mixture.
6. The Ca^{+2}/calmodulin-dependent ATPase activity can be calculated by subtracting the fitted slopes. **Figure 18.2** described representative trace with control microsomes and **Fig. 18.3** described representative trace with microsomes from cells overexpressing PMCA4.

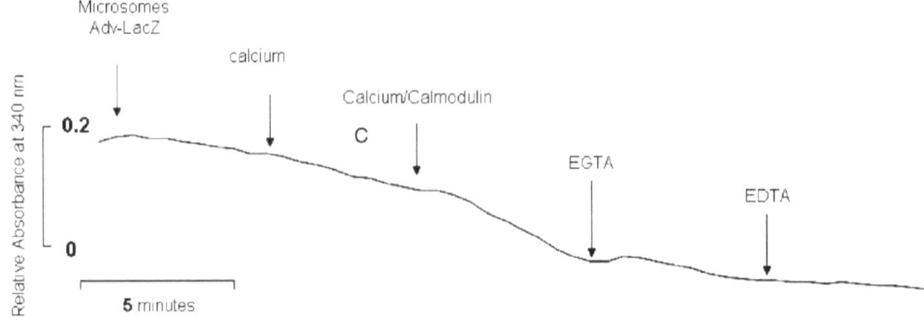

Fig. 18.2. Representative trace of ATPase assay using microsomal particles from cells infected with LacZ overexpressing virus as a negative control. The decrease of absorbance at 340 nm was mainly due to spontaneous degradation of NADH. No significant changes were observed after the addition of either calcium or calcium–calmodulin.

3.6. Adaptation of the Assay in 96-Well Format

We have optimized this assay to be conducted in a 96-well format (*see* **Fig. 18.4**). A microplate reader with 340 nm filter is needed. For controls, prepare wells containing microsomes only, microsomes + calcium, and microsomes + calcium and calmodulin.

Protocol:

1. All the measurements are carried out at 37°C in an Ascent microplate reader.
2. Add 250 μl of coupled enzyme assay reaction mixture in each well (pre-warm to 37°C). Record the baseline activity for 30 min after addition of 10 μg microsomal preparation

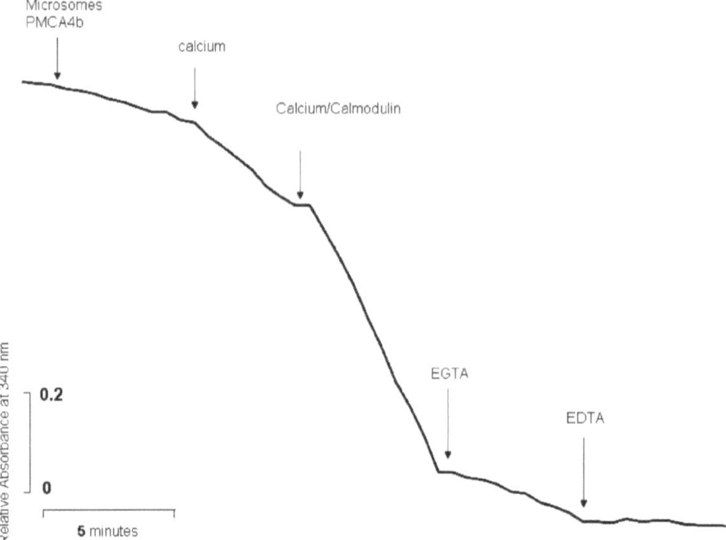

Fig. 18.3. Representative trace of ATPase assay using microsomal particles from cells infected with PMCA4 overexpressing virus. The decrease of absorbance at 340 nm after addition of calcium–calmodulin represented the calcium–calmodulin-dependent ATPase, which was mainly due to the PMCA4 activity.

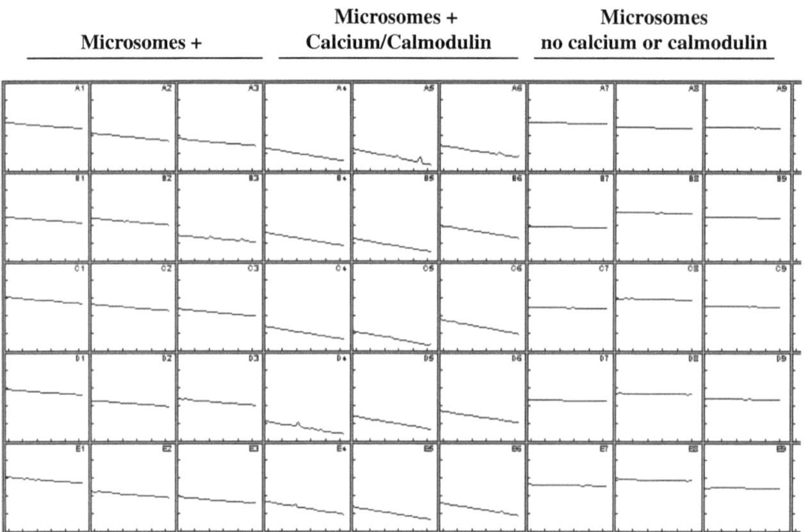

Fig. 18.4. Representative trace from 96-well format PMCA assay. About 20 μg of microsomes from cells overexpressing PMCA4b was added to each well. In columns 1–3, PMCA4b was activated with 4 μM free calcium. In columns 4–6, the pump was activated with 4 μM free calcium, and 5 μg calmodulin. Columns 7–9 served as negative controls.

to 250 μl of the reaction mixture. Monitor the reduction of NADH by automatic scanning setup at 30 s intervals.

3. To start the ATPase activation, add calcium and calmodulin mixture to the reaction and then follow the 340 nm absorbance for another 30 min.

4. Add substance (inhibitor/activator) to be tested and record the absorbance at 340 nm for another 30 min.
5. To stop the Ca^{+2}-dependent ATPase activity, add 2 mM EGTA and continue the reaction for another 15 min.
6. Calculate the Ca^{+2}/calmodulin-dependent ATPase activity using the equation below. Use the fitted slopes in each step to determine the effects of each substance added.

3.7. Data Analysis

Enzyme activity is expressed as the rate of NADH reduction per milligram microsomal protein. It can be calculated using the following equation:

$$E = \left(\frac{-S}{6.22 \times P}\right) \times 250$$

where E = enzyme activity (mM NADH/min/mg protein)

S = fitted slope (per minute) at the last 10 min of each measurement

6.22 is millimolar extinction coefficient of NADH at 340 nm and 37°C

P = microsomal protein added (in milligram protein)

250 is the reaction volume in μl

4. Notes

1. The amount of adenovirus added in the experiments depends on the number of cells plated. For example, addition of similar number of adenovirus particles to the cells will give a multiplicity of infection (MOI) of 1. In most experiments, we infected cells with an MOI of 25, meaning the number of adenovirus particles added was 25 times more than the number of cells plated. Cells were infected for 48 h before harvesting.
2. Free calcium concentration in the solution is calculated using Fabiato equation. A free software to calculate the free calcium (MaxChelator) can be downloaded from the following website: http://www.stanford.edu/~cpatton/downloads.htm.
3. Purified adenovirus tertiary stock needs to be aliquoted in small volume (30 μl) and stored at –80°C. We have found

that repeated freeze and thaw of the virus stock will reduce the virus viability.

4. The microsomes are resuspended in the final solution at the concentration of 2 mg/ml and should be stored as 200 μl aliquots in liquid nitrogen. Storing the microsomes in liquid nitrogen will preserve the activity better than storing in 80°C freezer.

Acknowledgments

This work was supported by Medical Research Council (MRC) Programme Grant (G0500025) to L.N. and British Heart Foundation (BHF) Project Grant (PG/05/082) to D.O and E.C. The laboratory is also supported by the NIHR Biomedical Research Funding Scheme. TM is a lecturer in Biochemistry Dept., Zagazig University, Egypt. Address for correspondence to: Dr Delvac Oceandy (delvac.oceandy@manchester.ac.uk).

References

1. Strehler, E.E. and Zacharias, D.A. (2001) Role of alternative splicing in generating isoform diversity among plasma membrane calcium pumps. *Physiol. Rev.* **81**, 21–50.
2. Bers, D.M. (2000) Calcium fluxes involved in control of cardiac myocyte contraction. *Circ. Res.* **87**, 275–281.
3. Oceandy, D., Buch, M.H., Cartwright, E.J., and Neyses, L. (2006) The emergence of plasma membrane calcium pump as a novel therapeutic target for heart disease. *Mini. Rev. Med. Chem.* **6**, 583–588.
4. Fujimoto, T. (1993) Calcium pump of the plasma membrane is localized in caveolae. *J. Cell Biol.* **120**, 1147–1157.
5. Schuh, K., Uldrijan, S., Telkamp, M., Rothlein, N., and Neyses, L. (2001) The plasmamembrane calmodulin-dependent calcium pump: a major regulator of nitric oxide synthase I. *J. Cell Biol.* **155**, 201–205.
6. Williams, J.C., Armesilla, A.L., Mohamed, T.M., Hagarty, C.L., McIntyre, F.H., Schomburg, S., Zaki, A.O., Oceandy, D., Cartwright, E.J., Buch, M.H., Emerson, M., and Neyses, L. (2006) The sarcolemmal calcium pump, alpha-1 syntrophin, and neuronal nitric-oxide synthase are parts of a macromolecular protein complex. *J. Biol. Chem.* **281**, 23341–23348.
7. Oceandy, D., Cartwright, E.J., Emerson, M., Prehar, S., Baudoin, F.M., Zi, M. Alatwi, N., Schuh, K. Williams, J.C. Armesilla, A.L., and Neyses, L. (2007) Neuronal nitric oxide synthase signaling in the heart is regulated by the sarcolemmal calcium pump 4b. *Circulation* **115**, 483–492.
8. Mohamed, T.M., Oceandy, D., Prehar, S., Alatwi, N., Hegab, Z., Baudoin, F.M., Pickard, A., Zaki, A.O., Nadif, R., Cartwright, E.J., and Neyses, L. (2009) Specific role of neuronal nitric-oxide synthase when tethered to the plasma membrane calcium pump in regulating the beta-adrenergic signal in the myocardium. *J. Biol. Chem.* **284**, 12091–12098.
9. Pande, J., Mallhi, K.K., Sawh, A., Szewczyk, M.M., Simpson, F., and Grover, A.K. (2006) Aortic smooth muscle and endothelial plasma membrane Ca2+ pump isoforms are inhibited differently by the extracellular inhibitor caloxin 1b1. *Am. J. Physiol. Cell. Physiol.* **290**, C1341–C1349.

Chapter 19

Assessment of the Contribution of the Plasma Membrane Calcium ATPase, PMCA, Calcium Transporter to Synapse Function Using Patch Clamp Electrophysiology and Fast Calcium Imaging

Chris J. Roome and Ruth M. Empson

Abstract

The plasma membrane calcium ATPase, or PMCA, functions to extrude calcium out of cells as a key component necessary for adequate calcium homeostasis in all cells. However, calcium is particularly important at synapses between neurons, where communication relies on the controlled rise and fall in presynaptic calcium that precedes the release of neurotransmitter. Here we show how to infer the real-time contribution of PMCA-mediated calcium extrusion to this presynaptic calcium dynamic and how this influences the properties of the synapse. To do this we have taken advantage of a well-studied synapse in the cerebellum. We use electrophysiology to assess the timing of short-term facilitation at this synapse in the presence and absence of PMCA2 using PMCA2 knockout mice and pharmacology and fast calcium imaging to measure the presynaptic calcium dynamics. These approaches are all highly applicable to other synapses and can help determine the contribution of PMCA, and other transporters or exchangers, to the calcium dynamics that underpin reliable synaptic transmission.

Key words: Membrane, ATPase, calcium transporter, synapse, patch clamp, electrophysiology, fast calcium imaging.

1. Introduction

The plasma membrane Ca^{2+} ATPases, PMCAs (there are four subtypes, 1–4, and these can also be alternatively spliced), are membrane-bound transporters, or pumps, that extrude Ca^{2+} across the plasma membrane, against the inward Ca^{2+} concentration gradient using energy derived from the hydrolysis of ATP (1)

Q. Yan (ed.), *Membrane Transporters in Drug Discovery and Development*, Methods in Molecular Biology 637,
DOI 10.1007/978-1-60761-700-6_19, © Springer Science+Business Media, LLC 2010

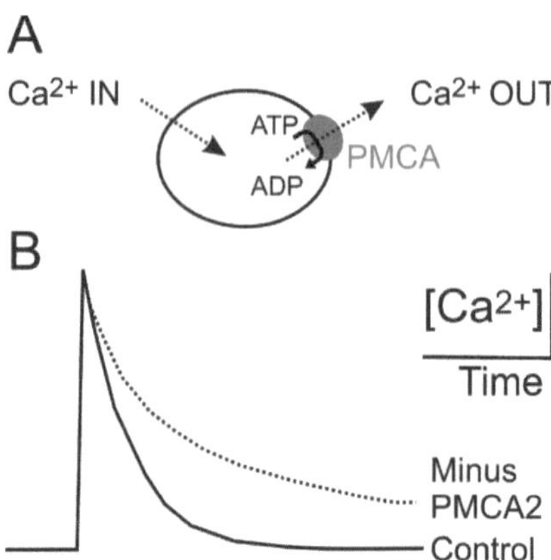

Fig. 19.1. The PMCA pump influences the time taken for cytosolic calcium levels to recover in a cell, neuron, or neuronal compartment. (**a**) A schematic of the PMCA pump and how it acts to extrude calcium out of the cytosol at the expense of ATP is shown. (**b**) A schematic peak rise in calcium concentration within the cytosol under normal conditions, filled line, and how the recovery of calcium levels slows when PMCA is removed, *dotted line*, minus PMCA, as, for example, in cells from the PMCA2 knockout mouse is shown.

(*see* **Fig. 19.1**). As such these transporters are vital for clearing calcium from the cytosolic space as one of the important "off" mechanisms in the calcium signaling toolkit (2). Importantly, many of the PMCAs express a calcium–calmodulin (Ca-CaM) binding domain within their C-terminus. Normally this domain of the pump remains in a state that auto-inhibits the pump and prevents wasteful expenditure of ATP. It is only when this domain of the pump has the opportunity to interact with calcium bound to calmodulin that the pump turns on and starts to extrude calcium. The Ca-CaM domain allows the PMCAs to "sense" low concentrations of intracellular calcium (consistent with the high affinity with which calmodulin binds calcium) and means that the PMCA calcium pumps are exquisitely designed to respond to small fluctuations in basal or baseline calcium. Importantly following a large calcium rise within the cell, the PMCAs also remain active to return the cytosolic calcium fully back to resting or basal levels.

Much of our knowledge regarding the functional biochemistry of the PMCAs came from the pioneering work of the laboratories of Carafoli, Strehler, and Penniston labs (3) and relied upon the development of some important assays to measure PMCA activity in vitro. These methods included fast cuvette-based measurements of phosphate production (based upon the phosphorolysis of 2-amino-6 mercapto-7-methylpurine as an absorbance change at 360 nm) following the activation of PMCA expressed in microsomal fractions by the addition of calcium–calmodulin and more traditional ^{45}Ca uptake measurements (4).

Unfortunately neither of the traditionally used biochemical methods to measure PMCA activity are applicable to determine its contribution to real-time calcium dynamics in living cells. It is however possible to measure calcium dynamics in living cells with calcium-sensitive fluorescent dyes and to use changes in the kinetics of the fluorescence changes when the PMCA is inhibited to infer its contribution. Indeed this has proven a successful strategy to demonstrate the active presence and contribution of the pump, in sympathetic neurons, snail neurons, retina, Drosophila motor neurons, the calyx of Held, cerebellar parallel fibers, and hippocampal CA3 synapses (5–11). Removal of PMCA activity results in a slower return of the calcium to baseline levels, following a calcium challenge (*see* **Fig. 19.1b**, dotted line). In some cases using ratiometric calcium dyes such as FURA and Indo it has been possible to observe raised baseline cytosolic calcium when the PMCA is inhibited (5). More recently this approach has successfully detected the influence of naturally occurring human PMCA mutations (associated with familial deafness) on pump efficiency and kinetics in CHO cells (12) and cochlear hair cells.

1.1. The Contribution of PMCA Activity to Calcium Dynamics During Synaptic Transmission at the Cerebellar Parallel Fiber to Purkinje Neuron Synapse

Within the nervous system, one of the most important roles of calcium is to trigger neurotransmitter release from vesicles present within the presynaptic terminal. Although this calcium rise is traditionally thought of as being very large, there are several arguments to suggest that release can still occur at calcium levels much closer to baseline levels (13), where the PMCA may have an important functional role.

To investigate the role of the PMCA for the control of presynaptic calcium we took advantage of a synapse in the cerebellum where PMCA2 is highly expressed, the parallel fiber to Purkinje neuron synapse (*see* **Fig. 19.2**). The Purkinje neurons are the main output cells of the cerebellum and the control of their activity is critical for the correct integration of sensory and motor information within the whole animal. The peak rise in calcium that occurs within the presynaptic parallel fibers can be measured during the initiation of synaptic transmission (14), and this calcium then decays back to baseline by virtue of the activation of Ca^{2+} "off" mechanisms within the terminal. This approach allowed us to investigate how the PMCA contributed to synapse function by measuring the calcium dynamics at the presynaptic site and how they change when the PMCA was removed. Importantly by also recording from the postsynaptic Purkinje neuron we could at the same time investigate the consequence of any changes for the functional outcome of the synapse. This we did by recording the timing of the synaptic response, the excitatory postsynaptic current, and EPSC in the Purkinje neuron (*see* lower panel **Fig. 19.2d**).

Fortunately the parallel fiber to Purkinje neuron synapse is a low-probability synapse, so that when the synapse is activated,

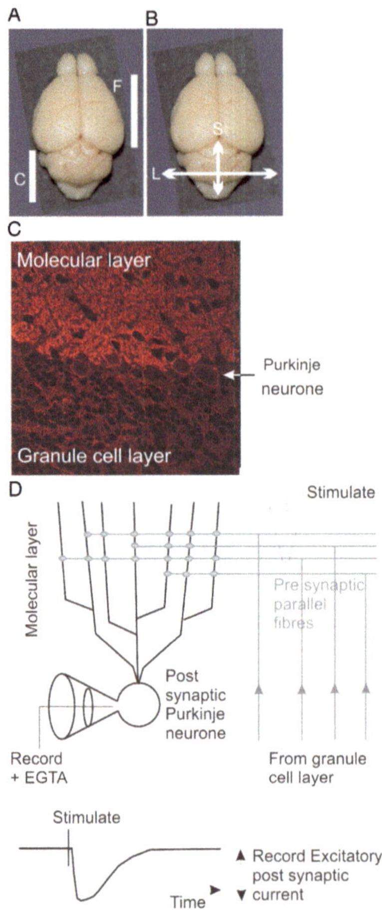

Fig. 19.2. The cerebellum, its circuitry, and the slicing and electrophysiology recording configurations. (**a**) The cerebellum, labeled C, is located to the rear of the mouse brain and slices of the cerebellum can be prepared in either a saggital, S, or L, longitudinal, or coronal, orientation, as shown by the *white arrows* in (**b**). *F* represents the forebrain. (**c**) A sagittally oriented cerebellar slice that has been labeled to show the expression of PMCA2 using immunohistochemistry is shown. The molecular layer located above the granule cell layer contains the dendrites of the Purkinje neurons, the main output cells of the cerebellum, where PMCA2 can be seen in the membrane of the cell body and richly expressed in the dendrites. The molecular layer also contains the axons from the granule cells that form the longitudinal parallel fibers and associated synapses, seen diagrammatically in (**d**). (**d**) The whole cell recording configuration from the cell body of the Purkinje neuron and the approximate position of the stimulation electrode are shown. The parallel fibers receive the stimulus applied in the molecular layer that leads to a rise in presynaptic calcium levels and release of glutamate from the parallel fiber terminals; this in turn leads to the excitatory postsynaptic current, EPSC, lower part of (**d**), seen as an inward current when the Purkinje neuron is held at –65 mV in voltage clamp. The whole cell electrode also contains a high concentration of EGTA to adequately buffer intracellular calcium and prevents any complication from raised intracellular calcium levels in the Purkinje neuron (compared with the presynaptic parallel fiber terminal compartment) when PMCA is removed either in the PMCA2 knockout mouse or by pharmacological means.

not all the vesicles release their contents into the synaptic cleft. This means that if the synapse receives another stimulus at the presynaptic side there are still vesicles available within the terminal ready to release their contents into the synaptic cleft. Actually, if a second stimulus is given to the terminals the response of the synapse is greater, and this is termed facilitation (*see* **Fig. 19.3a**). Following on from pioneering work by Katz and Miledi at the neuromuscular junction (15) this facilitation at the parallel fiber synapse is considered to be a result of build up of calcium within the presynaptic terminal, the so-called residual calcium (16). Since the PMCA is best suited at controlling the rate of decay of calcium and baseline levels of calcium we raised the hypothesis that PMCA should influence facilitation at this synapse.

The powerful combination of electrophysiology and calcium imaging allows us to measure the functional contribution of the PMCA2 transporter at synapses while also making use of the PMCA2 transgenic knockout mouse as a way of effectively removing PMCA. The methods we describe here can be applied to other synapses and importantly to investigate the role of the PMCA and other transporters and exchangers in shaping the kinetics of presynaptic calcium and the influence this brings to bear upon the timing and shaping of the postsynaptic response.

2. Materials

2.1. PMCA2$^{-/-}$ Transgenic Knockout Mice

1. The best way to remove PMCA2 from synapses is to delete it genetically. The rich expression of PMCA2 in the cerebellum provides an ideal opportunity to use the transgenic PMCA2 knockout mouse. Although we cannot be certain that other PMCAs exist at the synapses, for example, PMCA3 is thought to exist at cerebellar synapses (17), the fact that they are PMCA2 enriched makes both the cerebellum and the PMCA2 knockout mouse a good model. The PMCA2 transgenic knockout mouse (18) exhibits some abnormalities including hearing loss, vestibular disturbances, and ataxia, the latter being associated with a cerebellar phenotype including reduction in the size of the cerebellum. Interestingly there are no compensatory adjustments in other PMCAs in the cerebellum of the PMCA2$^{-/-}$ mouse (19).
2. The mice are available on two backgrounds, black B6 and white FVBN, and all studies should interchangeably use the mice on both backgrounds. PMCA2$^{-/-}$ mice are generated by PMCA2$^{-/+}$ crosses, where the numbers of

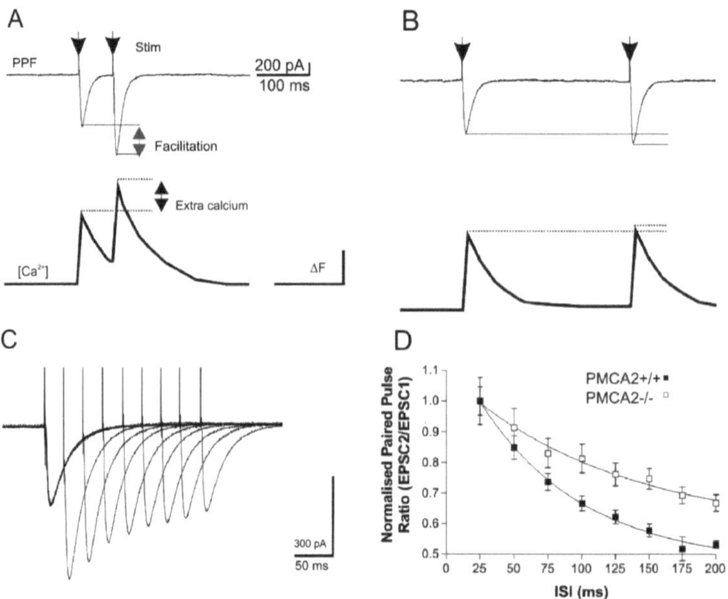

Fig. 19.3. Paired pulse facilitation of Purkinje neuron excitatory postsynaptic responses and how this phenomenon is influenced by altered levels of PMCA2. Using the schematic recording configuration in **Fig. 19.2d**, the postsynaptic Purkinje neuron, patch clamped at the cell body, receives a stimulus, *downward arrow* left-hand trace, that evokes an EPSC. A second stimulation, a few tens of milliseconds later, also shown on the left-hand trace (**a**) before calcium levels in the presynaptic terminal have had a chance to recover fully, see lower trace in (**a**), leads to an extra amount of calcium delta F is change in fluorescence of fluorescent calcium indicator in the presynaptic terminal that drives the release of more glutamate from readily available vesicles in the presynaptic terminal, the result being that the second EPSC is now larger than the first. This enhancement is termed paired pulse facilitation, PPF. If, however, the time between stimuli is increased, *see* (**a**) right-hand panel, the first rise in calcium has longer to decay more fully back to baseline, presumably as a consequence of activation of the so-called off mechanisms (including the extrusion of calcium out of the presynaptic compartment by the PMCA). Under these circumstances the extent of paired pulse facilitation is greatly reduced since there is less "additional" calcium available in the presynaptic terminal, *see* lower trace right-hand trace (**b**). In fact, as can be seen in (**c**), the extent of paired pulse facilitation decays exponentially as the inter-stimulus interval increases and is consistent with the decay of intracellular calcium levels. In (**d**), the PPF for a number of cells, normalized to maximum PPF, in both the double knockout mouse, PMCA2$^{-/-}$, and the wild-type PMCA2$^{+/+}$ is shown and the decay of PPF fitted with a single exponential over inter-stimulus intervals up to 200 ms. Genetic removal of PMCA2 leads to a significant slowing ($p < 0.001$, *t* test) of the recovery of PPF compared with the wild type; the mean halftime for decay of PPF in the PMCA2$^{-/-}$ cells was 87 ± 8.3 ms ($n = 12$) compared with 53.1 ± 3.3 ms ($n = 15$), fitted with a single exponential for all individual cells in Prism.

PMCA2$^{-/-}$, PMCA2$^{-/+}$, and PMCA2$^{+/+}$ follow an approximate Mendelian inheritance. The mice breed well, although the PMCA2$^{-/-}$ offspring are often smaller and sometimes require careful husbandry, especially if the litter is large.

2.2. Cerebellar Brain Slices

1. Living brain tissue slices retain neurons and synapses in their most naturally and easily accessible form as an in vitro preparation and provide the best conditions to test the influence of PMCA on synapse function. Slices of the cerebellum (*see* **Fig. 19.2**) allow the cells and circuitry of the cerebellum to be kept intact for up to 8 h (*see* (20) for an excellent review on cerebellar circuitry).
2. Living brain slices from mice are best prepared from young 3–4-week-old mice especially for use with patch clamp electrophysiology. Slices can be prepared with a vibratome, such as the Leica VT1000S, that renders as little damage to the surface of the cut slice as possible. This helps to ensure clean access of the recording patch clamp pipette to the Purkinje neurons within the slice. Slices can be prepared in a sagittal (S) or coronal (also called Longitudinal, L) direction depending upon their use (*see* **Fig. 19.2b**), whether for paired pulse electrophysiology, where the Purkinje neuron dendrites remain intact (sagittal), or coronal where the presynaptic parallel fibers remain intact for calcium imaging but where the dendrites of the Purkinje neurons are excessively cut.
3. Ice-cold artificial cerebrospinal fluid (aCSF): 126 mM NaCl, 2.5 mM KCl, 1.25 NaH_2PO_4, 1.3 mM $MgCl_2$, 2.4 mM $CaCl_2$, 10 mM glucose, and 26 mM $NaHCO_3$. It should be chilled to 4–6°C and oxygenated with a 95% O_2 and 5% CO_2 mix (carbogen), for 20–30 min prior to use.
4. A holding chamber containing oxygenated aCSF where the slices lie on a nylon mesh prior to recording is also required.

2.3. Electrophysiology – Whole Cell Patch Clamp Synaptic Recordings

1. A dedicated whole cell patch clamp electrophysiology setup is needed for these experiments comprising an upright microscope with infra red differential interference contrast (DIC) optics, with a Teflon-coated water immersion objective (Nikon, Olympus, and Zeiss all provide excellent models) and camera for best slice visualization. Manual or motorized manipulators (Narishige, Burleigh, Luigs and Neumann, Sutter all provide different versions) are needed to precisely position the stimulating and recording electrodes within the slice. Electrophysiological voltage clamp recordings should use a low-noise amplifier such as the Axopatch 200B or the Multiclamp, but we have also successfully used the switched voltage clamp amplifier the Axoclamp 2A (all except the Axoclamp 2A are provided by Molecular Devices). The output from the amplifier should be recorded digitally, using a commercially available analog

to digital converter and associated software of which there are a variety of options.

(The Axon Guide http://www.moleculardevices.com/pages/instruments/axon_guide.html provides a comprehensive guide to all aspects of electrophysiological recording techniques.)

2. We used an intracellular patch solution that contained 10 mM of the calcium chelator EGTA to strongly buffer calcium in the postsynaptic neuron (*see* **Fig. 19.2d**). This reduces any complications that might arise if, when the PMCA is inhibited, calcium rises not only in the presynaptic compartment but also in the postsynaptic cell, in this case the Purkinje neuron (where PMCA is rather heavily expressed) (*see* **Fig. 19.2c**). EGTA, as a rather low-affinity slow calcium buffer, present at a high concentration will heavily buffer all intracellular postsynaptic calcium changes. Other contents of the patch clamp solution are standard: 4.5 mM KCL, 20 mM KOH, 3.48 mM $MgCl_2$, 4 mM NaCl, 120 mM K^+ gluconate, 10 mM HEPES, 8 mM sucrose, 10 mM EGTA, 4 mM Na_2ATP, 0.4 mM Na_2GTP. Glass electrodes are filled with the solution and their resistance should be 3–5 MΩ.
3. Calcium increases in the presynaptic parallel fibers are activated using a brief electrical shock applied to the surface of the fibers using a focally placed stimulating electrode (*see* **Figs. 19.3** and **19.4**). The stimulating electrode is a silver wire inside a glass electrode filled with aCSF and with a resistance of approximately 500 kΩ and a stimulus isolator such as a Digitimer DS2A is required.

2.4. Fast Calcium Imaging from Presynaptic Parallel Fiber Terminals

1. Filling the presynaptic terminals with the fluorescent calcium indicator should be achieved as non-invasively as possible so as not to greatly perturb the calcium rises during the physiological synaptic response. The single wavelength Ca^{2+}-sensitive dye Calcium Green dextran (10,000 molecular weight) is particularly suitable, as the large dextran group prevents its exit from cells, so once in, it will remain, and also be actively transported along the living parallel fibers, but not by dead ones!
2. Calcium Green is a single wavelength dye that requires a peak wavelength of excitation light for its excitation before it can emit light (in the visible range) that is easily detected by all fluorescent microscopes through the appropriate filters. A sensitive and fast CCD camera capable of capture rates greater than 40 frames/s (e.g., Hamamatsu EM-CCD, but there are a variety of good products available) is required to detect the fluorescence emission. Capture of the emission

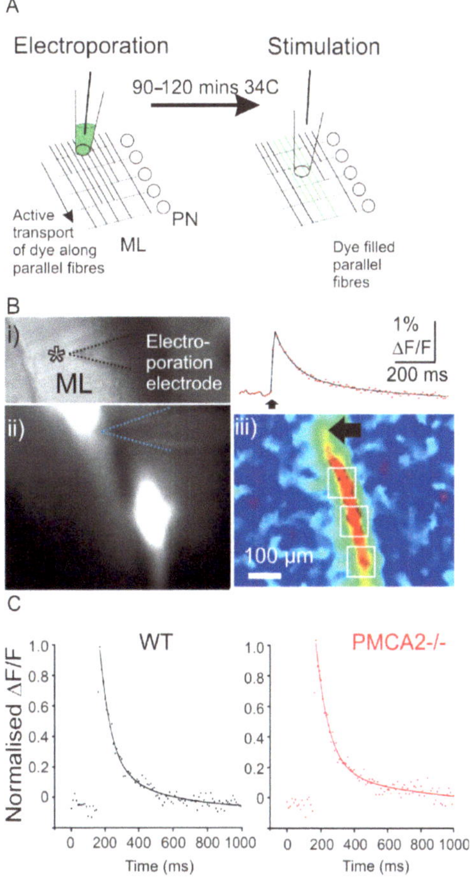

Fig. 19.4. Measurement of presynaptic calcium transients at the parallel fibers and how the calcium recovery is slowed in the PMCA2$^{-/-}$ double knockout mouse. (**a**) A schematic of the experimental approach to first label the fibers with calcium-sensitive dye before recording depolarization evoked calcium responses is shown. Longitudinal slices (*see* **Fig. 19.2b**) where longer lengths of the parallel fibers remain intact are labeled using a local electroporation of calcium-sensitive dye (Calcium Green dextran) and temperature-enhanced active transport of the dye along the parallel fiber beams. In (**b**, *ii*), depots of fluorescent dye can be observed following placement of the electroporation electrode into the molecular layer, gray scale image (*i*) *above*, with fluorescent beams emanating from the depots clearly visible. Placement of a stimulating electrode, (*ii*) in the beam and application of a stimulus, leads to a rise in presynaptic intracellular calcium within the beam, seen as an increased in fluorescence. Using regions of interest (*iii*) placed in the beam, recovery of calcium dynamics over time, represented by a change in fluorescence in response to the stimulus can be determined, trace above. Comparison of calcium recovery in wild type, where PMCA2 is present in the presynaptic terminals, (**c**) with the dynamics of the responses in the absence of PMCA2, (**c**) right-hand panel, indicate a slower recovery of the calcium when the pump is deleted. This slower recovery of calcium and therefore enhanced residual calcium in the PMCA2 double knockout accords with the slower decay of PPF that is visible in **Fig. 19.3d** and also entirely consistent with a role for PMCA2 in regulating presynaptic short-term plasticity at this synapse.

from the Calcium Green must also be synchronized with the stimulation of the presynaptic parallel fibers using a hardware interface such as a Master8 (AMPI) and software (such as Image-Pro).

3. Methods

3.1. PMCA2$^{-/-}$ Transgenic Mouse Breeding Program

1. All offspring from the PMCA2$^{+/-}$ mice are routinely PCR genotyped before weaning and used at 3–4 weeks of age for electrophysiology and Ca^{2+} imaging. Litter mates are routinely used with their ages differing by no more than 1 day. Sometimes it is prudent to conduct littermate experiments on the same day when experimental conditions should be as near identical as is possible.

3.2. Preparation of Cerebellar Slices

1. Slices for electrophysiology, paired pulse facilitation experiments are prepared in a sagittal orientation so that the Purkinje neuron dendrites are retained intact, whereas experiments for imaging the parallel fiber calcium transients use slices prepared in a coronal or longitudinal direction (*see* **Fig. 19.2b**).
2. Mice are anaesthetized prior to slice preparation (according to international standards within our local Ethical Guidelines) and the speed and care with which the brain is removed from the skull is critical for the successful preparation of the slices. Brain removal into ice-cold aCSF, preoxygenated with carbogen should occur in less than 1 min. As explained in **Note 1**, healthy slices and cells require fast and smooth preparation of slices that takes some practice.
3. The cerebellum is then placed upon a slicing platform, constructed from a small square block of 1% agar (standard agar or agarose, not necessarily molecular biology grade, is fine) placed on the Vibratome stage; this is achieved with a small drop of superglue to hold the tissue in place on the agar block, and the agar block is then glued to the Vibratome stage; in the case of sagittal slices the cerebellar vermis is sectioned flat; in the case of coronal slices the whole cerebellum is sectioned at an angle of approximately 40° from horizontal created by cutting the agar block at an appropriate angle. Slices are cut at 270–300 μm thickness and then transferred to an oxygenated aCSF-containing bath on a small platform comprising of a net made from stretched nylon stockings. This permits diffusion of nutrients, oxygen, and carbon dioxide to both sides of the slice and maintains

their viability for 6–8 h. Slices are best used approximately 90 min after preparation.

3.3. Whole Cell Electrophysiological Recordings to Assess Paired Pulse Facilitation (PPF)

1. Slices are transferred to the recording chamber where they are perfused with oxygenated aCSF at a rate of 3–4 ml/min; this solution also contains one of a variety of GABA-A receptor antagonists, e.g., gabazine, bicuculline, or picrotoxin, to prevent all synaptic inhibition within the cerebellar circuit. The slice must be secured in the chamber using a small weight, in the form of either a platinum wire or a small stone. Naturally, this small weight should not damage the tissue that is to be recorded from. The recording temperature is most often at room temperature, but if required the temperature of the recordings can be increased to 34–35°C if an appropriate in-line heater (e.g., SHM-6, Harvard Apparatus) is available.
2. Infrared-based optics to visualize the neurons helps identify the healthy neurons that are to be patched, *see* **Note 1**. Once the neuron has been selected, the stimulating electrode is placed in the molecular layer (*see* **Fig. 19.2d**), touching the surface of the slice. Once the whole cell configuration is obtained the cell is held at −65 mV in voltage clamp for 10 min to allow stabilization of the recording and diffusion of the patch pipette contents into the cell.
3. Following the stabilization period the cell can be stimulated to evoke an excitatory postsynaptic current, EPSC, and this should be between 200 and 400 pA in amplitude. A paired pulse protocol is used with an inter-stimulus interval (ISI) of between 25 and 200 ms (*see* **Fig. 19.3**), delivered every 30 seconds, so at 0 seconds the cell is stimulated with an ISI of 25 ms, 30 seconds later with an ISI of 50 ms, 60 seconds later with an ISI of 75 ms, and so on until reaching a 200 ms interval. If this family of interstimulus interval sequence produces facilitation of the second EPSC that also decays in a stable and predictable manner (*see* **Note 1**) then it should be repeated at least three times. If there is variability to the PPF or even depression of the second EPSC this points to a contamination of the response from activation of the climbing fiber input; this input is known to be a high-release probability synapse that readily depresses in paired pulse experiments (21). It is imperative to keep the stimulus protocol going at a 30 second interval and to minimize variability to the size of the first EPSC; any random alterations to the interval can badly influence the variability to the extent of measured paired pulse facilitation.

4. The extent of paired pulse facilitation is calculated by dividing the amplitude of the second EPSC by the amplitude of the first. In this case a PPF ratio of 2 would be obtained in the case where the second EPSC was two times the size of the first. The PPF is calculated for each family of EPSC traces (25–200 ms responses) and averaged. The decay of the PPF over time is then fitted with a mono or dual exponential, depending upon the data, to generate either one or two decay time constants, *T1* and *T2* for each cell. This procedure can be achieved with most statistical packages such as Prism or Origin, and if necessary constraints to the fits, such as the plateau level, based upon the complete recovery of the PPF can be applied. The values of *T1* and *T2* can then be compared across cells from different experimental groups. To compare average PPF decay profiles the families of PPF data can also be normalized and compared across different experimental groups (*see* **Fig. 19.3d**).

As also shown in **Fig. 19.3d** the decay of PPF was significantly faster in parallel fiber to Purkinje neuron synapses from wild-type PMCA2$^{+/+}$ compared with PMCA2$^{-/-}$ slices. We also obtained similar PPF results when we treated PMCA2$^{+/+}$ cells with pharmacological inhibitors of the PMCA. There are a variety of inhibitors available and these are outlined in the Notes **Section 5**. Our result strongly supports the proposal that removal of the PMCA2 pump from the presynaptic parallel fiber terminals alters the availability of residual calcium that contributes to facilitation at this synapse. Without PMCA2 the removal of calcium from the presynaptic terminal slows down allowing the residual calcium to build up so that the decay of paired pulse facilitation slows down, or takes longer to recover.

3.4. Calcium Imaging at Parallel Fiber Presynaptic Terminals

1. Although the decay of paired pulse facilitation gives an indirect measure of the contribution of the PMCA to the decay of presynaptic calcium at this parallel fiber to Purkinje neuron synapse (and also other synapses) a direct measurement of calcium greatly aids interpretation. We therefore used calcium imaging of the presynaptic parallel fiber terminals to directly measure the contribution of PMCA2 to the calcium decay following natural synaptic stimulation of the parallel fibers.
2. Calcium Green dextran (Invitrogen, Molecular Probes) can be readily electroporated into the parallel fibers where it is then transported along the axons to fill the presynaptic terminals (*see* **Fig. 19.4a**). By cutting the cerebellar slices in the longitudinal or coronal direction, long lengths of the parallel fibers are retained intact.

   After placing the longitudinal slice into the same recording chamber as used for the electrophysiology recordings above, a glass stimulating electrode, of a similar type to that used to stimulate the parallel fibers for PPF (*see* **Section 2.3**, but now 2–3 MΩ resistance) filled with 5% Calcium Green dextran in saline, is placed into the molecular layer of the slice (*see* **Fig. 19.4a,b(i)**). A minimal volume of dye is required, 2–5 μl, but be sure to rid the electrode of any bubbles and also ensure that the silver wire touches the saline and dye within the glass electrode. Using a programmable stimulator such as the Master8, deliver 900 short 0.01 ms duration current pulses of 50–70 μA at 5 Hz (five times per second, 200 ms interval). The Calcium Green dextran containing electrode will be visible in the fluorescent excitation light, and the exit of the dye from the electrode into the slice during the application procedure can be reliably followed. A depot of fluorescent dye will appear as a large spot within the molecular layer, as shown in **Fig. 19.4b (ii)**. If it does not, then check for bubbles in the application electrode and also ensure that it is in the slice, and repeat. It is also possible to re-use the application electrode for multiple depot placements within the same longitudinal slice. Typically we injected two or three regions, as shown in **Fig. 19.4b (ii)**. It is useful to identify the shape of the slice and any landmarks as well as the positions of the dye depots with a laboratory book sketch. Be careful not to over-inject a region, as outlined in Notes, **Section 5**.

Once the depot of dye is placed, the dye must be given the opportunity to transport along the parallel fibers, and this process is greatly facilitated by raising the temperature of the slice (*see* **Fig. 19.4a**). Gently remove the slice from the recording chamber and return it to the holding chamber (*see* **Section 3.2**) immersed in a water bath to raise its temperature to 34°C, for 90–120 min. After this time return the slice to the recording chamber for the calcium imaging and use the "landmarks" to re-orientate it correctly (and the same way up). If, for example, the depots cannot be relocated after returning the slice then try flipping the slice over, as it is probably depot side down!

3. If the Calcium Green dye has been reliably transported then a "beam" of low-intensity fluorescence can be observed exiting from the often brightly fluorescent (and therefore probably high calcium) depot (*see* **Fig. 19.4b (ii)**). Once the beam has been located, place a stimulating aCSF-filled electrode (500 kΩ) onto the position of the beam, as far away from the depot as possible.
4. Using similar stimulus intensities as those used to evoke the EPSCs in the electrophysiology PPF experiments stimulate the beam and simultaneously record the fluorescence within

the beam. Typically, we recorded the calcium transients at a temperature of 35°C (compared with a room temperature of 25°C) when the calcium transients are smaller, but decay faster. The changes in fluorescence should be recorded at a frequency of at least 40 Hz, and preferably at 100 Hz, for a period of 1 second, whereupon the calcium should have decayed back to baseline. To achieve this time resolution it will be necessary to bin pixels on the CCD camera, but it is perfectly acceptable to increase time resolution at the expense of some spatial resolution. Typically an 8 × 8 binning will sufficiently enhance the acquisition performance of the camera. A single stimulation is delivered ten times at an interval of 30 seconds. Typically, since the changes in fluorescence are small, in the order of 1–2%, the fluorescence responses must be averaged. An average of 10 responses is appropriate to generate a calcium fluorescence transient with sufficient signal to noise ratio (*see* **Fig. 19.4b (iii)**).

5. The raw fluorescence can be extracted using regions of interest within the area of the Calcium Green labeled parallel fiber beam that responds, see white boxes **Fig. 19.4b** (iii), and changes can be expressed as the change in absolute fluorescence values divided by the baseline fluorescence, so-called deltaF/F. If necessary a linear bleaching correction should also be applied.
6. In order to compare the stimulus-evoked calcium transients across different slices and different animals the deltaF/F responses can be normalized and the calcium decay fitted with a two-phase exponential decay, in much the same way as the decay of PPF was treated, *see* **Section 3.3**. As shown in **Fig. 19.4c**, the absence of PMCA2 enhanced the time taken for the calcium to decay fully back to baseline, and the second, slower time constant of the decay was significantly enhanced in the parallel fiber responses from PMCA2$^{-/-}$ mice. Ideally, the same slowing of the presynaptic calcium signal should be observed following pharmacological inhibition of the PMCA, *see* Notes, **Section 3**. However, some of the inhibitors of the PMCAs have an intrinsic fluorescence that can perturb single wavelength calcium measurements; therefore an alternative inhibitor, listed in Notes, **Section 5**, is preferable.

4. Future Perspectives

The combination of electrophysiology and calcium imaging together with the advantages offered by the cerebellar circuitry in the PMCA2 knockout mouse has proved a powerful approach to assess the contribution of the PMCA2 transporter to synapse function. However, the application of these techniques should not stop here; future studies based upon manipulations of PMCA2 structure and function and how this influences presynaptic calcium dynamics will further advance the field. Furthermore the same approaches can identify the contribution of the PMCA at other synapses, such as hippocampal excitatory and inhibitory presynaptic terminals (10).

The methods used here also have application to assess the contribution of other transporters, such as the Na^{+}/Ca^{2+} exchanger, NCX, in particular NCX1 and NCX3, and even their contribution at other synapses; indeed PPF was enhanced and slowed in the hippocampus of the NCX2 knockout mouse (22) presumably as a result of enhanced presynaptic calcium (although this was not measured). Like PMCA2, both these exchangers are present within the cerebellar molecular layer (23) and like the PMCA, pharmacological inhibition of these exchangers lacks specificity. However, the combination of transgenic knockout animals for NCX1 and 3, together with the methods described here, could represent a powerful approach to better understand the interplay of NCX and PMCA for calcium homeostasis during synapse function.

5. Notes

1. Reliable and consistent paired pulse facilitation experiments require long-lasting stable recordings from the Purkinje neurons. For this it is vital that the cerebellar slices are healthy; critical indicators include a fast and smooth preparation of the slices, "plump"-looking Purkinje cell bodies (but not swollen) with little evidence of dead (shrunken, dark-looking Purkinje neuron cell bodies) within the slice and little evidence of dark dendrites at the surface of the slice. If slices are unhealthy, practice in "slick" slice preparation is needed, also check quality of water used to make all solutions (MilliQ water or similar with a resistance of approximately 18 MΩ is needed).
2. Depot injections of Calcium Green should not use more than two applications; the most critical stage is the active transport. Indeed the beginning of active transport should be observable during the application process, but it is further enhanced by the longer incubation at higher temperature. Do not leave slices for more than 2 h at the higher temperature; return the water bath back to ambient temperature as soon as possible.
3. Pharmacological Tools to Prevent PMCA Action

All studies on a transgenic knockout mouse require controls to ensure that observed changes represent the consequence of the knockout and not an adaptive change. Adequate controls require that pharmacological blockade of the PMCA should produce similar outcomes to the genetic deletion, and there are several different ways to do this.

Carboxyeosin. The membrane permeable ester analog of carboxyeosin is taken up across the membrane and cleaved by intracellular esterases to deliver carboxyeosin inside the cell. Carboxyeosin specifically inhibits P-type ATPases, the family of ATPases (24) that PMCA belongs to, but other P-type ATPases including the sarcoplasmic endoplasmic reticulum calcium ATPase (SERCA) will also be inhibited in a living cell, although this should be minimal if carboxyeosin is used at 10 μM or less (25). Carboxyeosin should be used at less than or equal to 10 uM and is best made up in dry DMSO and stored frozen. Carboxyeosin is fluorescent and will readily render polythene tubings a bright pink color unless assiduous washing with 25% ethanol followed by distilled water is employed.

Lanthanum chloride, at 100 μM, can also inhibit the PMCAs with a strong affinity of approximately 3 μM making its use in neuronal tissue a good option when combined with other approaches (26, 27)

High pH (8–8.5) also inhibits the PMCA; by virtue of inhibiting the pump's exchange of H^+ for Ca^{2+} as it crosses the membrane (28, 29), a potentially confounding effect, however, is the possibility that the high pH influences synapse physiology through alternative mechanisms (30).

Calmodulin inhibitors have also been used to inhibit the PMCAs (31).

More recently the caloxins, short-peptide inhibitors of the PMCAs, have also been developed (32). These bind to the extracellular domains (or "mouth") of the pump and are advantageous since they can be applied to the outside of cells and caloxin 3A1 exhibits selectivity for PMCA versus SERCA. Efforts to engineer the caloxins to give PMCA subtype specificity, for example, in different tissues (33) and even toward splice variant specificity are ongoing (34). Although caloxins are yet to find extensive use in neuronal preparations, caloxin 2A1 was used to demonstrate PMCA involvement during domoic acid-induced intracellular acidification (23).

Acknowledgments

We acknowledge the support of a University of Otago Research Grant, The Neurological Foundation of New Zealand (to RME), and a University of Otago PhD Scholarship to CJR.

References

1. Strehler, E.E. and Zacharias, D.A. (2001) Role of alternative splicing in generating isoform diversity among plasma membrane calcium pumps. *Physiol. Rev.* **81**, 21–50.
2. Berridge, M.J., Bootman, M.D., and Roderick, H.L. (2003) Calcium signalling: dynamics, homeostasis and remodelling. *Nat. Rev. Mol. Cell Biol.* **4**, 517–529.
3. Strehler, E.E., Heim, R., and Carafoli, E. (1991) Molecular characterization of plasma membrane calcium pump isoforms. *Adv. Exp. Med. Biol.* **307**, 251–261.
4. Caride, A.J., Penheiter, A.R., Filoteo, A.G., Bajzer, Z., Enyedi, A., Penniston, J.T. (2001) The plasma membrane calcium pump displays memory of past calcium spikes. Differences between isoforms 2b and 4b. *J. Biol. Chem.* **276**, 39797–39804.
5. Wanaverbecq, N., Marsh, S.J., Al-Qatari, M., and Brown, D.A. (2003) The plasma membrane calcium-ATPase as a major mechanism for intracellular calcium regulation in neurons from the rat superior cervical ganglion. *J. Physiol.* **550**, 83–101.
6. Thomas, R.C. (2008) The plasma membrane calcium ATPase (PMCA) of neurons is electroneutral and exchanges 2 H^+ for each Ca^{2+} or Ba^{2+} ion extruded. *J. Physiol.* **587**(2) 315–327.
7. Kreitzer, M.A., Collis, L.P., Molina, A.J., Smith, P.J., and Malchow, R.P. (2007) Modulation of extracellular proton fluxes from retinal horizontal cells of the catfish by depolarization and glutamate. *J. Gen. Physiol.* **130**, 169–82.
8. Kim, M.H., Korogod, N., Schneggenburger, R., Ho, W.K., and Lee, S.H. (2005) Interplay between Na^+/Ca^{2+} exchangers and mitochondria in Ca^{2+} clearance at the calyx of Held. *J. Neurosci.* **25**, 6057–6065.
9. Lnenicka, G.A., Grizzaffi, J., Lee, B., and Rumpal, N. (2006) Ca^{2+} dynamics along identified synaptic terminals in Drosophila larvae. *J. Neurosci.* **26**, 12283–12293.
10. Jensen, T.P., Filoteo, A.G., Knopfel, T., and Empson, R.M. (2007) Pre synaptic plasma membrane Ca^{2+} ATPase isoform 2a regulates excitatory synaptic transmission in rat hippocampal CA3. *J. Physiol.* **579**, 85–99.
11. Empson, R.M., Garside, M.L., and Knopfel, T. (2007) Plasma membrane Ca^{2+} ATPase 2 contributes to short-term synapse plasticity at the parallel fiber to Purkinje neuron synapse. *J. Neurosci.* **27**, 3753–3758.
12. Ficarella, R., Di Leva, F., Bortolozzi, M., Ortolano, S., Donaudy, F., Petrillo, M., Melchionda, S., Lelli, A., Domi, T., Fedrizzi, L., Lim, D., Shull, G.E., Gasparini, P., Brini, M., Mammano, F., and Carafoli, E. (2007) A functional study of plasma-membrane calcium-pump isoform 2 mutants causing digenic deafness. *Proc. Natl. Acad. Sci. USA.* **104**, 1516–1521.
13. Lou, X., Scheuss, V., and Schneggenburger, R. (2005) Allosteric modulation of the presynaptic Ca^{2+} sensor for vesicle fusion. *Nature.* **435**, 497–501.
14. Sabatini, B.L. and Regehr, W.G.(1997) Control of neurotransmitter release by presynaptic waveform at the granule cell to Purkinje cell synapse. *J. Neurosci.* **17**, 3425–3435.
15. Katz, B. and Miledi, R.(1968) The role of calcium in neuromuscular facilitation. *J. Physiol.* **195**, 481–492.
16. Chen, C. and Regehr, W.G. (1999) Contributions of residual calcium to fast synaptic transmission. *J. Neurosci.* **19**, 6257–6266.
17. Burette, A. and Weinberg, R.J. (2007) Perisynaptic organization of plasma membrane calcium pumps in cerebellar cortex. *J. Comp. Neurol.* **500**, 1127–1135.
18. Kozel, P.J., Friedman, R.A., Erway, L.C., et al. (1998) Balance and hearing deficits in mice with a null mutation in the gene encoding plasma membrane Ca^{2+}-ATPase isoform 2. *J. Biol. Chem.* **273**, 18693–18696.
19. Kurnellas, M.P., Lee, A.K., Li, H., Deng, L., Ehrlich, D.J., and Elkabes, S. (2007) Molecular alterations in the cerebellum of the plasma membrane calcium ATPase 2 (PMCA2)-null mouse indicate abnormalities in Purkinje neurons. *Mol. Cell Neurosci.* **34**, 178–188.
20. Choi, H.S. and Eisner, D.A. (1999) The role of sarcolemmal Ca^{2+}-ATPase in the regulation of resting calcium concentration in rat ventricular myocytes. *J. Physiol.* **515**, 109–118.
21. Gatto, C. and Milanick, M. A. (1993). Inhibition of the red blood cell calcium pump by eosin and other fluorescein analogues. *Am. J. Physiol.* **264**, C1577–C1586.
22. Herscher, C.J. and Rega, A.F. (1996) Pre-steady-state kinetic study of the mechanism of inhibition of the plasma membrane $Ca^{(2+)}$-ATPase by lanthanum. *Biochemistry* **35**, 14917–14922.
23. Vale González, C., Alfonso, A., Suñol, C., Vieytes, M.R., and Botana, L.M. (2006) Role of the plasma membrane calcium adenosine triphosphatase on domoate-induced

intracellular acidification in primary cultures of cerebellar granule cells. *J. Neurosci. Res.* **84**, 326–337.
24. Willoughby, D., Thomas, R., and Schwiening, C. (2001). The effects of intracellular pH changes on resting cytosolic calcium in voltage-clamped snail neurons. *J. Physiol.* **530**, 405–416.
25. Benham, C.D., Evans, M.L., and McBain, C.J. (1992). Ca^{2+} efflux mechanisms following depolarisation evoked calcium transients in cultured rat sensory neurons. *J. Physiol.* **455**, 567–583.
26. Taira, T., Smirnov, S., Voipio, J., and Kaila, K. (1993). Intrinsic proton modulation of excitatory transmission in rat hippocampal slices. *Neuroreport* **4**, 93–96.
27. Scheuss, V., Yasuda, R., Sobczyk, A., and Svoboda, K. (2006) Nonlinear [Ca^{2+}] signaling in dendrites and spines caused by activity-dependent depression of Ca^{2+} extrusion. *J. Neurosci.* **26**, 8183–8194.
28. Szewczyk, M.M., Pande, J., and Grover, A.K. (2008) Caloxins: a novel class of selective plasma membrane Ca^{2+} pump inhibitors obtained using biotechnology. *Pflugers Arch.* **456**, 255–266.
29. Pande, J., Mallhi, K.K., and Grover, A.K. (2005) Role of third extracellular domain of plasma membrane Ca^{2+}-Mg^{2+}-ATPase based on the novel inhibitor caloxin 3A1. *Cell Calcium.* **37**, 245–250.
30. Pande, J., Mallhi, K.K., Sawh, A., Szewczyk, M.M., Simpson, F., and Grover, A.K. (2006) Aortic smooth muscle and endothelial plasma membrane Ca^{2+} pump isoforms are inhibited differently by the extracellular inhibitor caloxin 1b1. *Am. J. Physiol. Cell Physiol.* **290**, C1341–C1349.
31. Ito, M. (2006) Cerebellar circuitry as a neuronal machine. *Prog. Neurobiol.* **78**, 272–303.
32. Hashimoto, K. and Kano, M. (1998) Presynaptic origin of paired-pulse depression at climbing fibre-Purkinje cell synapses in the rat cerebellum. *J. Physiol.* **506**, 391–405.
33. Jeon, D., Yang, Y.M., Jeong, M.J., Philipson, K.D., Rhim, H., and Shin, H.S. (2003) Enhanced learning and memory in mice lacking Na^{+}/Ca^{2+} exchanger 2. *Neuron.* **38**, 965–976.
34. Canitano, A., Papa, M., Boscia, F., Castaldo, P., Sellitti, S., Taglialatela, M., and Annunziato, L. (2002) Brain distribution of the Na^{+}/Ca^{2+} exchanger-encoding genes NCX1, NCX2, and NCX3 and their related proteins in the central nervous system. *Ann. NY. Acad. Sci.* **976**, 394–404.

SUBJECT INDEX

Q. Yan (ed.), *Membrane Transporters in Drug Discovery and Development*, Methods in Molecular Biology 637, DOI 10.1007/978-1-60761-700-6, © Springer Science+Business Media, LLC 2010

E

F

G

H

Q

R

S

FSC
www.fsc.org
MIX
Papier aus verantwortungsvollen Quellen
Paper from responsible sources
FSC® C105338

If you have any concerns about our products,
you can contact us on
ProductSafety@springernature.com

In case Publisher is established outside the EU,
the EU authorized representative is:
Springer Nature Customer Service Center GmbH
Europaplatz 3, 69115 Heidelberg, Germany

Printed by Libri Plureos GmbH
in Hamburg, Germany